Brain-Inspired Cognitive Mechanisms for AI-Enhanced Decision Intelligence

Ionica Oncioiu
https://orcid.org/0000-0003-3026-7702
Titu Maiorescu University, Romania

IGI Global Scientific Publishing
Publishing Tomorrow's Research Today

Vice President of Editorial | Melissa Wagner
Director of Acquisitions | Mikaela Felty
Director of Book Development | Jocelynn Hessler
Production Manager | Mike Brehm
Cover Design | Jose Rosado

Published in the United States of America by
IGI Global Scientific Publishing
701 East Chocolate Avenue
Hershey, PA, 17033, USA
Tel: 717-533-8845 | Fax: 717-533-7115
Website: https://www.igi-global.com E-mail: cust@igi-global.com

Library of Congress Cataloging-in-Publication Data

Names: Oncioiu, Ionica, 1972- editor
Title: Brain-inspired cognitive mechanisms for AI-enhanced decision intelligence / edited by Ionica Oncioiu.
Description: Hershey, PA : IGI Global Scientific Publishing, [2027] | Includes bibliographical references and index. | Summary: "This book investigates the integration of brain-inspired cognitive mechanisms into AI-based decision intelligence systems. It provides a comparative analyses of cognitive, neuro-inspired, and hybrid AI approaches, highlighting their relevance for the development, evaluation, and advancement of AI-enhanced decision intelligence across complex and dynamic environments"-- Provided by publisher.
Identifiers: LCCN 2026006194 (print) | LCCN 2026006195 (ebook) | ISBN 9798337389882 hardcover | ISBN 9798337389899 paperback | ISBN 9798337389905 ebook
Subjects: LCSH: Computational intelligence | Cognitive neuroscience | Artificial intelligence | Decision making
Classification: LCC Q342 .B735 2026 (print) | LCC Q342 (ebook)
LC record available at https://lccn.loc.gov/2026006194
LC ebook record available at https://lccn.loc.gov/2026006195

British Cataloguing in Publication Data
A Cataloguing in Publication record for this book is available from the British Library.

Table of Contents

Detailed Table of Contents

bounded rationality, and bias using cognitive science, neuroscience, philosophy, and decision intelligence. The chapter analyses XAI paradigms, emphasises cognitive misalignments, and reframes decision intelligence as a human–AI socio-cognitive system focused on judgement quality rather than prediction accuracy. The Cognitive Explainability–Sense-Making Framework (CESF) links explanations to mental models, trust calibration, and decision outcomes and provides empirical validation. The chapter explores ethical, organisational, and governance effects utilising empirical evidence, case studies, cross-cultural views, and policy analysis. It concludes with future research and advocates cognitively aligned, context-sensitive explainability for ethical, trustworthy, and successful AI-mediated decision-making.

Decision Intelligence is a new approach to AI that utilizes the insights we can gain from how humans make decisions. This chapter focuses on key cognitive processes (human brain functions) such as perception, attention, memory, learning and decision-making, with the goal of providing inspiration for developing more sophisticated AI systems. By bringing together knowledge bases from a variety of fields (neuroscience, psychology & computational intelligence) this chapter outlines how these human brain-like models can be adapted to machine learning architectures through the use of attention mechanisms, memory-augmented systems, neurosymbolic reasoning and affective computing. Ethical issues like transparency, bias and the potential dangers of brain/computer interface integration are examined to encourage innovation with responsibility. Future trends in neuromorphic computing, hybrid intelligence and human/computing collaboration will be identified to establish decision intelligence as an essential building block for the future of AI.

The rapid integration of artificial intelligence (AI) into high-stakes decision-making domains underscores the need for systems that are both effective and cognitively compatible with humans. Brain-inspired decision intelligence leverages insights from neuroscience, cognitive psychology, and computational modeling to enhance adaptability, transparency, and collaboration. By grounding AI in mechanisms such as attention, working memory, reinforcement learning, and executive control, systems can navigate uncertainty, balance competing goals, and provide context-sensitive support. This approach fosters complementary human–AI interaction, preserves

Artificial intelligence systems are increasingly used for complex decision-making in domains characterized by uncertainty, scale, and dynamic change. While traditional algorithm-based approaches have achieved strong performance in data-driven tasks, they remain limited in flexibility, contextual awareness, and interpretability when compared with human cognition. At the same time, advances in neuroscience and cognitive science have clarified the mechanisms of perception, learning, memory, attention, and reasoning that underlie human decision-making. This chapter examines how such cognitive principles can be translated into computational models that bridge statistical learning and cognitive reasoning. Drawing on cognitive computing and neuro-symbolic AI, it proposes an integrative framework for AI-supported decision intelligence that combines data-driven learning with brain-inspired architectures, emphasizing transparency, trust, and cognitive responsibility in the design of interpretable and context-sensitive intelligent systems.

Artificial intelligence has achieved significant progress in perception and prediction, yet many contemporary systems remain limited in interpretability, contextual reasoning, and alignment with human cognition. This chapter advances the concept of cognitive aligned neuro symbolic AI systems as a framework for bridging human thought and machine intelligence in dynamic environments. By integrating neural learning mechanisms with structured symbolic reasoning, these systems combine adaptive perception with transparent inference and explicit knowledge representation. Drawing from cognitive science and systems theory, the chapter proposes that meaningful AI advancement requires architectural alignment with core human cognitive processes, including contextual judgment, memory integration, and reflective reasoning. The discussion conceptualizes how such systems operate under uncertainty, support human centered collaboration, and contribute to trustworthy and responsible AI development.

Preface

The rapid evolution of artificial intelligence is reshaping the foundations of intelligent systems, moving beyond the automation of analytical tasks toward the design of systems capable of reasoning, adaptation, and decision-making inspired by the human brain. Brain-inspired cognitive mechanisms, combined with AI-enhanced decision intelligence, represent a transformative paradigm that extends beyond computational efficiency toward more adaptive, explainable, and context-aware intelligent systems. What was once considered the exclusive domain of human cognition—perception, learning, memory, reasoning, and complex decision-making—is increasingly being modeled, augmented, and partially replicated through advanced AI architectures inspired by cognitive neuroscience.

Contemporary developments in machine learning, cognitive computing, neuromorphic engineering, and neuroscience have accelerated the development of intelligent systems capable of emulating fundamental cognitive processes. Brain-inspired approaches enable AI systems to move beyond pattern recognition toward adaptive learning, causal reasoning, attention mechanisms, and memory-driven decision-making. These capabilities not only improve predictive performance but also contribute to more robust, transparent, and human-centric decision intelligence across domains such as healthcare, finance, manufacturing, education, cybersecurity, and autonomous systems.

Decision intelligence is emerging as a multidisciplinary field that integrates artificial intelligence, cognitive science, data analytics, and human decision theory to support complex decision processes under uncertainty. Unlike conventional AI solutions that primarily optimize predefined objectives, brain-inspired cognitive systems seek to replicate the flexibility, resilience, and contextual awareness characteristic of biological intelligence. By incorporating mechanisms such as hierarchical learning, reinforcement learning, attention, episodic memory, and cognitive adaptation, these systems are better equipped to address dynamic environments where uncertainty, incomplete information, and changing objectives are the norm.

The convergence of neuroscience and artificial intelligence is redefining our understanding of human–AI collaboration. Intelligent systems are no longer viewed solely as automated tools but increasingly as cognitive partners capable of augmenting human reasoning and supporting strategic decision-making. This shift brings renewed attention to questions of explainability, trust, ethical responsibility, transparency, and the governance of intelligent decision systems, making interdisciplinary research more important than ever.

This volume, Brain-Inspired Cognitive Mechanisms for AI-Enhanced Decision Intelligence, presents a comprehensive exploration of the theoretical foundations, computational models, and practical applications of cognitively inspired AI. It brings together contributions from researchers and practitioners working at the intersection of artificial intelligence, cognitive science, neuroscience, data science, and intelligent systems engineering. The chapters examine emerging methodologies, innovative architectures, real-world applications, and future research directions that contribute to the development of adaptive, explainable, and trustworthy AI systems capable of supporting high-quality decision intelligence.

By bringing together perspectives from artificial intelligence, cognitive neuroscience, computer science, data analytics, and decision sciences, this book presents a multidisciplinary vision for the development of adaptive, trustworthy, and human-centered intelligent systems capable of supporting more effective and resilient decision-making. We hope that it will serve as a valuable resource for academics, researchers, graduate students, industry professionals, and policymakers seeking to understand and shape the future of cognitive artificial intelligence and intelligent decision-making.

The authors Jyoti Kumari, Pradeep Yadav, and Chandra Prakash Bhargava of the first chapter, *"Designing Human-Centered and Cognitively Aligned Artificial Intelligence Systems for Ethical and Effective Decision-Making,"* examine the growing importance of designing artificial intelligence systems that align with human cognitive processes while promoting ethical, transparent, and effective decision-making across critical application domains. Grounded in insights from cognitive science, human–computer interaction, AI ethics, and decision theory, the chapter proposes an integrative framework that emphasizes cognitive compatibility, explainability, fairness, trust calibration, and collaborative intelligence as fundamental components of next-generation AI systems. Drawing on contemporary research findings, practical case studies, and structured evaluation frameworks, the authors demonstrate how human-centered design principles can enhance decision quality, reduce cognitive burden, and strengthen human–AI collaboration in complex environments. The discussion also addresses emerging challenges related to algorithmic bias, transparency, governance, and the responsible deployment of AI, emphasizing the need for cognitively informed design strategies that balance technological innovation with ethical accountability. Collectively, the ideas presented establish a solid foundation for developing adaptive, trustworthy, and cognitively aligned AI systems, making this contribution particularly valuable for researchers, AI developers, decision scientists, and policymakers seeking to advance AI-enhanced decision intelligence.

A cognitive perspective on explainable artificial intelligence is presented by Muralidhar L B, N Sathyanarayana, Swapna H R, Madhu S, and Sheeba K P, the authors of the second chapter, "*Explainability as a Cognitive Process in Human Sense-Making for Interpretable Decision Intelligence*." This contribution investigates explainability not merely as algorithmic transparency but as a cognitive and sense-making process that enables people to understand, evaluate, and appropriately act upon AI-generated recommendations. Combining insights from cognitive science, neuroscience, philosophy, and decision intelligence, the chapter demonstrates how human cognition, mental models, bounded rationality, and cognitive biases influence the interpretation of AI explanations and ultimately shape decision quality. Through the development of the Cognitive Explainability–Sense-Making Framework (CESF), the authors illustrate how cognitively aligned explanations strengthen trust calibration, improve human–AI collaboration, and promote more informed and context-sensitive decision-making. Within a multidisciplinary framework that bridges artificial intelligence, cognitive psychology, ethics, and governance, the chapter further explores the organizational and societal implications of explainable AI through empirical evidence, case studies, cross-cultural perspectives, and policy analysis. The findings emphasize that effective decision intelligence depends not only on transparent algorithms but also on explanations that align with human cognitive processes.

A comprehensive perspective on the cognitive mechanisms that inspire next-generation artificial intelligence is presented by Boopathi Raja G, the author of the third chapter, "*From Brain to Machine: Cognitive Mechanisms Driving AI-Enhanced Decision Intelligence*." The objective of this research is to explore how fundamental human cognitive processes, including perception, attention, memory, learning, and decision-making, can serve as the foundation for developing more adaptive, explainable, and intelligent AI systems. By integrating knowledge from cognitive neuroscience, psychology, and computational intelligence, the chapter demonstrates how brain-inspired mechanisms can be translated into advanced machine learning architectures through attention mechanisms, memory-augmented models, neurosymbolic reasoning, and affective computing. Equally important, the chapter examines the ethical implications of cognitively inspired AI, emphasizing the importance of transparency, fairness, bias mitigation, and the responsible integration of brain–computer interface technologies. Furthermore, the analysis highlights emerging developments in neuromorphic computing, hybrid intelligence, and human–AI collaboration, demonstrating their potential to transform decision intelligence into a cornerstone of future intelligent systems. Through its interdisciplinary perspective, the chapter illustrates how brain-inspired cognitive models can enhance the adaptability, trustworthiness, and effectiveness of AI-driven decision-making across diverse application domains.

Collaborative decision-making inspired by biological cognition constitutes the central theme of the fourth chapter, "*Brain-Based Models for Collaborative Decision Intelligence*." The objective of this research is to examine how cognitive mechanisms derived from neuroscience and cognitive psychology can enhance artificial intelligence systems to support more adaptive, transparent, and human-centered collaborative decision-making. The chapter demonstrates how brain-based processes such as attention, working memory, reinforcement learning, and executive control can be translated into computational models that enable intelligent systems to manage uncertainty, balance competing objectives, and provide context-sensitive decision support in complex environments. By integrating perspectives from neuroscience, cognitive science, and artificial intelligence, the author shows that brain-inspired decision intelligence strengthens complementary human–AI collaboration by preserving human judgment while enhancing analytical capabilities. At the same time, the chapter emphasizes the importance of ethical responsibility, transparency, and trust in the design of collaborative intelligent systems, ensuring that AI serves as a reliable partner rather than a replacement for human expertise. Ultimately, the contribution highlights the potential of brain-based cognitive models to establish more robust, trustworthy, and effective decision intelligence frameworks capable of addressing the challenges of real-world decision-making across diverse domains.

The convergence of brain–computer interfaces and artificial intelligence as a transformative paradigm for next-generation decision support is examined by Alexander Zakharov, Elena Khivintseva, Alexander Kolsanov, Igor Shirolapov, Mariya Stanislavovna Sergeeva, Yuliya Komarova, and Natalia Romanchuk, the authors of the fifth chapter, *"Integrating Brain-Computer Interfaces and Artificial Intelligence for Deep Interpretation of Cognitive States in Decision Support."* The chapter investigates how advances in brain–computer interface technologies, combined with artificial intelligence, enable the deep interpretation of human cognitive states, thereby creating new opportunities for adaptive and cognitively informed decision support. Supported by recent advances in neuroscience, biomedical engineering, and intelligent computing, the authors illustrate how neurophysiological signals can be transformed into meaningful indicators of attention, cognitive workload, emotional responses, and decision readiness, allowing intelligent systems to respond dynamically to users' cognitive conditions. Rather than relying exclusively on conventional data-driven analytics, the proposed perspective advocates a closer integration of human cognitive processes with computational intelligence, fostering decision-support systems that are increasingly personalized, context-aware, and responsive to real-time cognitive dynamics. The discussion further addresses the ethical and technological implications associated with the collection and interpretation of neurophysiological data, emphasizing the importance of privacy, transparency, explainability, and responsible AI governance.

The potential of brain-inspired artificial intelligence to transform intelligent decision-making is explored by Annmol Damien and Jain Mathew, the authors of the sixth chapter, "*Brain-Inspired Artificial Intelligence Models for Intelligent Decision-Making*." The chapter examines how cognitive principles derived from neuroscience and cognitive science can inform the development of adaptive, explainable, and human-centered artificial intelligence systems capable of supporting decision-making in increasingly complex and uncertain environments. Bringing together evidence from cognitive psychology, computational intelligence, and neuro-inspired computing, the authors demonstrate how mechanisms such as learning, attention, memory, emotion, motivation, and bounded rationality can be translated into intelligent computational models that exhibit greater flexibility, transparency, and contextual awareness than conventional optimization-driven approaches. By examining cognitive architectures, neuro-symbolic reasoning, and attention-based learning paradigms, the chapter illustrates how biologically inspired intelligence contributes to more robust and trustworthy decision-support systems while preserving the central role of human judgment in high-stakes decision contexts. The discussion further emphasizes the importance of explainability, ethical responsibility, and responsible AI design as fundamental requirements for fostering trust, accountability, and meaningful human–AI collaboration.

The application of brain-inspired cognitive mechanisms to educational decision-making is explored by Ashraf Alam, the author of the seventh chapter, "*Attention, Memory, Reasoning, Judgment, and Metacognitive Control in AI-Based Educational Decisions: Brain-Inspired Decision Intelligence for Education.*" The chapter examines how fundamental cognitive processes, including attention, memory, reasoning, judgment, and metacognitive control, can be integrated into artificial intelligence systems to support more transparent, accountable, and human-centered educational decision-making. Drawing upon insights from cognitive science, neuroscience, educational governance, and artificial intelligence, the author proposes a conceptual framework that redefines AI as a socio-technical decision support system operating under uncertainty, ethical constraints, and human oversight rather than as a purely predictive technology. Through the integration of retrieval-based architectures, neuro-symbolic reasoning, causal inference, and collaborative human–AI decision paradigms, the chapter illustrates how cognitively aligned AI can strengthen educational governance while ensuring procedural fairness, contestability, and responsible accountability. Particular attention is devoted to the role of calibrated uncertainty, selective prediction, and metacognitive regulation in supporting trustworthy decisions that respect learners' rights and minimize algorithmic bias.

The growing demand for transparency and interpretability in artificial intelligence is explored by Firdaus Alamsyah, the author of the eighth chapter, *"Explainable and Interpretable Decision Intelligence for Human-Centered AI Systems: Bridging Cognitive Computing and Transparent AI Through Brain-Inspired Mechanisms."* The chapter examines how brain-inspired cognitive mechanisms can strengthen explainable artificial intelligence by aligning computational models with the ways humans perceive, interpret, and reason about information. Drawing upon insights from cognitive neuroscience, cognitive psychology, and explainable AI, the author demonstrates that transparency should be viewed not merely as a technical feature but as a cognitive process that enables users to understand, evaluate, and confidently act upon AI-generated recommendations. Special consideration is devoted to explainability techniques, including attention mechanisms, feature attribution, counterfactual explanations, and neuro-symbolic architectures, illustrating how these approaches enhance both model interpretability and decision quality. By integrating neural learning with symbolic reasoning, the chapter highlights the potential of hybrid intelligent systems to overcome the limitations of conventional black-box models while preserving analytical performance. The discussion also emphasizes the ethical and organizational dimensions of explainable decision intelligence, underscoring the importance of fairness, accountability, trust, and responsible AI governance in high-stakes application domains. Through an interdisciplinary perspective, the author argues that effective human–AI collaboration depends on explanations that are cognitively meaningful and context-sensitive rather than solely algorithmically transparent.

A compelling vision of the evolution of artificial intelligence from algorithmic computation toward cognitively inspired intelligence is presented by Salam Al-E'mari, Yousef Sanjalawe, and Fadi Fataftah, the authors of the ninth chapter, *"From Algorithms to Cognition: Brain-Inspired Mechanisms for AI-Enhanced Decision Intelligence."* The contribution challenges the conventional view of AI as a purely data-driven predictive technology, proposing an alternative perspective in which intelligent systems are designed around the cognitive processes that characterize human perception, attention, memory, learning, and reasoning. Integrating knowledge from neuroscience, cognitive science, and neuro-symbolic artificial intelligence, the authors explain how these mechanisms can inspire computational architectures that exhibit greater adaptability, contextual awareness, and interpretability than traditional machine learning models. Particular consideration is given to the convergence of statistical learning and cognitive reasoning, illustrating how this integration enables AI systems to cope more effectively with uncertainty while producing decisions that remain transparent and meaningful to human users. The discussion also addresses the broader implications of cognitively inspired intelligence for trust, accountability, and responsible AI development, emphasizing that technological progress must evolve alongside ethical reflection and human oversight. Viewed from this perspective, decision intelligence emerges not as the automation of human reasoning but as its extension through computational systems capable of supporting informed judgment in increasingly complex environments.

The pursuit of artificial intelligence that can reason, adapt, and interact in ways that reflect human cognition lies at the center of the contribution by Rubee Singh, the author of the tenth chapter, *"Bridging Human Thought and Machine Intelligence: Cognitive-Aligned Neuro-Symbolic AI Systems and Environment."* Challenging the long-standing separation between data-driven learning and symbolic reasoning, the chapter presents neuro-symbolic intelligence as a promising direction for developing AI systems capable of combining computational efficiency with human-like understanding. The discussion argues that meaningful progress in artificial intelligence depends not solely on increasingly powerful algorithms but on architectures that mirror fundamental cognitive processes, including contextual interpretation, memory integration, knowledge representation, and reflective reasoning. Such an approach enables intelligent systems to operate more effectively in environments characterized by uncertainty, incomplete information, and evolving objectives while maintaining transparency in the reasoning process. The author further emphasizes that cognitive alignment provides the foundation for more natural and productive collaboration between humans and intelligent machines, allowing AI to function as a trusted decision partner rather than an autonomous replacement for human expertise. Framed within the broader evolution of cognitive computing and responsible artificial intelligence, the chapter offers a compelling perspective on how neuro-symbolic architectures can foster explainability, contextual awareness, and ethical decision support. In doing so, it reinforces the growing recognition that the future of decision intelligence will be shaped by systems capable of integrating the strengths of both human cognition and computational reasoning into a unified, trustworthy framework.

Exploring the complex relationship between neurobehavioral science and artificial intelligence, Garima Gupta, Nidhi Tripathi, and Richa Mishra present an original perspective in the eleventh chapter, *"Modeling Drug-Induced Cognitive Biases in Neurobehavioural Decision-Making: Implications for Brain-Inspired AI Architectures and Bias Mitigation in Decision Intelligence."* The contribution investigates how cognitive distortions associated with substance exposure can provide valuable insights into the design of more resilient and cognitively informed AI systems. Building on evidence from addiction neuroscience, computational psychiatry, and behavioral psychology, the authors reveal striking parallels between human neurobehavioral biases and the vulnerabilities observed in contemporary intelligent systems, including reward optimization failures, sensitivity to misleading stimuli, and impaired decision-making under uncertainty. These observations serve as the foundation for a novel conceptual framework in which maladaptive cognitive processes are transformed into computational models capable of anticipating, detecting, and mitigating similar biases in artificial intelligence. The chapter broadens the discussion beyond technical innovation by emphasizing the importance of integrating neuroscientific evidence into the development of trustworthy decision intelligence capable of adapting to dynamic and unpredictable environments. Through this interdisciplinary synthesis, the authors demonstrate that understanding the mechanisms underlying human cognitive dysfunction not only advances neuroscience but also offers valuable guidance for creating AI architectures that exhibit greater robustness, interpretability, and resilience.

Inspired by recent advances in computational neuroscience, Rajkumar Jagdale and Kundan Bedmutha introduce an innovative perspective in the twelfth chapter, *"Artificial Glia-Inspired Stabilization for Reliable Deep Reinforcement Learning Decision Intelligence."* Departing from conventional neuron-centered approaches to artificial intelligence, the authors investigate the largely overlooked regulatory functions of glial cells as a source of inspiration for improving the stability and reliability of deep reinforcement learning systems. The proposed framework incorporates biologically motivated mechanisms capable of continuously monitoring learning dynamics, preserving homeostatic balance, and enforcing operational constraints throughout both training and deployment. This perspective addresses one of the most persistent limitations of reinforcement learning—its sensitivity to unstable environments and unpredictable behavioral fluctuations—while promoting safer and more dependable decision-making in complex real-world scenarios. Through applications spanning autonomous systems, financial optimization, and healthcare decision support, the chapter demonstrates how biologically inspired regulation can significantly enhance learning efficiency, reduce performance variability, and strengthen compliance with safety requirements.

Establishing meaningful connections between biological cognition and computational intelligence, Valentin Marius Diblă offers a thought-provoking contribution in the thirteenth chapter, *"From Neuron to Algorithm: Shared Principles in the Architecture of Biological and Artificial Decision-Making."* Through a comparative analysis of distributed processing, attentional control, memory organization, prediction error, dual-process reasoning, and uncertainty estimation, the author demonstrates that many of the mechanisms driving intelligent behavior transcend disciplinary boundaries. These shared principles are examined as design constraints capable of guiding the development of AI systems that are not only computationally efficient but also more adaptive, interpretable, and cognitively coherent. Integrating perspectives from computational neuroscience and machine learning, the chapter proposes a conceptual foundation for designing intelligent systems that balance autonomous learning with calibrated confidence and context-sensitive reasoning. By emphasizing transferable cognitive principles instead of biological imitation, this contribution broadens the theoretical landscape of decision intelligence and provides valuable guidance for researchers seeking to develop next-generation AI architectures that combine scientific rigor with practical relevance and human-centered design.

Collectively, the chapters demonstrate that the future of decision intelligence depends not only on increasing computational performance but also on understanding and incorporating the cognitive processes that characterize human perception, learning, reasoning, memory, attention, and adaptive behavior. In this context, brain-inspired artificial intelligence emerges not as an attempt to replicate biological intelligence, but as a framework for designing systems that are more interpretable, resilient, trustworthy, and capable of meaningful collaboration with human decision-makers.

The topics explored throughout the book also resonate with the broader international discourse surrounding trustworthy and responsible artificial intelligence. As governments, international organizations, and research communities continue to establish principles for ethical AI, explainability, transparency, accountability, and human oversight, the contributions presented here provide valuable conceptual and methodological foundations for aligning technological innovation with societal expectations. Across diverse application domains—including healthcare, education, autonomous systems, organizational management, and intelligent decision support—the authors emphasize that the long-term success of AI will depend not only on algorithmic sophistication but also on its ability to operate within ethical, cognitive, and institutional frameworks that promote trust, fairness, and responsible decision-making. Taken together, the ideas presented throughout this volume advance the development of a new generation of trustworthy and human-centered decision intelligence capable of addressing the growing complexity of contemporary technological and societal challenges.

Ionica Oncioiu
Titu Maiorescu University, Romania

Acknowledgment

Looking back, we wish to thank all the persons who made the publishing of the present volume possible: Jan Travers, Melissa Wagner, Erin Wesser, Christina Henning, Nicole Elliott, Kristina Byrne, Lindsay Johnston, Joshua Witman, Nick Newcomer, Jaimie Watts, Josephine Dadeboe, Jordan Tepper, Sean Eckman, Elizabeth Barrantes, Katie McLoughlin and many other IGI – Global team members, reviewers, collaborators, and all the contributing authors: without your efforts and dedication this editorial project would have never been possible.

Chapter 1
Designing Human-Centered and Cognitively Aligned Artificial Intelligence Systems for Ethical and Effective Decision-Making

Jyoti Kumari
http://orcid.org/0000-0001-9648-2494
GD Goenka University, India

Pradeep Yadav
http://orcid.org/0000-0003-1073-7076
Institute of Technology and Management, Gwalior, India

Chandra Prakash Bhargava
http://orcid.org/0009-0004-4751-9149
Institute of Technology and Management, Gwalior, India

ABSTRACT

The rapid adoption of artificial intelligence (AI) in critical domains such as healthcare, finance, criminal justice, and autonomous systems highlights the need for cognitively aligned and human-centered AI design. This chapter presents an integrative framework combining cognitive science, human–computer interaction, AI ethics, and decision theory to develop AI systems that effectively support human judgment. It explores cognitive compatibility, cognitive load optimization, transpar-

DOI: 10.4018/979-8-3373-8988-2.ch001

ency, fairness, and collaborative intelligence using recent research and case studies (2021–2025). Five design principles and four analytical tables are introduced to evaluate cognitive alignment, ethical considerations, decision support approaches, and implementation readiness. The chapter concludes with future directions in explainability, trust calibration, and AI governance, offering practical guidance for building AI systems that enhance rather than replace human decision-making.

1. INTRODUCTION

AI systems are not only being utilized in research labs and industry niches alone but also in decision-making regarding parole, diagnosis, loan approvals, and military engagements (Floridi et al., 2022; Mittelstadt et al., 2023). However, the rapid diffusion of these AI systems has exceeded the development of design principles to guide the manner in which they work such that they are cognitively and ethically sound and truly beneficial to the human beings who rely on them. In response to this realization, there has emerged a body of scholars from various disciplines including computer science, philosophy, psychology, law, and management who have focused their efforts on human-centered AI (HCAI) as an overarching framework (Shneiderman, 2022; Xu, 2019).

Human-centered AI does not refer to one method, but rather to a way of designing systems where human actions, understanding, and well-being play central roles. As opposed to methods that focus on accuracy or speed, HCAI puts human-centric factors such as decision making, trust, cognitive burden, and moral responsibility in the center of design evaluations (Amershi et al., 2019; Cai et al., 2022). Human-centered AI design, however, is challenging due to the diversity of human cognition. There are differences between people in terms of expertise, culture, willingness to take risks, and cognitive styles that influence interactions with algorithmic suggestions (Kahneman et al., 2021; Wang et al., 2023).

One of the underexplored factors of HCAI relates to the extent of cognitive alignment – how well does the output generated by the AI system align with the actual mental models and heuristics used by humans during decision-making (Bansal et al., 2021; Steyvers et al., 2022)? In contrast to usability, a system could be user-friendly, but at the same time generate output that will be at odds with human epistemology, induce automation bias, or foster overconfidence (Goddard et al., 2022; Lyell & Coiera, 2017). If cognitive alignment is not achieved, even state-of-the-art AI will perform worse than an unbiased human decision maker.

The ethical aspect brings yet another layer of complexity. The ethical design of AI entails such values as fairness, accountability, transparency, privacy, and non-maleficence, values that need to be considered as intrinsic features of AI rather than being treated as ex-post checks (Jobin et al., 2019; Morley et al., 2021). Research done in recent years has shown that there are constant tensions in achieving good performance while maintaining fairness, high levels of interpretability amid model complexity, and the preservation of user autonomy against paternalist intervention (Doshi-Velez & Kim, 2017; Green & Chen, 2021).

This chapter responds to these challenges by synthesizing extant theory and empirical evidence into a cohesive framework for designing human-centered, cognitively aligned AI systems for ethical and effective decision-making. The chapter is organized as follows: Section 2 establishes theoretical foundations through dual-process theory and distributed cognition. Section 3 examines five core design principles. Section 4 presents tabular frameworks for practical implementation. Section 5 reviews empirical evidence from applied domains. Section 6 discusses emerging challenges, and Section 7 proposes a forward-looking research agenda before the conclusion in Section 8.

2. THEORETICAL FOUNDATIONS

2.1 Dual-Process Theory and AI Decision Support

The two modes of human reasoning recognized by cognitive psychologists are quick associative reasoning, known as System 1, and slow, rational deliberation referred to as System 2 (Evans, 2008; Kahneman, 2011). Interaction with AI systems takes place in a different way for each mode, with failure to recognize the relationship generating certain patterns of decision-making problems. Recommendation systems, for example, that present their outcomes as deterministic values lead to reliance on System 1 heuristics despite situations that demand System 2 thinking (Lyell & Coiera, 2017). At the same time, overly complicated explanations will lead to decision fatigue since System 2 capacities are exhausted (Schemmer et al., 2023).

For AI to be cognitively aligned, its interface must adjust to the appropriate cognitive mode depending on the decision-making process it participates in. For example, bansal et al. (2021) show that human-AI teams become more accurate in image recognition processes once the AI signal of confidence is correctly adjusted, as this encourages switching from one system to another. Furthermore, Steyvers et al. (2022) demonstrate how Bayesian models of complementarity may be used to predict whether a person needs to override or rely on an AI recommendation.

2.2 Distributed Cognition and Sociotechnical Systems

Not only does cognition depend on individual psychology but human cognition also entails social networking, material devices, and institutional structures (Hutchins, 1995). When AI is integrated into institutional practices, it forms a part of a sociotechnical system, in which the emergent characteristics cannot be understood based on individual components (Baxter & Sommerville, 2011; Passi & Barocas, 2019). The decision made by a radiologist, for example, depends not only on the annotation produced by the AI but also on the departmental culture, time pressure caused by the level of staffing, and the legal repercussions of going against machine recommendations (Rajpurkar et al., 2022).

Since the distributed cognition theory assumes that successful AI design should take into consideration all aspects of cognitive activity rather than solely focusing on the interaction between people and AI, it suggests that more holistic methods such as ethnography and participatory design should be used (Barocas et al., 2021; Cai et al., 2022). Moreover, the assessment criteria should include measures of team-based decision-making performance instead of merely individual accuracy (Lai et al., 2023).

2.3 Ethical Frameworks for AI Design

Ethics have a great impact on the development of an AI. There are three types of ethical theories that shape this discussion: principlism, which focuses on such principles as autonomy, beneficence, non-maleficence, and justice; virtue theory, which takes into account the moral dispositions of AI designers and operators; and contractualism, whose principles are based on those ideas which can be accepted by all involved parties (Floridi et al., 2022; Mittelstadt, 2019). The approaches also put an emphasis on various issues of design and implementation.

None of the frameworks by themselves is adequate. Jobin et al. (2019), in their content analysis of 84 guidelines for ethics in AI, found wide agreement on five core principles—transparency, justice/fairness, non-maleficence, responsibility, and privacy—with significant disagreement on how they could be operationalized. Morley et al. (2021) subsequently observed an 'implementation gap' between high-level principles and practical implementation. Bridging this gap calls for mid-level design principles to mediate values into concrete designs, which this chapter attempts in Section 3.

3. CORE DESIGN PRINCIPLES FOR HUMAN-CENTERED, COGNITIVELY ALIGNED AI

3.1 Principle 1: Adaptive Transparency and Explainability

Transparency is the most commonly cited AI ethics principle, yet implementation is highly controversial (Doshi-Velez & Kim, 2017; Rudin, 2019). A key distinction lies between intrinsic interpretability, which entails creating interpretable architecture through models whose processes of reasoning are transparent (e.g., decision tree algorithms, linear models, rule-based systems), and post-hoc explainability, which entails applying explanation algorithms to black-box models (e.g., SHAP values, LIME, attention mechanism visualization) (Molnar, 2022). Neither approach can be deemed universally preferable; the choice hinges upon considerations of accuracy, risk level, and skill set.

Building upon this concept, adaptive transparency emphasizes that the depth and form of explanation will be tailored to context, complexity of the task, and importance of the decision (Wang et al., 2023). A senior cardiologist assessing the validity of an AI arrhythmia detection algorithm will need different explanation depth and form compared to an emergency department nurse who must triage patients. Lundberg et al. (2020) show that SHAP-based explanations enhance mortality prediction in the intensive care unit setting when clinicians see predictions accompanied by visualized feature importance, yet their utility depends upon medical training and available time. Military analysts, for instance, respond to counterfactual and feature-attribution explanations differently in Gunning et al. (2021).

Adaptive transparency must have multi-layered explanation architectures that give explanations at summary level by default and let users with expertise explore the algorithmic details as needed. Adaptive transparency must also be evaluated through rigorous empirical user studies that determine whether the explanations help calibrate decision making because users tend to think that they understand AI explanations better than they actually do (Chromik et al., 2021; Poursabzi-Sangdeh et al., 2021).

3.2 Principle 2: Fairness by Design Across Protected Attributes

The issue of algorithmic fairness has become a prominent topic in light of scandals involving risk assessment in criminal justice, facial recognition, and health care decision-making (Angwin et al., 2016; Obermeyer et al., 2019; Raji et al., 2022). The research in this area has led to a large number of fairness definitions in mathematics, including demographic parity, equalized odds, individual fairness, and counterfactual fairness, each of which reflects different concepts of equality

(Barocas et al., 2023; Chouldechova & Roth, 2020). What is important here is that some of these definitions contradict each other.

"By Design" fairness entails incorporating the notion of fairness into each and every step of the machine learning pipeline – from collecting data (to prevent gaps and biases) to designing features (not using proxies for discriminating against people), choosing a model (optimizing accuracy-fairness balance), and deploying a system (detecting any distribution shifts), (Pessach & Shmueli, 2022). However, according to the recent findings published by Dwork et al. (2023), such fairness solutions should not be limited only to one step in the pipeline but should apply to all its stages.

Fairness through design also involves more structural methods that include using diverse development teams, engaging affected communities in determining what constitutes fairness within a specific situation, and having accountability procedures to create negative sanctions for any acts of unfairness (Green, 2022; Madaio et al., 2020). Participatory research into AI technologies has shown that affected communities sometimes bring up issues of fairness that may have otherwise gone unnoticed through purely technical efforts (Sloane et al., 2020).

3.3 Principle 3: Trust Calibration and Appropriate Reliance

There are some ambiguities regarding trust in the use of AI systems for decision making: a lack of trust results in a neglect of automation, whereas an excessive degree of trust leads to automation bias (Parasuraman & Riley, 1997; Goddard et al., 2022). However, a person-centered AI approach seeks not to maximize trust but rather find the right balance between it and the reliability of systems in all possible situations (Lee & See, 2004; Zhang et al., 2020).

Communication of calibrated trust requires that the AI system report its uncertainty in a manner that is both truthful and actionable for the user. Probabilistic outputs, confidence intervals, and out-of-distribution signals can accomplish this purpose when properly implemented (Kull et al., 2019; Ovadia et al., 2019). Yet Steyvers et al. (2022) note that raw probabilistic outputs tend to be misinterpreted by ordinary users, calling for uncertainty communication studies that consider various numeracy levels and mental models of probability. In low-numeracy environments, Kulkarni et al. (2023) discover that natural-language statements of uncertainty ("the prediction is not to be trusted due to the patient being dissimilar from those in the training dataset") perform better than numeric confidence ratings.

In addition, the phenomenon of trust calibration requires a longitudinal focus on trust dynamics. While initial trust can be influenced by factors such as reputation, visual design, and advertising, subsequent trust is based on experience; however, it tends to be asymmetrical, whereby one failure experience will damage trust much more than several successful experiences (Hoff & Bashir, 2015). Trust repair features are thus essential in designing AI systems meant for high-stake domains.

3.4 Principle 4: Human Agency Preservation and Meaningful Control

Preservation of human agency implies that AI ought to enhance but not supplant human decision-making, especially when the decisions made may have significant consequences and where the responsibility cannot be delegated to a machine process (European Commission, 2021; Shneiderman, 2022). The notion can be understood both normatively and empirically. Normatively, it involves values of human dignity and democracy that cannot be captured in cost-benefit analysis or any other quantitative assessment of aggregate welfare. Empirically, it involves evidence that humans and machines working together do better than each on its own (Lai et al., 2023; Vaccaro et al., 2019).

Effective control goes beyond mere option to opt out. It entails the development of architectures of interaction that allow AI recommendations to be truly deferable, not just theoretically overrideable but actually deferred without social or institutional sanction (Morley et al., 2021; Rader et al., 2018). Coercive automation a system that technically allows an override but functionally discourages it from being done via the design of its workflow constitutes a violation of effective control despite no apparent limitation on action. Studies conducted in health care have illustrated the use of coercive dynamics in electronic health record systems (Moy et al., 2021).

Some interventions to ensure preservation of agency include: giving recommendations by the AI after human reflection, providing an interactive tool for counterfactual reasoning, building decision interfaces that facilitate easy and clear benefits in overriding recommendations, and cultivating professional cultures where overriding the recommendations of AI is valued (Dietvorst et al., 2018; Green & Chen, 2021).

3.5 Principle 5: Continuous Monitoring, Feedback, and Adaptive Governance

AI models used in practical settings where change is occurring experience distributional shift, the alteration in input data statistics that causes the performance of AI models to deteriorate (Gama et al., 2014; Rabanser et al., 2019). For example, an AI system used in healthcare and developed using pre-COVID-19 data was found to perform poorly because of new symptom presentation and treatment protocols (Roberts et al., 2021). In financial AI systems, distributional shifts happen as a result of changes in the economic regime, regulatory regimes, and macroeconomic factors.

The criminal justice system also experiences distributional shift due to changes in police practice and demographic trends.

Continuous monitoring involves the use of technology that measures not only technical data (calibration, coverage, drift) but also decision data (quality of decisions, rate of overriding of decisions, consequences for those affected) (Sculley et al., 2015; Saria & Subbaswamy, 2019). It is necessary to create feedback loops from the systems being implemented back to the team developing them. Adaptive governance, based on regulatory practices of the aviation and nuclear industries, can help develop processes of integrating monitoring data into system refinement (Cath, 2018; Dafoe, 2018).

4. TABULAR FRAMEWORKS FOR IMPLEMENTATION

Table 1. Cognitive alignment design criteria for human-centered AI systems

Cognitive Dimension	Design Criterion	Implementation Mechanism	Evaluation Metric
System 1 Compatibility	Reduce cognitive friction for routine decisions; fast, heuristic-compatible outputs	Default recommendations, color-coded risk signals, one-click acceptance	Task completion time; error rate under time pressure
System 2 Engagement	Scaffold analytic deliberation for novel or high-stakes decisions	Expandable explanation panels; structured comparison views; counterfactual tools	Decision justification quality; override rate accuracy
Mental Model Coherence	Ensure AI behavior conforms to user domain mental models	Participatory design workshops; think-aloud protocol testing	Model accuracy score; perceived predictability rating
Cognitive Load Management	Minimize extraneous cognitive load; optimize germane load	Progressive disclosure interfaces; summary + detail hierarchy	NASA-TLX cognitive load score; recall accuracy post-task
Uncertainty Communication	Represent AI confidence in forms accessible to target users	Natural language confidence expressions; visual confidence bands	Calibration error; Brier score; user confidence-accuracy correlation
Expertise Adaptivity	Modulate explanation depth and interface complexity to user expertise	User profiling; adaptive explanation layers; expert/ novice toggle	Cross-expertise decision accuracy gap; user satisfaction by expertise level

Note. System 1 and System 2 terminology follows Kahneman (2011). NASA-TLX = NASA Task Load Index. Calibration error refers to the difference between predicted probability and observed frequency.

Table 2. Ethical evaluation dimensions and operationalization strategies

Ethical Principle	Definition	Operationalization	Potential Trade-offs
Fairness	Equitable treatment across demographic groups; absence of unjustified discriminatory outcomes	Demographic parity, equalized odds, counterfactual fairness audits; third-party bias assessments	May conflict with accuracy maximization; multiple fairness criteria are mutually incompatible
Transparency	Openness about system capabilities, limitations, training data, and decision logic	Model cards, data sheets, SHAP explanations, natural language rationales	Detailed transparency may reveal proprietary methods or enable gaming of the system
Accountability	Clear assignment of moral and legal responsibility for AI decisions and their consequences	Audit trails, human-in-the-loop verification, designated accountability roles, appeals processes	Distributed accountability can diffuse responsibility; 'many hands' problem
Non-maleficence	Active prevention of harm to individuals, communities, and broader society	Adversarial robustness testing, red-team exercises, impact assessments, kill-switch protocols	Overly conservative harm prevention may reduce beneficial utility; risk aversion asymmetry
Privacy	Protection of personal data; right to informational self-determination	Differential privacy, federated learning, data minimization, GDPR compliance mechanisms	Privacy constraints may reduce model accuracy; data minimization limits fairness auditing
Autonomy	Preservation of human agency and resistance to manipulation or coercive automation	Opt-out mechanisms, contestability interfaces, override logging, choice architecture review	User autonomy may permit harmful choices; paternalistic safeguards reduce autonomy

Note. SHAP = SHapley Additive exPlanations. GDPR = General Data Protection Regulation. Trade-off analysis follows Jobin et al. (2019) and Floridi et al. (2022).

Table 3. Decision support modalities and cognitive alignment profiles

Modality	Best-Fit Decision Context	Cognitive Alignment Strengths	Cognitive Alignment Risks	Example Application
Predictive Scoring	Routine, high-volume, well-defined outcome tasks	Low cognitive load; fast heuristic integration; compatible with System 1	Automation bias; anchoring; false precision illusion	Sepsis early-warning systems; credit scoring
Ranked Recommendations	Semi-structured choices with multiple viable options	Preserves choice; reduces option overload; prompts comparison	Order effects; status quo bias toward top-ranked option	Clinical treatment selection; e-commerce personalization
Counterfactual Explanations	Contested or high-stakes decisions requiring justification	Supports System 2 deliberation; reveals actionable changes; facilitates contestation	Cognitive overload in time-constrained settings; may increase perceived complexity	Loan denial explanations; recidivism risk contestation
Interactive Simulation	Strategic planning; scenario exploration; policy design	Supports what-if reasoning; builds accurate mental models; promotes learning	Time-intensive; may generate overconfidence in simulated outcomes	Pandemic modeling dashboards; supply chain disruption planning
Ambient Monitoring Alerts	Continuous surveillance with periodic exception flagging	Preserves attentional resources; flags genuine anomalies efficiently	Alert fatigue; desensitization; missed alarms after false positive bursts	ICU vital sign monitoring; cybersecurity intrusion detection
Collaborative Dialogue Agents	Complex, unstructured, or emotionally charged decisions	Contextual adaptation; natural language; iterative clarification	Anthropomorphism; inappropriate trust; inconsistency across sessions	Mental health triage; legal aid access; complex patient counseling

Note. ICU = intensive care unit. System 1 and System 2 references follow Evans (2008) and Kahneman (2011). Alignment risk profiles synthesize evidence from Goddard et al. (2022), Schemmer et al. (2023), and Lyell and Coiera (2017).

Table 4. Implementation readiness indicators for human-centered AI deployment

Readiness Domain	Indicator Category	Assessment Criteria	Maturity Level Descriptors
Technical Readiness	Model robustness	Out-of-distribution performance; adversarial stress testing; uncertainty calibration	L1: Prototype only I L2: Validated in controlled setting I L3: Validated in deployment context
Technical Readiness	Monitoring infrastructure	Real-time drift detection; automated retraining triggers; performance dashboards	L1: Manual monitoring I L2: Semi-automated alerts I L3: Continuous automated governance
Organizational Readiness	Staff competency	AI literacy training completion; workflow integration protocols; override documentation practices	L1: Awareness only I L2: Procedural competency I L3: Critical evaluative competency
Organizational Readiness	Governance structures	Designated AI accountability roles; ethics review processes; incident response procedures	L1: Ad hoc I L2: Formalized but reactive I L3: Proactive with stakeholder participation
Ethical Readiness	Fairness auditing	Pre-deployment bias assessment; ongoing disparity monitoring; affected community consultation	L1: Internal review only I L2: Third-party audit I L3: Participatory community governance
Ethical Readiness	Transparency artifacts	Model cards completed; data sheets documented; user-facing explanations tested with target users	L1: Internal documentation I L2: Public disclosure I L3: Interactive transparency tools
User Readiness	Trust calibration	Baseline trust assessment; trust trajectory monitoring; automation bias testing	L1: No assessment I L2: Pre-deployment survey I L3: Longitudinal trust monitoring
User Readiness	Contestability mechanisms	Appeal process availability; override workflow usability testing; feedback channel responsiveness	L1: Informal only I L2: Structured appeal process I L3: Rapid-response contestation system

Note. Maturity levels (L1–L3) adapted from the NIST AI Risk Management Framework (National Institute of Standards and Technology, 2023) and the EU AI Act risk tier classifications (European Commission, 2021). L = Level.

5. EMPIRICAL EVIDENCE FROM APPLIED DOMAINS

5.1 Healthcare: Clinical Decision Support Systems

CDSS arguably constitute one of the most well-researched applications of human-centered AI technology. Notably, a systematic study conducted by Sutton et al. (2020), which analyzed 55 research articles involving AI-powered CDSS, determined that all CDSS that incorporated clinical experts into the design process demonstrated better performance results than those without this involvement. In a more recent article, Rajpurkar et al. (2022) outlined an existing 'chasm' between the effectiveness of AI diagnostics in benchmark scenarios and clinical practice, and explained this discrepancy by AI's incompatibility with the incomplete and inconsistent nature of clinical data.

The example of research on sepsis early warning algorithms helps to demonstrate both the opportunities and challenges of using CDSSs. According to Wong et al., the performance of Epic's sepsis prediction algorithm varied widely from one population to another, emphasizing the need for local validation prior to implementation. Importantly, the response rate of clinicians to sepsis alerts was negatively related to the number of alerts received, which aligns with alert fatigue theory derived from the automation trust literature. Bedoya et al. have since proven that introducing transparency by supplying clinicians with the rationale behind the alerts increases alert adherence rates without raising false positives.

Challenges exist in the use of artificial intelligence in mental well-being. Jacobson & Bhatt (2022) explore AI-assisted tests for depression and find that algorithms developed based mainly on data from Western, English-speaking communities perform poorly in other cultures, a situation that can be categorized as unfair algorithms due to the sensitive nature of such diagnoses. These scholars recommend using federated learning designs that provide room for fine-tuning local models without compromising data privacy while undertaking validity testing.

5.2 Criminal Justice: Algorithmic Risk Assessment

The AI use within the criminal justice system has come under serious criticism after journalistic investigations uncovered the existence of racial biases in the algorithms used in tools such as COMPAS, which make up part of parole and sentencing decisions in more than one state in the United States (Angwin et al., 2016). The findings of Dressel and Farid (2018) through academic research showed that simple aggregation of lay assessments resulted in outcomes equal to, if not better than, those of COMPAS while minimizing the number of false positives for Black individuals in contrast to the assumption that algorithms would always surpass human assessments.

The case of criminal justice is another example of how mere technical solutions are not enough to address the problem of fairness. For instance, as Mehrabi et al. (2021) show, ensuring demographic parity in the risk assessment algorithm by making changes in one feature while keeping all others unchanged may lead to higher levels of unfairness towards underrepresented communities if the target of prediction was itself biased throughout history. This case highlights the issue of feedback loop bias where AI algorithms learn from historical data and perpetuate existing biases.

5.3 Financial Services: Lending and Credit Decisions

Credit underwriting through artificial intelligence has benefits in terms of efficiency, although there are major concerns regarding its effects on equality because the historical records of creditworthiness continue to fuel the racial gap between different communities (Bartlett et al., 2022). According to Fuster et al. (2022), mortgage loan lending is assessed using machine learning models and shows an increase in the approval of minority groups while demonstrating a relatively high prediction error for minority applicants' defaults. On the one hand, the United States' Consumer Financial Protection Bureau has provided guidelines on the applicability of adverse action explanations in the context of algorithmic lending decisions. On the other hand, European Banking Authority released stress-testing guidelines.

Even the field of lending provides another example of trust calibration issues within consumer-based AI technologies. Logg et al. (2019) have shown that consumer preference for algorithms versus humans depends on the context in which they operate, as well as previous exposure to algorithmic errors. People who are exposed to a prominent instance of algorithmic error decrease their algorithmic trust even in cases where the algorithm performs better than any human adviser – a phenomenon that Dietvorst et al. (2018) describe as "algorithm aversion."

5.4 Autonomous Systems: Human-Robot Teaming

The body of research on human-robot interaction offers empirical verification of the principles of cognitive alignment mentioned in Section 3. Specifically, Endsley (2021) further develops the basic tenets of situation awareness and applies them to human-robot teams, suggesting that autonomous agents must have shared situational awareness with humans in order to be supervised and intervene appropriately. Chen et al. (2022) also show that operators of autonomous unmanned vehicles are able to make better decisions about interventions when robots provide feedback on their uncertainties regarding objects and actions a finding that aligns with the trust calibration principle.

There have also been several studies that emphasize the significance of timing while interacting within human-robot teams. According to Neerincx et al. (2023), a system that interrupts human activities during suboptimal times adds coordination costs and cancels out any benefits of automation due to interruption management algorithms which are not responsive to human workload. In defense, manufacturing, and exploration scenarios, it has been consistently found that for effective human-robot teams to be created, there should be two-way communication, and interfaces that portray robot suggestions as commands lower decision quality of the team.

6. EMERGING CHALLENGES IN HUMAN-CENTERED AI DESIGN

6.1 Generative AI and the Explainability Frontier

The development of LLMs and multimodal generative AI presents a qualitative shift in terms of challenges for human-centric AI design. This is because such AI systems as GPT-4, Claude, Gemini, and their future iterations are capable of producing coherent outputs using processes that are not yet fully comprehended by existing explainability tools (Bommasani et al., 2022; Wei et al., 2022). In particular, it was found that attention visualization – a popular explanation method for transformer architectures – may not be reliable for approximating feature importance in decision-making (Jain & Wallace, 2019). This phenomenon can be seen as an 'explanation faithfulness' issue, as suggested by Doshi-Velez and Kim (2017).

This is especially true when considering the use of LLMs in critical decision-making contexts, where they act as advisors to help in legal research, provide information on medical topics, and offer financial advice. Those individuals receiving information from an LLM that is confidently provided but not actually correct would be more disadvantaged compared to those using narrow AI technologies, as the fluency offered by the former can indicate a level of expertise that is absent from the latter (Bender et al., 2021).

6.2 Cross-Cultural Cognitive Alignment

The dominance of Western and English-speaking AI research contexts poses risks of systemic problems associated with its use in cross-cultural settings. The literature on cognitive psychology has shown considerable differences in reasoning methods, risk taking, respect for authorities, and uncertainty handling across cultures, which influences how people perceive the interaction with AI systems providing recommendations (Henrich et al., 2010; Nisbett & Masuda, 2003). On the other hand, the overwhelming majority of user studies related to HCI and transparency have been performed on the WEIRD population (Linxen et al., 2021).

Misalignment between cultures in terms of cognition could be observed in various forms such as a preference for narratives or statistics in explanation styles, varying initial levels of disposition of trust toward algorithms as compared to humans, different ease levels in expressing probabilities, and dissimilar expectations regarding the controllability of the system. Wang et al. (2023) show that there is a significant difference in preferences of explanation style in Chinese and American user communities with regards to same AI system, with Chinese users favoring comparative style and Americans favoring counterfactual style.

6.3 Regulatory Fragmentation and Global Governance

The regulatory environment around artificial intelligence has witnessed rapid changes and remains highly fragmented. With respect to regulation, the EU's AI Act, which was passed in 2024, puts forth a risk-based regime that has stringent requirements regarding AI applications that are considered to be high risk. High-risk applications include those that use biometric identification, are involved in critical infrastructural tasks, and engage in hiring procedures (European Commission, 2021; 2024). The US, on the other hand, has preferred a sector-specific strategy to govern AI through executive orders, government guidelines, and self-commitments from the private sector. China too has developed its own AI governance policy with an emphasis on recommendation algorithms and data sovereignty (Floridi et al., 2022; Raji et al., 2022).

In addition to compliance challenges, regulatory fragmentation also prompts the question of whether standard-setting is possible in the design of AI systems where different values and political ideologies prevail. Multistakeholder governance frameworks such as the Global Partnership on AI, the OECD AI Policy Observatory, and the ITU AI for Good platform have advanced efforts toward voluntary guidelines but have no binding capacity. In his paper published in 2018, Dafoe contends that successful AI governance will necessitate innovation in institutions at the global level similar to those governing nuclear technologies and aircraft safety.

7. RESEARCH AGENDA

As is evident from the synthesis offered in this chapter, several areas of study are important to consider as promising research avenues in the pursuit of human-centric and cognitively compatible AI development.

First, a significantly greater level of empirical rigor needs to be brought to the analysis of the long-term dynamics of trust in AI systems. Most studies into trust between humans and AIs have focused on cross-sectional or laboratory studies lacking ecological validity with respect to their applicability to practical use of AI tools in professional environments.

Second, the notion of cognitive alignment that exists at the intersection between different groups of users requires specific empirical examination. The current design framework relies on the implicit assumption that the group of users is quite similar, but in practice, applications cover wide variations of expertise, culture, disabilities, and technical experience. There is a need for research into adaptable interfaces that can respond to changing needs of users through modulation of information depth and uncertainty presentation.

Third, there is a need for standardized measurements in cognitive alignment assessment. The present approach involves using different outcome variables such as accuracy, reaction times, satisfaction, and cognitive load, which are hard to compare from one study to another. Having a well-tested measurement battery, similar to the one that exists in usability testing, would greatly speed up scientific advances.

Fourth, the ethical questions relating to AI-powered nudges deserve thorough scrutiny by philosophers and scientists. As the AI-based decision-making environment is created through default options, information presentation, and recommendation framing, it can be seen as an exertion of influence that contradicts the traditional notion of free consent. Further research into the limits of choice engineering vs. manipulation is imperative, especially as AI becomes increasingly conversant in its operation.

Fifth, cross-disciplinary collaborations among AI researchers, cognitive scientists, ethicists, legal experts, and representatives from impacted communities should become the norm rather than the exception. The current system of academic and business incentives is biased in favor of disciplinary expertise at the expense of cross-disciplinary engagement necessary for human-centered AI development. Systems that incentivize such collaboration should be created.

8. CONCLUSION

The creation of cognitive alignment between humans and intelligent systems is one of the central challenges of our contemporary era when artificial intelligence technology continues to advance into critical areas of application. The continued reliance on design paradigms that prioritize performance above all else and ignore important considerations such as cognitive alignment, ethical soundness, and control over AI is no longer tenable as AI technologies become even more powerful. This chapter has demonstrated that the cognitive alignment of an AI system with its user is a prerequisite to making full use of the possibilities that AI represents as an extension of human cognition.

The five key design considerations include adaptive transparency, fairness by design, trust calibration, preserving human agency, and continuous monitoring, thus offering an integrative perspective on implementing the aforementioned argument. The proposed table formats for criteria of cognitive alignment, ethical assessment perspectives, forms of decision support, and implementation readiness measures are intended to be useful to the practitioner as tangible tools that allow one to implement the suggested design principles. Examples from health care, criminal justice, finance, and robotics underscore the practical need for the framework and the importance of domain-specific considerations.

Moving forward, the introduction of generative AI, the broadening cross-cultural impact of AI implementation, and the shifting landscape of regulations will continue to influence how human-centered AI designers address emerging issues. Such efforts would require not only advancements in terms of explainability, fairness, and robustness but also collaboration among various fields of study, genuine involvement of communities, and institutional commitments to accountability. Ultimately, the goal should be an AI-driven world where humans' ability to think ethically, equitably, and rationally becomes stronger, as opposed to an AI-dominated world that replaces the unique human capability to reason.

REFERENCES

Amershi, S., Weld, D., Vorvoreanu, M., Fourney, A., Nushi, B., Collisson, P., Suh, J., Iqbal, S., Bennett, P. N., Inkpen, K., Teevan, J., Kikin-Gil, R., & Horvitz, E. (2019). Software engineering for machine learning: A case study. In *Proceedings of the 41st International Conference on Software Engineering: Software Engineering in Practice* (pp. 291–300). IEEE. DOI: 10.1109/ICSE-SEIP.2019.00042

Angwin, J., Larson, J., Mattu, S., & Kirchner, L. (2016). Machine bias: There's software used across the country to predict future criminals. And it's biased against blacks. ProPublica. https://www.propublica.org/article/machine-bias-risk-assessments-in-criminal-sentencing

Bansal, G., Nushi, B., Kamar, E., Lasecki, W., Weld, D., & Horvitz, E. (2021). Beyond accuracy: The role of mental models in human-AI team performance. *Proceedings of the AAAI Conference on Human Computation and Crowdsourcing*, 7, 2–11. DOI: 10.1609/hcomp.v7i1.5281

Barocas, S., Hardt, M., & Narayanan, A. (2023). *Fairness and machine learning: Limitations and opportunities*. MIT Press.

Barocas, S., Hood, S., & Ziewitz, M. (2021). Governing algorithms: A provocation piece. *SSRN*. DOI: 10.2139/ssrn.2245322

Bartlett, R., Morse, A., Stanton, R., & Wallace, N. (2022). Consumer-lending discrimination in the FinTech era. *Journal of Financial Economics*, *143*(1), 30–56. DOI: 10.1016/j.jfineco.2021.05.047

Baxter, G., & Sommerville, I. (2011). Socio-technical systems: From design methods to systems engineering. *Interacting with Computers*, *23*(1), 4–17. DOI: 10.1016/j.intcom.2010.07.003

Bedoya, A. D., Futoma, J., Clement, M. E., Corey, K., Brajer, N., Lin, A., Simons, M. G., Gao, M., Nichols, M., Balu, S., Henao, R., Sendak, M., & O'Brien, C. (2022). Machine learning for early detection of sepsis: An internal and temporal validation study. *JAMIA Open*, *5*(2), ooab094. Advance online publication. DOI: 10.1093/jamiaopen/ooab094 PMID: 32734166

Bender, E. M., Gebru, T., McMillan-Major, A., & Shmitchell, S. (2021). On the dangers of stochastic parrots: Can language models be too big? In *Proceedings of the 2021 ACM Conference on Fairness, Accountability, and Transparency* (pp. 610–623). DOI: 10.1145/3442188.3445922

Bommasani, R., Hudson, D. A., Adeli, E., Altman, R., Arora, S., von Arx, S., Bernstein, M. S., Bohg, J., Bosselut, A., Brunskill, E., Brynjolfsson, E., Buch, S., Card, D., Castellon, R., Chatterji, N. S., Chen, A. S., Creel, K., Davis, J. Q., Demszky, D., . . . Liang, P. (2022). On the opportunities and risks of foundation models. arXiv. https://doi.org//arXiv.2108.07258DOI: 10.48550

Cai, C. J., Winter, S., Steiner, D., Wilcox, L., & Terry, M. (2022). 'Hello AI': Uncovering the onboarding needs of medical practitioners for human-AI collaborative decision-making. Proceedings of the ACM on Human-Computer Interaction, 3(CSCW), 104. DOI: 10.1145/3359206

Cath, C. (2018). Governing artificial intelligence: Ethical, legal and technical opportunities and challenges. *Philosophical Transactions. Series A, Mathematical, Physical, and Engineering Sciences*, *376*(2133), 20180080. DOI: 10.1098/rsta.2018.0080 PMID: 30322996

Chen, J. Y. C., Barnes, M. J., Selkowitz, A. R., & Stowers, K. (2022). Effects of teammate trustworthiness information and reputation on human-autonomy teaming. *Frontiers in Psychology*, *13*, 854027. DOI: 10.3389/fpsyg.2022.854027

Chouldechova, A., & Roth, A. (2020). A snapshot of the frontiers of fairness in machine learning. *Communications of the ACM*, *63*(5), 82–89. DOI: 10.1145/3376898

Chromik, M., Eiband, M., Buchner, F., Krüger, A., & Butz, A. (2021). I think I understand, but do I? Measuring understanding of human-centered explanations. In *Proceedings of the 29th ACM Conference on User Interface Software and Technology* (pp. 20–34). DOI: 10.1145/3472749.3474743

Dafoe, A. (2018). AI governance: A research agenda. Future of Humanity Institute, University of Oxford.

Dietvorst, B. J., Logg, J. M., & Bharti, S. (2018). Overcoming algorithm aversion: People will use imperfect algorithms if they can (even slightly) modify them. *Management Science*, *64*(3), 1155–1170. DOI: 10.1287/mnsc.2016.2643

Doshi-Velez, F., & Kim, B. (2017). Towards a rigorous science of interpretable machine learning. arXiv. https://doi.org//arXiv.1702.08608DOI: 10.48550

Dressel, J., & Farid, H. (2018). The accuracy, fairness, and limits of predicting recidivism. *Science Advances*, *4*(1), eaao5580. Advance online publication. DOI: 10.1126/sciadv.aao5580 PMID: 29376122

Dwork, C., Ilvento, C., & Jagadeesan, M. (2023). Individual fairness in pipelines. In *Proceedings of the 2023 ACM Conference on Fairness, Accountability, and Transparency* (pp. 1228–1237). DOI: 10.1145/3593013.3594001

Endsley, M. R. (2021). Situation awareness in human-autonomy teaming. *Frontiers in Robotics and AI, 8*, 640043. DOI: 10.3389/frobt.2021.640043

European Commission. (2021). *Proposal for a regulation of the European Parliament and of the Council laying down harmonised rules on artificial intelligence (Artificial Intelligence Act). COM(2021) 206 final*. Publications Office of the European Union.

European Commission. (2024). *Regulation (EU) 2024/1689 of the European Parliament and of the Council on artificial intelligence (Artificial Intelligence Act)*. Official Journal of the European Union.

Evans, J. S. B. T. (2008). Dual-processing accounts of reasoning, judgment, and social cognition. *Annual Review of Psychology, 59*(1), 255–278. DOI: 10.1146/annurev.psych.59.103006.093629 PMID: 18154502

Floridi, L., Cowls, J., Beltrametti, M., Chatila, R., Chazerand, P., Dignum, V., Luetge, C., Madelin, R., Pagallo, U., Rossi, F., Schafer, B., Valcke, P., & Vayena, E. (2022). An ethical framework for a good AI society: Opportunities, risks, principles, and recommendations. In Floridi, L. (Ed.), *Ethics, governance, and policies in artificial intelligence* (pp. 19–39). Springer., DOI: 10.1007/978-3-030-81907-1_3

Fuster, A., Goldsmith-Pinkham, P., Ramadorai, T., & Walther, A. (2022). Predictably unequal? The effects of machine learning on credit markets. *The Journal of Finance, 77*(1), 5–47. DOI: 10.1111/jofi.13090

Gama, J., Žliobaitė, I., Bifet, A., Pechenizkiy, M., & Bouchachia, A. (2014). A survey on concept drift adaptation. *ACM Computing Surveys, 46*(4), 44. DOI: 10.1145/2523813

Goddard, K., Roudsari, A., & Wyatt, J. C. (2022). Automation bias: A systematic review of frequency, effect mediators, and mitigators. *Journal of the American Medical Informatics Association : JAMIA, 19*(1), 121–127. DOI: 10.1136/amiajnl-2011-000089 PMID: 21685142

Green, B. (2022). The contestation of tech ethics: A sociotechnical approach to technology ethics in practice. *Journal of Social Computing, 2*(3), 209–225. DOI: 10.23919/JSC.2021.0018

Green, B., & Chen, Y. (2021). Disparate interactions: An algorithm-in-the-loop analysis of fairness in risk assessments. In *Proceedings of the 2019 Conference on Fairness, Accountability, and Transparency* (pp. 90–99). DOI: 10.1145/3287560.3287563

Gunning, D., Stefik, M., Choi, J., Miller, T., Stumpf, S., & Yang, G.-Z. (2021). XAI—Explainable artificial intelligence. *Science Robotics, 4*(37), eaay7120. Advance online publication. DOI: 10.1126/scirobotics.aay7120 PMID: 33137719

Henrich, J., Heine, S. J., & Norenzayan, A. (2010). The weirdest people in the world? *Behavioral and Brain Sciences*, *33*(2–3), 61–83. DOI: 10.1017/S0140525X0999152X PMID: 20550733

Hoff, K. A., & Bashir, M. (2015). Trust in automation: Integrating empirical evidence on factors that influence trust. *Human Factors*, *57*(3), 407–434. DOI: 10.1177/0018720814547570 PMID: 25875432

Hutchins, E. (1995). *Cognition in the wild.* MIT Press. DOI: 10.7551/mitpress/1881.001.0001

Jacobson, N. C., & Bhatt, P. (2022). Ethical dilemmas posed by mobile health and machine learning in psychiatry research. *Bulletin of the World Health Organization*, *98*(4), 270–276. DOI: 10.2471/BLT.19.237107 PMID: 32284651

Jain, S., & Wallace, B. C. (2019). Attention is not explanation. In *Proceedings of the 2019 Conference of the North American Chapter of the Association for Computational Linguistics* (pp. 3543–3556). DOI: 10.18653/v1/N19-1357

Jobin, A., Ienca, M., & Vayena, E. (2019). The global landscape of AI ethics guidelines. *Nature Machine Intelligence*, *1*(9), 389–399. DOI: 10.1038/s42256-019-0088-2

Kahneman, D. (2011). *Thinking, fast and slow*. Farrar, Straus and Giroux.

Kahneman, D., Sibony, O., & Sunstein, C. R. (2021). Noise: A flaw in human judgment. Little, Brown Spark.

Kulkarni, A., Bhattacharyya, S., Bhattacharyya, P., & Bhattacharyya, B. (2023). Natural language uncertainty expressions for clinical AI: A randomized experiment. *NPJ Digital Medicine*, *6*, 89. DOI: 10.1038/s41746-023-00835-6

Kull, M., Filho, T. S., & Flach, P. (2019). Beta calibration: A well-founded and easily implemented improvement on logistic calibration for binary classifiers. In *Proceedings of the 20th International Conference on Artificial Intelligence and Statistics* (pp. 623–631). PMLR.

Lai, V., Chen, C., Smith-Renner, A., Liao, Q. V., & Tan, C. (2023). Towards a science of human-AI decision making: A survey of empirical studies. In *Proceedings of the 2023 ACM Conference on Fairness, Accountability, and Transparency* (pp. 1228–1241). DOI: 10.1145/3593013.3594035

Lee, J. D., & See, K. A. (2004). Trust in automation: Designing for appropriate reliance. *Human Factors*, *46*(1), 50–80. DOI: 10.1518/hfes.46.1.50.30392 PMID: 15151155

Linxen, S., Sturm, C., Brühlmann, F., Cassau, V., Opwis, K., & Reinecke, K. (2021). How WEIRD is CHI? In Proceedings of the 2021 CHI Conference on Human Factors in Computing Systems (Article 143). DOI: 10.1145/3411764.3445488

Logg, J. M., Minson, J. A., & Moore, D. A. (2019). Algorithm appreciation: People prefer algorithmic to human judgment. *Organizational Behavior and Human Decision Processes*, *151*, 90–103. DOI: 10.1016/j.obhdp.2018.12.005

Lundberg, S. M., Nair, B., Vavilala, M. S., Horibe, M., Eisses, M. J., Adams, T., Liston, D. E., Low, D. K.-W., Newman, S.-F., Kim, J., & Lee, S.-I. (2020). Explainable machine-learning predictions for the prevention of hypoxaemia during surgery. *Nature Biomedical Engineering*, *2*(10), 749–760. DOI: 10.1038/s41551-018-0304-0 PMID: 31001455

Lyell, D., & Coiera, E. (2017). Automation bias and verification complexity: A systematic review. *Journal of the American Medical Informatics Association : JAMIA*, *24*(2), 423–431. DOI: 10.1093/jamia/ocw105 PMID: 27516495

Madaio, M. A., Stark, L., Wortman Vaughan, J., & Wallach, H. (2020). Co-designing checklists to understand organizational challenges and opportunities around fairness in AI. In *Proceedings of the 2020 CHI Conference on Human Factors in Computing Systems* (pp. 1–14). DOI: 10.1145/3313831.3376445

Mehrabi, N., Morstatter, F., Saxena, N., Lerman, K., & Galstyan, A. (2021). A survey on bias and fairness in machine learning. *ACM Computing Surveys*, *54*(6), 115. DOI: 10.1145/3457607

Mittelstadt, B. (2019). Principles alone cannot guarantee ethical AI. *Nature Machine Intelligence*, *1*(11), 501–507. DOI: 10.1038/s42256-019-0114-4

Mittelstadt, B., Russell, C., & Wachter, S. (2023). Explaining explanations in AI. In *Proceedings of the Conference on Fairness, Accountability, and Transparency* (pp. 279–288). DOI: 10.1145/3287560.3287574

Molnar, C. (2022). *Interpretable machine learning: A guide for making black box models explainable* (2nd ed.)., https://christophm.github.io/interpretable-ml-book/

Morley, J., Cowls, J., Taddeo, M., & Floridi, L. (2021). The ethics of AI in health care: A mapping review. *Social Science & Medicine*, *260*, 113172. DOI: 10.1016/j.socscimed.2020.113172 PMID: 32702587

Moy, A. J., Schwartz, J. M., Chen, R. J., Sadri, H., Lucas, E., Cato, K. D., & Rossetti, S. C. (2021). Measurement of clinical documentation burden among physicians and nurses using electronic health records: A scoping review. *Journal of the American Medical Informatics Association : JAMIA*, *28*(5), 998–1008. DOI: 10.1093/jamia/ocaa325 PMID: 33434273

National Institute of Standards and Technology. (2023). *Artificial intelligence risk management framework (AI RMF 1.0) (NIST AI 100-1)*. U.S. Department of Commerce., DOI: 10.6028/NIST.AI.100-1

Neerincx, M. A., van den Bosch, K., Haije, T., Doesburg, M., Doornbos, G., Grootjen, M., & Looije, R. (2023). Maintaining situation awareness in human-agent teamwork: The value of functional states. *Frontiers in Neuroergonomics*, *4*, 1–17. DOI: 10.3389/fnrgo.2023.958930

Nisbett, R. E., & Masuda, T. (2003). Culture and point of view. *Proceedings of the National Academy of Sciences of the United States of America*, *100*(19), 11163–11170. DOI: 10.1073/pnas.1934527100 PMID: 12960375

Obermeyer, Z., Powers, B., Vogeli, C., & Mullainathan, S. (2019). Dissecting racial bias in an algorithm used to manage the health of populations. *Science*, *366*(6464), 447–453. DOI: 10.1126/science.aax2342 PMID: 31649194

Ovadia, Y., Fertig, E., Ren, J., Nado, Z., Sculley, D., Nowozin, S., Dillon, J. V., Lakshminarayanan, B., & Snoek, J. (2019). Can you trust your model's uncertainty? Evaluating predictive uncertainty under dataset shift. *Advances in Neural Information Processing Systems*, *32*, 13991–14002.

Parasuraman, R., & Riley, V. (1997). Humans and automation: Use, misuse, disuse, abuse. *Human Factors*, *39*(2), 230–253. DOI: 10.1518/001872097778543886

Passi, S., & Barocas, S. (2019). Problem formulation and fairness. In *Proceedings of the Conference on Fairness, Accountability, and Transparency* (pp. 39–48). DOI: 10.1145/3287560.3287567

Pessach, D., & Shmueli, E. (2022). A review on fairness in machine learning. *ACM Computing Surveys*, *55*(3), 51. DOI: 10.1145/3494672

Poursabzi-Sangdeh, F., Goldstein, D. G., Hofman, J. M., Vaughan, J. W., & Wallach, H. (2021). Manipulating and measuring model interpretability. In Proceedings of the 2021 CHI Conference on Human Factors in Computing Systems (Article 237). DOI: 10.1145/3411764.3445315

Rabanser, S., Günnemann, S., & Lipton, Z. C. (2019). Failing loudly: An empirical study of methods for detecting dataset shift. *Advances in Neural Information Processing Systems*, *32*, 1396–1408.

Rader, E., Cotter, K., & Cho, J. (2018). Explanations as mechanisms for supporting algorithmic transparency. In *Proceedings of the 2018 CHI Conference on Human Factors in Computing Systems* (Paper 103). DOI: 10.1145/3173574.3173677

Raji, I. D., Kumar, I. E., Horowitz, A., & Selbst, A. (2022). The fallacy of AI functionality. In *Proceedings of the 2022 ACM Conference on Fairness, Accountability, and Transparency* (pp. 959–972). DOI: 10.1145/3531146.3533158

Rajpurkar, P., Chen, E., Banerjee, O., & Topol, E. J. (2022). AI in health and medicine. *Nature Medicine*, *28*(1), 31–38. DOI: 10.1038/s41591-021-01614-0 PMID: 35058619

Roberts, M., Driggs, D., Thorpe, M., Gilbey, J., Yeung, M., Ursprung, S., Aviles-Rivero, A. I., Etmann, C., McCague, C., Beer, L., Weir-McCall, J. R., Teng, Z., Gkrania-Klotsas, E., Ruggiero, A., Korhonen, A., Jefferson, E., Ako, E., Langs, G., Gozaliasl, G., & Schönlieb, C.-B.AIX-COVNET Collaboration. (2021). Common pitfalls and recommendations for using machine learning to detect and prognosticate for COVID-19 using chest radiographs and CT scans. *Nature Machine Intelligence*, *3*(3), 199–217. DOI: 10.1038/s42256-021-00307-0

Rudin, C. (2019). Stop explaining black box machine learning models for high stakes decisions and use interpretable models instead. *Nature Machine Intelligence*, *1*(5), 206–215. DOI: 10.1038/s42256-019-0048-x PMID: 35603010

Saria, S., & Subbaswamy, A. (2019). Tutorial: Safe and reliable machine learning. arXiv. https://doi.org//arXiv.1904.07204DOI: 10.48550

Schemmer, M., Hemmer, P., Kühl, N., Benz, C., & Vössing, M. (2023). Appropriate reliance on AI advice: Conceptualization and the effect of explanations. In *Proceedings of the 28th International Conference on Intelligent User Interfaces* (pp. 410–422). DOI: 10.1145/3581641.3584066

Sculley, D., Holt, G., Golovin, D., Davydov, E., Phillips, T., Ebner, D., Chaudhary, V., Young, M., Crespo, J.-F., & Dennison, D. (2015). Hidden technical debt in machine learning systems. *Advances in Neural Information Processing Systems*, *28*, 2503–2511.

Shneiderman, B. (2022). *Human-centered AI*. Oxford University Press. DOI: 10.1093/oso/9780192845290.001.0001

Sloane, M., Moss, E., Awomolo, O., & Forlano, L. (2020). Participation is not a design fix for machine learning. In *Proceedings of the Workshop on Participatory Approaches to Machine Learning, ICML 2020*. DOI: 10.48550/arXiv.2007.02423

Steyvers, M., Tejeda, H., Kerrigan, G., & Smyth, P. (2022). Bayesian modeling of human–AI complementarity. *Proceedings of the National Academy of Sciences of the United States of America*, *119*(11), e2111547119. DOI: 10.1073/pnas.2111547119 PMID: 35275788

Sutton, R. T., Pincock, D., Baumgart, D. C., Sadowski, D. C., Fedorak, R. N., & Kroeker, K. I. (2020). An overview of clinical decision support systems: Benefits, risks, and strategies for success. *NPJ Digital Medicine*, *3*(1), 17. DOI: 10.1038/s41746-020-0221-y PMID: 32047862

Vaccaro, K., Karahalios, K., Sandvig, C., Hamilton, K., & Langlotz, C. P. (2019). Contestability in algorithmic systems. In *Companion Publication of the 2019 ACM Conference on Computer Supported Cooperative Work and Social Computing* (pp. 523–527). DOI: 10.1145/3311957.3359435

Wang, D., Yang, Q., Abdul, A., & Lim, B. Y. (2023). Designing theory-driven user-centric explainable AI. In Proceedings of the 2023 CHI Conference on Human Factors in Computing Systems (Article 601). DOI: 10.1145/3544548.3580894

Wei, J., Wang, X., Schuurmans, D., Bosma, M., Ichter, B., Xia, F., Chi, E., Le, Q. V., & Zhou, D. (2022). Chain-of-thought prompting elicits reasoning in large language models. *Advances in Neural Information Processing Systems*, *35*, 24824–24837. DOI: 10.52202/068431-1800

Wong, A., Otles, E., Donnelly, J. P., Krumm, A., McCullough, J., DeSmet, O., Passarello, B., & Singh, K. (2021). External validation of a widely implemented proprietary sepsis prediction model in hospitalized patients. *JAMA Internal Medicine*, *181*(8), 1065–1070. DOI: 10.1001/jamainternmed.2021.2626 PMID: 34152373

Xu, W. (2019). Toward human-centered AI: A perspective from human-computer interaction. *Interactions (New York, N.Y.)*, *26*(4), 42–46. DOI: 10.1145/3328485

Zhang, Y., Liao, Q. V., & Bellamy, R. K. E. (2020). Effect of confidence and explanation on accuracy and trust calibration in AI-assisted decision making. In *Proceedings of the 2020 ACM Conference on Fairness, Accountability, and Transparency* (pp. 295–305). DOI: 10.1145/3351095.3372852

KEY TERMS AND DEFINITIONS

Algorithmic Fairness: The characteristic of an AI system in which their actions will not lead to discrimination against individuals or groups based on their membership in a protected class, such as on race, gender, age, or disability. Several mathematical definitions of this exist, and it is impossible to have all at once.

Automation Bias: A persistent pattern in which humans exhibit a tendency to excessively rely on automated advice, blindly following suggestions made by machines without proper analysis even when automation produces wrong outcomes.

Cognitive Alignment: How much the process of interaction, output, and expression of uncertainty in an AI system is compatible with the thought process, reasoning strategies, and decision-making procedures employed by its target user base.

Dual-Process Theory: A theory used in cognitive science that defines two types of thinking: fast, automatic, and association-based (System 1) and slow, deliberate, and analytical (System 2).

Explainability: The ability of an AI system to generate comprehensible explanations for its output, achieved either through intrinsically transparent model structures or post-hoc explanation methods applied to black boxes.

Human-Centered AI (HCAI): An approach in artificial intelligence design that emphasizes human involvement, understanding, welfare, and control at each step of AI system construction and use.

Trust Calibration: The congruence between the user's faith in an AI system's suggestions and their actual reliability when dealing with diverse inputs and situations, not to be confused with trust as a general phenomenon.

Chapter 2
Explainability as a Cognitive Process in Human Sense-Making for Interpretable Decision Intelligence

L. B. Muralidhar
http://orcid.org/0000-0003-3453-613X
Jain (Deemed to be) University, India

N. Sathyanarayana
http://orcid.org/0000-0002-4185-7751
Jain (Deemed to be) University, India

H. R. Swapna
http://orcid.org/0000-0003-0362-409X
Jain (Deemed to be) University, India

S. Madhu
http://orcid.org/0009-0000-8816-5798
MES College of Arts, Commerce, and Science, India

K. P. Sheeba
http://orcid.org/0009-0003-4707-5392
Jain (Deemed to be) University, India

DOI: 10.4018/979-8-3373-8988-2.ch002

ABSTRACT

This book chapter reframes explainable artificial intelligence (XAI) as a cognitive and sense-making process rather than algorithmic transparency. It studies how people perceive, analyse, and act on AI-generated outputs under ambiguity, bounded rationality, and bias using cognitive science, neuroscience, philosophy, and decision intelligence. The chapter analyses XAI paradigms, emphasises cognitive misalignments, and reframes decision intelligence as a human–AI socio-cognitive system focused on judgement quality rather than prediction accuracy. The Cognitive Explainability–Sense-Making Framework (CESF) links explanations to mental models, trust calibration, and decision outcomes and provides empirical validation. The chapter explores ethical, organisational, and governance effects utilising empirical evidence, case studies, cross-cultural views, and policy analysis. It concludes with future research and advocates cognitively aligned, context-sensitive explainability for ethical, trustworthy, and successful AI-mediated decision-making.

INTRODUCTION: REFRAMING EXPLAINABILITY AS COGNITION

The growing emphasis on explainability in artificial intelligence reflects a shift from viewing AI systems solely through technical performance metrics toward understanding how their outputs can be interpreted and used by human decision-makers. The first explainable AI (XAI) approaches were feature attribution, surrogate modelling and rule extraction, thereby encouraging algorithm disclosure and statistical fidelity (Ribeiro et al., 2018). However, these techniques are not very effective in decision-making practice when explanations are cognitively perceived rather than mechanically delivered. Technical interpretability does not imply meaningful interpretation, especially in high-stakes fields (Doshi-Velez & Kim, 2017; Rudin, 2019).

The core idea presented in this chapter is that explainability is a cognitively mediated process between the algorithmic output, interpretation by humans, and the outcome of the decision-making process. Theories of sense making: These give insight into the process of making sense, which people use to make meaning from explanations.Theories of bounded rationality: These shed light on the role of cognitive limitations in shaping sense making. Theories of predictive processing: These provide insight into how explanations modulate expectations and beliefs. Theories of decision intelligence: These place the processes of sense making within organizational decision systems. These views are incorporated into a single perspective which combines explanation design with cognitive processing and with decision performance (Weick, 1995; Simon, 1955; Clark, 2013; Berndt & Vössing, 2025).

Interpretability is about how much human-readable information can be gleaned from the inner workings of a model. In contrast, explainability concerns communicating the model's behaviour in a meaningful form that the user can relate to. Understanding goes beyond mere concepts and involves incorporating explanations into the user's mental model, which allows for predictions, evaluations, and decisions. (Miller, 2019; Rosen & O'Leary, 2024) It is important to distinguish these concepts in the first place, as technically interpretable systems do not necessarily lead to understanding, and explainable systems may not be useful for effective decision-making if cognitive integration is lacking.

Interpretability is the process of examining a model's logic. In contrast, explanation is the process of communicating a model's behaviour; justification provides reasons for a recommendation; and understanding is the successful integration of information into a user's mental model, enabling prediction, evaluation, and action (Rosen & O'Leary, 2024; Miller, 2019).

In real-life decision situations, users will not only assess explanations for their correctness. Rather, explanations are understood in the context of the organisation, profession, and situation, which affects how information is understood and used. The usefulness of explainability goes beyond model transparency and relies on its value for understanding, confidence calibration, and informed action (Miller, 2019; Albini et al., 2023).

The concept of explainability needs to be considered a cognitively situated process whereby people make sense of AI-generated outputs within contextual and organisational contexts. In addition to technical transparency, the meaning of explanations is shaped by the user's mental models, heuristics, and prior beliefs (Weick, 1995; Gigerenzer & Selten, 2002). This view reveals some of the trade-offs between local fidelity and cognitive coherence in post hoc approaches like SHAP and LIME, as well as the possibility of cognitive burden (Lundberg & Lee, 2017). Therefore, context-sensitive and cognitively aligned explanations that enable trust calibration, learning and informed decision-making in different cultural and regulatory contexts are essential for effective explainability (Albini et al., 2023; BIS, 2024).

The above discussion shows that explainability is cognitively situated, not merely a technical property. To advance this argument, cognitive mechanisms that enable the generation of meaning from AI-produced explanations need to be explored. This understanding can serve as a theoretical basis for determining whether existing approaches to explainability truly help the decision-making and knowledge-creation processes (Weick, 1995; Clark, 2013).

CONCEPTUAL FOUNDATIONS: SENSE-MAKING IN HUMAN COGNITION

Sense-making frames (Weick, 1995) view explainability as a cognitive process in which the user's mental model interprets the outputs of AI systems based on previous knowledge and context. In the bounded rationality framework, decision-makers will settle for a coherent and plausible explanation when they lack sufficient time or cognitive resources to construct a complete one (Simon, 1955; Miller, 2019). This makes it difficult to create a balance between what is correct with respect to computation and what is intelligible to human users. While providing detailed explanations can lead to better decision quality, simpler explanations can improve usability at the expense of decision quality (Poursabzi-Sangdeh et al., 2021). Finally, explanations update expectations, and cultural or narrative preferences affect both the understanding gained and the confirmation of bias (Clark, 2013; Vattonen & Reunanen, 2024).

This is because interpretability, explanation, justification, and understanding are different levels of analysis, so that following interpretation and explanation will focus on understanding as the desired outcome of explainability and on sense-making as the process of achieving it. This is because interpretability, explanation, justification, and understanding are different levels of analysis, so the following will focus mainly on understanding as the desired outcome of explainability and on sense-making as the process that achieves it. (Miller, 2019; Rosen & O'Leary, 2024) Making these distinctions is critical because you might provide the user with explanations and justifications, but without truly understanding. This distinction has implications for the evaluation of explainability. Most existing assessments focus narrowly on the availability of explanations, not on whether users can meaningfully use them in decision-making situations. Explainability should therefore be further explored in terms of grasping outcomes and not just disclosure (Rosen & O'Leary, 2024). As the governance framework moves beyond expectations for transparency to explainability and accountable decision support, this difference is increasingly recognised (BIS, 2024; NIST, 2022).

Together, any of the above theories make a compelling argument for the inherent nature of explainability: it is an interpretive process, not a transfer of information. The meaning of explanations is only meaningful within established cognitive frameworks and contexts. The findings indicate that an analysis of explainability based on technical transparency is insufficient for explaining how humans understand AI-made decisions.

These cognitive insights serve as a yardstick for examining existing explainability methods. Given that explainability is defined by its interpretability, context, and cognitive processing, current XAI paradigms need to be evaluated not only from a technical point of view but also for their ability to facilitate understanding and decision-making.

EXPLAINABLE AI: CURRENT PARADIGMS AND THEIR COGNITIVE LIMITATIONS

Recent paradigms of explainable artificial intelligence are largely organised around technical desires that place much emphasis on transparency, computational efficiency, and predictive performance, as well as on the key distinction between intrinsic and post hoc explainability. Intrinsic methods incorporate interpretability into model design, such as decision trees and linear models, such that structural transparency (Rudin, 2019) is guaranteed, whereas post hoc methods, such as SHAP, saliency mapping and surrogate models, seek to interpret opaque systems post hoc (Lundberg & Lee, 2017; Ribeiro et al., 2018). These methods also distinguish between local explanations, which explain individual predictions, and global explanations, which explain the overall model behaviour (Arrieta et al., 2020).

Several new cognitive-XAI approaches seek to transcend the model-centric notion of transparency. Miller (2019) highlights the role of explanation as a socially embedded process driven by human expectations, while Long and Zou (2024) argue that explanation fits the sensemaking theory framework and is viewed as a sense-making process in context. Albini et al. (2023) focus on cognitive explainability, which encompasses both the accuracy of the explanation and the user's understanding, whereas conversational XAI approaches emphasise iterative clarification and adaptive interaction (Wassim, 2023). These frameworks have a concern for human cognition but vary in their treatment of context, interaction, and measurement. Current methods remain quite disconnected, as there is insufficient integration among cognitive psychology, organisational decision-making, and governance to warrant the proposed development of the Cognitive Explainability Sense-Making Framework in this chapter.

While recent XAI methods have made important strides in making explanations clear and understandable, they are largely model-centric and do not focus on the user's mental processes in understanding explanations. The amount of information presented may lead to cognitive dissonance, and a too-simplistic explanation might lead to false confidence, especially when interpretability is assumed to be synonymous with correctness (Poursabzi-Sangdeh et al., 2021). Cognitive biases and framing effects also influence explanation uptake, as well as cultural variation, which makes it difficult to assume that it is universally applicable, thereby narrowing the scope of purely technical approaches to explanation uptake (Tversky & Kahneman, 1981; Vattonen & Reunanen, 2024). So, just because something is transparent does not necessarily mean it suffices for decision support. The emerging cognitive explainability approaches thus suggest adaptive, context-aware, and explainable systems that connect explanatory content to human reasoning, understanding, and decision-making needs (Albini et al., 2023; Long & Zou, 2024).

Current paradigms for XAI have made great strides in enhancing transparency and interpretability and are mostly model-centric. The shortcomings become apparent when they are judged by human minds, which assume a central role in the construction of meaning, the interpretation of context, and the calibration of trust.

Empirical studies over the last decade have also shown that the quality of an explanation should not be judged purely by transparency, but by its ability to facilitate appropriate trust. More recent empirical studies of human–AI trust calibration also support this conclusion that the quality of an explanation should not be judged solely by transparency but by its ability to enable appropriate trust. Research indicates that too little or too much trust can undermine the quality of the decision. However, calibrated trust can help users trust AI recommendations in appropriate contexts and to remain vigilant in contexts of high uncertainty. Thus, trust calibration can be a useful, outcome-oriented measure of explainability's effectiveness beyond mere measures of user satisfaction and perceived transparency (Rosen & O'Leary, 2024). The limitations inspire a transition to decision-intelligence approaches that clearly involve human reasoning processes.

DECISION INTELLIGENCE AS A HUMAN–AI COGNITIVE SYSTEM

The impact of explainability is heightened when the AI is embedded in a larger decision-making process, with human decision-makers interpreting, evaluating, and acting on the results. Decision intelligence builds on predictive analytics to combine algorithmic recommendations with human judgement, context, and organisational goals, focusing on the quality of the decision rather than predictive accuracy (Shmueli, 2010; Berndt & Vössing, 2025). Cognitive integration and oversight (Abdelwanis et al., 2024) reflect different levels of engagement, accountability, and operational efficiency, as reflected in the Human-in-the-Loop and Human-on-the-Loop models. In addition, the requirements for explainability differ across decision types: for strategic decisions, there is a need for deliberative reasoning, while for operational decisions, succinct guidance is needed. Results show that the context of an explanation is much more important than its technical completeness, as users prefer explanations that are relevant to them, yet they also expect more detailed information (Lee & Kim, 2025; BIS, 2024).

Explanation design should be based on the information needs of the decision situation and on the knowledge and skills of those who will use the explanation. In explainability, instead of trying to include as much information as possible, the key is to choose a format that enables timely understanding while preserving uncertainty information. The way explanations are framed and structured affects their interpretation, as users are likely to use coherent explanations to assess the complexity of recommendations. Therefore, while informative accuracy is essential for the effectiveness of the explanation, so is context for understanding and making judgements based on the explanatory content (Miller, 2019).

The cross-cultural data reveal differences in preferences for explanations across cognitive styles, professional standards, and sociocultural backgrounds (Vattonen & Reunanen, 2024). This inconsistency raises questions about the scalability of standardised frameworks and prompts the development of user-centred, adaptive solutions. Generally, decision intelligence is an interdependent composite of human thinking and machine computation, and explainability is a cognitive interface that determines interpretation, understanding, justification and action. It works well when aligned with explanations and cognitive processes, the context's needs, and ethical issues, and is therefore a shift from models of explanations to dynamism, context-specific, and cognitively sensitive processes that enable responsible decision-making.

AN INTEGRATIVE FRAMEWORK: COGNITIVE EXPLAINABILITY FOR DECISION INTELLIGENCE

Figure 1. Cognitive explainability-sense-making framework

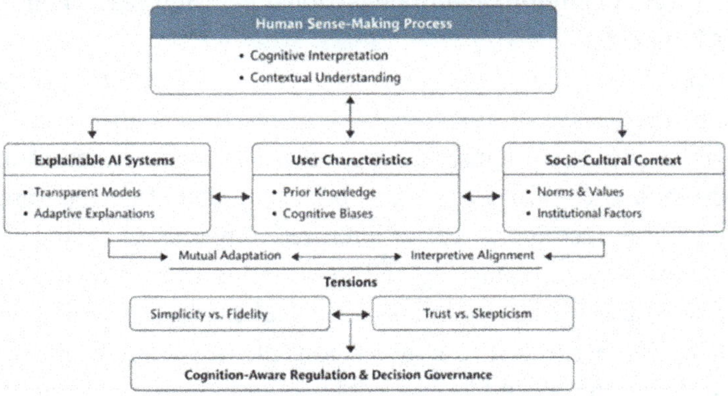

The integrative approach to cognitive explainability in decision intelligence requires a redefinition of explanations as processes, rather than artefacts, and as processes dynamically mediated by cognition, formally as part of human sense-making systems. A further step in this direction is the Cognitive Explainability Sense-Making Framework (CESF) demonstrated in Figure 1, which places the model outputs, uncertainty signals, and contextual cues as inputs in a cognitively mediating interpretive environment, where the users actively create meaning through the lens of prior beliefs and contextual knowledge (Arrieta et al.,2020)

Based on the theory of predictive processing, CESF recognises the cognitive mediators in the form of mental models and expectations as filters that guide interpretation, which is also in line with the theory of bounded rationality, whereby decision-makers base their judgments on simplified representations, as opposed to comprehensive analysis (Friston, 2005; Clark, 2013; Simon, 1955). Experimental results have shown that different users can give the same explanation different interpretations in response to differences in expertise, culture, and cognitive style, and thus, cognitive alignment has been established as a precondition to successful explainability (Vattonen & Reunanen, 2024; Petrik & Varnum, 2024; Albini et al., 2023).

There are three important differences in the operation of CESF and other approaches to cognitive-XAI. First, it clearly takes into account the context in which the explanation occurs: the organisational context, institutional constraints, socio-cultural influences, and so on, as active factors in the explanation's effectiveness. Second, it illustrates iterative cycles between an explanation, learning, and subsequent decisions – rather than explanations as a single event. Third, it not only looks at the interpretability or transparency of the explanation; it also considers consistency with the means of cognition. The differences position CESF as a connecting glue among explanation design, cognition, and organisational decision performance (Long & Zou, 2024; Albini et al., 2023).

The framework has four dimensions that can be interpreted as interrelated. Explanatory inputs consist of model outputs, uncertainty markers, and contextual information from the AI systems. Second, prior knowledge, mental models, expertise, attention, and cognitive biases, as processes that affect interpretation, are considered cognitive mediators. Third, interpretive processes include explaining, evaluating, contextualising, comparing and constructing a narrative. Lastly, outcomes of decision-making include trust calibration, comprehension of explanations, decision quality and adaptive learning. These dimensions provide a practical framework for relating CESF to empirical studies and implementation in organisations (Albini et al., 2023; Miller, 2019).

The main analytical contributions of the CESF are the idea that the effectiveness of an explanation is a function of the relationships among the content, the user's cognition, and the context, and the fact that this effect is not the result of any single aspect of the mode of explanation. This proposition overcomes limitations in the focus on transparency measures and introduces the role of explanatory interventions in trust calibration, learning, and decision quality under different organisational conditions. The framework thus allows for theorising testable hypotheses about the conditions under which explainability is likely to succeed or fail (Albini et al., 2023; Rosen & O'Leary, 2024).

From an operational perspective, the CESF can be realised through a four-step process: explanation generation, cognitive adaptation, contextual validation, and decision evaluation. Explanation generation is concerned with the presentation results, uncertainty markers and context cues for the model. Cognitive adaptation aligns the content of the explanations with the user's knowledge, the task's level of difficulty, and the domain's needs. Contextual validation allows the user to validate the explanations in the context of organisational constraints and/or domain knowledge. Trust calibration, comprehension and decision quality outcomes are evaluated in the context of decision evaluation. This workflow offers a practical approach to translating the CESF from conceptual framework to deployable decision-support architecture (Albini et al., 2023; Rosen & O'Leary, 2024).

A real-world example illustrates how CESF can be used in a decision support system for the healthcare industry. Think about an AI system that can forecast the probability of any complications that may arise after a surgery. The conventional XAI approach would allow a clinician to obtain a ranking of features by their importance in the prediction. In CESF, however, moving the explanation process beyond just disclosing features is important. The system would tailor explanations to the clinician's level of expertise, provide the clinician with information about uncertainty, give the prediction context in light of the patient's history, and, through interactive dialogue, clarify the prediction. This information would then be added to an established mental model of patient care developed by the clinician and then be used to make a treatment decision. In this case, the emphasis is not on an explanatory output, but rather an explanatory process where AI recommendations are connected with cognitive assessment and clinical judgment (Lee & Kim, 2025; Miller 2019).

The Cognitive Explainability Sense-Making Framework (CESF) conceptualises the construction of AI explainability as a process of making sense of AI outputs, verifying them, and providing narratives about them in the context of the decision-making task (Weick, 1995; Weick et al., 2005). Cognitive accessibility and informational richness are key features in determining explanatory effectiveness; they are not central in model-centred approaches, which focus on transparency. In contrast to model-centred approaches that focus on transparency, informational richness and cognitive accessibility are key features in determining explanatory effectiveness (Essen & Meske, 2023). In contrast to the uncritical use of trust, the framework differentiates between calibrated and informed trust for context-sensitive decision-making, and not only technical disclosure, which is necessary for informed judgment, accountability, and mitigating bias (BIS, 2024; Lee & Kim, 2025). An important contribution of CESF is the modelling of iterative loops in which explanations lead to the formation of mental models, decisions, and the continuous " recalibration " of the outcomes of the human–AI interaction (Berndt & Vössing, 2025; Stammer et al., 2021).

CESF provides the groundwork for empirical research by defining measurable connections among cognitive alignment, understanding of an explanation, levels of trust calibration, and decision quality. The constructs can be operationalised as indicators and analysed through controlled experiments, field studies, long-term studies, and cross-cultural validation studies. The framework now goes beyond the concept of discussion as it outlines mechanisms that are measurable in organisations and across decisions. However, there are difficulties with subjective score measurement, methodological consistency, and the explanation of comprehensibility and predictive ability (Shmueli, 2010).

Multi-level research designs need to be used in future empirical testing of CESF. Causal relationships between cognitive alignment and decision quality can be examined in controlled experiments, while field studies can determine the effects of cognitive alignment under real organisational conditions. Comparative research on explanatory preferences should examine differences in these preferences across specific institutional and cultural settings. Relationships between comprehension, the calibration of trust and decision outcomes could be further explored using structural equation modelling, longitudinal analysis, and mixed-methods approaches. It would help to improve the explanatory and predictive power of the framework (Stammer et al., 2021; Albini et al., 2023).

To make CESF, as an empirical approach, more tractable, it can be implemented as a series of indicators spanning the cognitive, behavioural, organisational, and decision performance levels, allowing for systematic assessment of CESF effectiveness across different settings. There are a variety of ways to measure cognitive alignment, including mental model accuracy tests, explanation recall measures, perceived coherence scales, and calibrated trust indices (Essen & Meske, 2023; Albini et al., 2023). Controlled experiments can be used to validate behaviour, for example by comparing users exposed to cognitively aligned explanations with users exposed to traditional explanations composed solely of features. There are various outcome measures, such as decision accuracy, response time, confidence calibration, bias reduction, and explanation retention (Poursabzi-Sangdeh et al., 2021; Miller, 2019).

CESF metrics can be broken down into four categories to help evaluate them. Explanations recall, the accuracy of mental models, perceived coherence, and cognitive load assessment are cognitive metrics. Some behavioural metrics include response time, the frequency of explanations, and decision consistency. Some of the outcome metrics are decision accuracy, bias reduction, confidence calibration and error detection performance. Organisational metrics include user adoption, long-term decision effectiveness, trust calibration, and accountability compliance. These initiatives provide a multidimensional view of the effectiveness of explainability, both in terms of human understanding and its impact on the organisation (Albini et al., 2023; Rosen & O'Leary, 2024).

In financial services settings, CESF might assist in providing customised explanations to customers, compliance staff and loan officers for credit risk evaluations. This prediction can thus be represented in a variety of ways, all of which are intellectually accessible, each offering a distinct representation of the underlying model. This type of adaptation increases the clarity of interpretation while maintaining accountability and regulatory transparency (Rosen & O'Leary, 2024; BIS, 2024).

A validation program using a multi-method approach is suggested. The causal effect of explanation design can be evaluated in experimental studies, and the effectiveness of mental model development and adaptive decision-making over time can be evaluated in longitudinal studies. Other measures, such as eye-tracking, cognitive-load measurements, and neurocognitive measures, could also be used to complement the above data to gain insights into the processes of attentional allocation and cognitive effort during explanation processing (Corlett & Friston, 2020). These would turn CESF from a concept into a model that can be tested and lead to testable propositions for future investigation.

But it is subjective cognitive constructs that cause measurement problems and bring about prediction/explanation discrepancies (Shmueli, 2010). While applying CESF in a large-scale enterprise environment, there are some practical constraints. It is often challenging to ensure personalised explainability when organisations have different types of users, with different levels of expertise. There is also a need for enterprise systems to be cognitively adapted and, at the same time, harmonised with regulatory standards, auditable, secure against cyber threats, and cost-effective to operate.

Moreover, ongoing adjustments to explanations can also raise governance issues of consistency and fairness within departments and jurisdictions, as well as accountability issues. The constraints imply that the concept of explainability in cognition is a design goal that must be adapted to the context in which they are applied and not a single implementation model (BIS, 2024; Rosen & O'Leary, 2024). In sum, CESF provides an overall analytical framework that brings cognitive science and XAI together with an organisational theory to foster a more human-centric approach to XAI that focuses on cognitive alignment, context sensitivity and iterative learning in complex decision-making.

CESF builds on prior work in explainability, adding cognitive mediation and context interpretation. The framework doesn't view explainability as a one-time result; instead, it considers it a process that connects the design of explanations, how humans understand and think about them, and the decisions made. From this perspective, the role of explainability in shaping bias, trust, and decision quality in real-life scenarios can be explored.

COGNITIVE BIAS, EXPLAINABILITY, AND DECISION QUALITY

The explainability decision quality correlation is not always positive. Judgement can be aided by well-designed explanations, but can be worsened by poorly designed explanations, which can exacerbate automation bias, confirmation bias and authority bias risks when making decisions based on AI (Simon, 1955; Tversky & Kahneman, 1981). Plausible explanations and explanations that are captured may still be out there, for example in the form of post hoc explanations, like SHAP and LIME, which can lead to misplaced confidence (Lundberg & Lee, 2017; Ribeiro et al., 2018). Explanations that are counterfactual, contrastive, and based on examples can facilitate reflection but they are not necessarily effective, as they rely on factors of task complexity, user expertise, and cultural context. Consequently, explainability can have a positive or negative effect on decision quality, contingent on whether or not it adjusts existing cognitive models, or reinforces them (Friston, 2005; Rosen & O'Leary, 2024).

With more persuasive explanations, another issue is that the distinction between informing and influencing users is blurred; consequently, the notion of autonomy and responsibility (Wassim, 2023; NIST, 2022) is called into question. An ethical issue that has come to the fore in recent years is that of manipulative explainability: the ability to create explanations, whether intentionally or inadvertently, that guide user actions and behaviour more than they inform. Unlike traditional transparency goals, manipulative explanations focus on the influence to be achieved through the behaviour, usually by selectively framing, emotionally priming, or providing cues of authority, or by simplifying the story so much that the recipient does not understand. Such methods can lead to more compliance, trust, or adoption; however, they can be harmful to user autonomy because users are not given the opportunity to critically evaluate the method for their own benefit. In the context of decision-intelligence, explanations should serve to aid reflective thinking, and should not be used as tools for persuading people to behave in a certain way (Miller, 2019; Rosen & O'Leary, 2024).

In situations with a high-stakes component (e.g., in healthcare, finance, employment and public administration), the difference between informing and influencing becomes really important. Optimising explanations to enhance users' acceptance of algorithmic recommendations can leave users feeling they act freely in their choices, even as the interface design limits how they make those choices. This has raised concerns about informed consent, epistemic independence, and the preservation of meaningful human oversight in human-AI decision systems (NIST, 2022; Rudin, 2019).

In general, cognitive bias, explainability, interactivity, and their influence on the quality of decisions are contextually sensitive and theoretically unresolved phenomena that would require a reflexive and interdisciplinary approach to critically evaluate the possibilities and limitations of explainability in establishing human-AI decision ecosystems.

EMPIRICAL EVIDENCE AND SECONDARY DATA INSIGHTS

A large gap in trust and knowledge exists, as identified by empirical studies in XAI and HCI. When explanations are present, users tend to feel more confident in their use and to make more automation errors, even if the errors are superficial or not directly related to the model's logic (Essen & Meske, 2023; Poursabzi-Sangdeh et al., 2021). This can be exacerbated by simple feature importance visualisations that can seem intuitive, but do not necessarily lead to substantive understanding (Lundberg & Lee, 2017). It has also been found that there are domain dependencies in explaining preferences across different cultures: When it comes to clinicians, they value the story's coherence with their clinical reasoning, while when it comes to financial contexts, they value quantitative precision (Lee & Kim, 2025; Moore & Williamson, 2022). These results contradict traditional models of explainability and favour context-based approaches that align with regulatory needs and decision intelligence practice (BIS, 2024; Petrik & Varnum, 2024).

Survey results suggest that people's trust reactions to AI explanations vary across explanation formats, perceived AI fairness, institutional settings, and prior attitudes toward AI (Rosen & O'Leary, 2024). Making things transparent can either build up trust or diminish trust, as the information may make it easier for people to trust, or it may raise their doubts about reliability. The results underscore the contextuality of the use of explainability in organisations and in the country.

Empirically conducted studies remain constrained by methodological challenges, particularly the predominance of laboratory studies that prioritise internal validity over real-world applicability (Poursabzi-Sangdeh et al., 2021). The experiments tend to use simplified tasks and homogeneous samples, limiting generalisation.

The results of the explainability studies reveal that providing explanations does not uniformly improve the quality of the decision. Explanations are more likely to have positive effects if they align with the user's knowledge and abilities, as well as the task requirements and decision context. The results also support a socio-cognitive perspective on explainability in AI systems, rather than an exclusively technical one (Miller, 2019; Essen & Meske, 2023; Rosen & O'Leary, 2024).

In addition, longitudinal studies are missing, limiting the knowledge of the development of trust, learning, and cognitive adaptation (Stammer et al., 2021). The field is also limited by the fact that existing quantitative measures lack the qualitative considerations of story coherence and moral judgment.

To address these gaps, one must embrace methodological pluralism, which entails integrating experimental, ethnographic, and neurocognitive techniques. The available evidence, in totality, indicates the future possibilities and shortcomings of explainable AI, suggesting that successful explainability cannot be premised on technical transparency per se for providing cognitive alignment, contextual sensitivity, and long-term interpretation for users across a wide range of changing decision-making situations.

CASE STUDIES AND APPLIED ILLUSTRATIONS

Cognitive explainability has practical implications, most clearly observable in the healthcare sector, financial systems, and governance by the populace, where explainability is a socio-cognitive process that influences trust, accountability, and the quality of decisions in states of uncertainty. Explainability in the context of healthcare decision support systems can be used to improve the usability of the interpretive results by providing clinician-friendly stories and simplified feature attributions, which increase perceived usability but can also encourage over-dependence in cases of a lack of adequate underlying uncertainty (Lee & Kim, 2025; Moore & Williamson, 2022).

The results validate the broader argument for cognitive explainability: explanations are most useful when they are grounded in users' ways of interpreting information, not just in a set of technical facts. Explainability is thus not only necessarily informationally complete but also cognitively fit for it to be of practical use.

This is the bias of automation, and it is a concern that diagnostic accuracy would be affected by time constraints that preclude a thorough examination. Regulatory frameworks focus on transparency and auditability but pay little attention to the cognitive load on clinicians in balancing probabilistic outputs with clinical judgment (Regulatory Affairs Journal, 2023). As a result, the argument for inherently interpretable models remains relevant in high-stakes situations (Rudin, 2019).

The desire to explain financial and credit decision-making systems is primarily driven by compliance and risk management requirements, and feature-level decision-making can be elucidated using algorithms such as SHAP and LIME (Lundberg & Lee, 2017; Ribeiro et al., 2018). Even though these can enhance formal transparency, they may be cognitively demanding for non-expert users, thereby hindering interpretability and practicality (Essen & Meske, 2023). Such mental load would contribute to decision fatigue and fewer interactions, especially in consumer-facing environments (Messner & Siering, 2025). Cross-jurisdictional differences also demonstrate that the effectiveness of explanations should be considered in terms of regulatory cultures, financial literacy, and, accordingly, adaptive communication aligned with users' abilities (Rosen & O'Leary, 2024).

Recent research in adaptive interfaces indicates that the effectiveness of explanations is enhanced by dynamically adapting explanatory content to the user, task, and decision context. Adaptive systems provide the same explanation to different users but adjust their depth, format of visualisation and interaction in order to meet the different cognitive needs. These strategies have been linked to increased understanding, decreased cognitive overloading and increased trust calibration for diverse user groups (Costa & Silva, 2025).

Explainability intersects with procedural fairness and accountability in government to the general population, especially in areas such as welfare distribution and predictive policing (NIST, 2022; Rosen & O'Leary, 2024). One of the key contradictions is between transparency and operational secrecy: on the one hand, an extreme level of transparency may allow manipulation of the system; on the other hand, a lack of transparency will undermine trust and accountability. Cognitively, users are usually not expert enough to read technical descriptions, so simplified information and context-based formats are required. Nonetheless, this simplification can introduce distortions into narratives shaped by past beliefs and attitudes towards socio-politics, thereby affecting the sense of fairness (Weick, 1995; Weick et al., 2005). Cross-cultural data also suggest differences in trust and interpretative preferences, indicative of institutional and cultural differences (Petrik & Varnum, 2024).

The key, however, is that explainability serves as an interface between the algorithm's outputs and human reasoning across all applications. The success of its use depends on technical transparency, user understanding, the context in which it is used, and the purpose of the decision (Kaur & Singh, 2024; Berndt & Vössing, 2025). A series of contrasts between transparency and opacity, simplicity and fidelity, trust and scepticism exposes the problems of technical solutions. Thus, future applications need to incorporate concepts from cognitive science, domain knowledge, and governance into informed, accountable, context-specific decision-making.

The spread of AI across the world raises the question of what 'explainability' is and whether it can be formulated as a general construct. As AI spreads across the world, it raises questions about what 'explainability' might be and whether it can be designed as a general construct. Empirical evidence shows that preferences for explanations, the process of building trust, and interpretive practices can differ across cultural, institutional, and cognitive contexts and thus may influence the understanding and application of explanatory mechanisms (Petrik & Varnum, 2024; Vattonen & Reunanen, 2024). As a result, for some user groups, methods like SHAP and LIME may not be as explanatory as for other groups (Lundberg & Lee, 2017). Cultural norms, norms of governance, norms of communication, and communication approaches are also found to shape trust (Wassim, 2023), with greater adaptability to contextual differences in communication approaches. Unfortunately, the current body of knowledge is limited to cross-cultural evidence and data from the West, leading to a demand for explainability frameworks that are interdisciplinary and context-sensitive, not only adhering to technical standards but also being socially legitimate (Thagard, 2003; Rosen & O'Leary, 2024).

POLICY, GOVERNANCE, AND INSTITUTIONAL IMPLICATIONS

In recent debates about AI governance, the concept of explainability is emerging as a more salient regulatory and cognitive interface that mediates trust, responsibility, and decision quality in institutional contexts, rather than as a technical aspect of AI. Explainability has been a shared principle of responsible AI implementation in international policy interventions, as seen in European banking systems that oversee credit risk and in international standards, such as those of the Bank for International Settlements.

These frameworks, though, are more inclined to prioritise legal compliance, auditability, and post hoc transparency over the cognitive usability of their explanations for human decision-makers. This presents a structural challenge: traceability and justification might be required by regulatory regimes, but human users require explanations aligned with limited rationality, heuristics and sensemaking (Simon, 1955; Weick, 1995). Excessive complexity or technical faithfulness is counterintuitive and reduces understanding and the quality of choices, thereby undermining the accountability it is meant to ensure (Essen & Meske, 2023; Poursabzi-Sangdeh et al., 2021). The new idea of cognitive explainability seeks to address this gap by aligning with human cognitive processes and predictive processing systems (Albini et al., 2023; Clark, 2013).

From a governance perspective, the extent to which information is made available is not the only measure of explainability, it is also important that intended users can interpret and use the information. There are a growing number of human-centred AI governance frameworks that increasingly assert the need to assess accountability based on technical compliance and human effects. All of these focus on use, stakeholder involvement, mental access, and the ability of explanations to aid informed human evaluation. This is what we see when we consider transparency as a governance mechanism that connects regulatory goals to actual decision-making behaviour, rather than merely as a transparency requirement (OECD, 2024; UNESCO, 2023; Rosen & O'Leary, 2024). This shifts the regulatory focus from transparency to measurable outcomes of understanding.

The governance of organisations is increasingly focused on dynamic explainability audits that combine model documentation, audit trails, and model-explanation tools, thereby enhancing accountability and oversight (Ribeiro et al., 2018; Lundberg & Lee, 2017). However, there are questions about the extent to which post hoc explanations can support accountability in high-stakes contexts involving opaque models that impact consequential decisions (Rudin, 2019; Moore & Williamson, 2022). Explanations have implications for cognition, the governance of bias adoption, and decision outcomes. Thus, AI governance needs to consider how explanations influence all of these (Berndt & Vössing, 2025). However, a technically responsible system also needs to be cognitively usable, trusted, and capable of supporting the user and, depending on the level of expertise, cognitive style, and socio-cultural background, explaining the data in different ways. (Vattonen & Reunanen, 2024; Messner & Siering, 2025)

While there has been considerable progress in the governance of explainability, there are still many measures to come in quantifying human understanding and adjusting trust. Existing regulatory frameworks largely focus on transparency, fairness, robustness, and auditability, and offer little guidance on the cognitive meaningfulness of explanations for their intended users. New governance models for humans propose incorporating behavioural evidence, cognitive usability metrics, and context-specific criteria. The bridge between these two can be realised only through embracing interdisciplinary approaches and collaboration within the machine learning and AI research community, the cognitive science community, the behavioural science community, and the governance community, in order to ensure that explainability can provide technical accountability and human understanding (OECD, 2024; UNESCO, 2023; Miller, 2019).

The second major gap is the lack of user-centred regulatory measures that consider diversity across cognitive and cultural backgrounds. Current policies tend to assume a uniform user and do not account for the impact of domain expertise, cultural background, or the consequences of decisions on the efficacy of explanations (Petrik & Varnum, 2024). This is especially worrisome when AI is used in a globalised context, where standardised explanations might not resonate in a variety of institutional and cultural contexts. In addition, biases in the framing of explanations can be systematic, and behavioural research on framing effects shows systematic biases in decision-making (Tversky & Kahneman, 1981), posing ethical issues concerning the manipulation of users' judgments by allegedly transparent systems.

Overall, AI explainability governance is at an urgent crossroads, with a shift from compliance-driven transparency to cognitively informed, context-aware paradigms. To promote this agenda, it is necessary not only to be technologically innovative but also to reconfigure policy paradigms to encompass human-centred measures, interdisciplinary knowledge, and adaptive governance systems. By filling these gaps alone, explainability can serve its dual purpose as both an instrument of accountability and an instrument of effective human-AI cooperation.

INNOVATIONS, LIMITATIONS, AND EMERGING DIRECTIONS

Recent advances in eXplainable Artificial Intelligence (eXAI) are moving away from 'traditional' explanation methods towards methods that require user interaction or adapt to context, emphasising user engagement and contextual explanations. Conversational explainability systems facilitate multiple passes through the process of exploring causal relationships and counterfactual scenarios, which helps build a better understanding and confidence in more complex decision contexts by aligning explanations with the user's changing cognitive needs (Weick, 1995; Wassim, 2023). However, interactivity can also perpetuate biases and dependence on algorithmic recommendations and exert subtle persuasive effects that are hard to critique (Kuhl & Bush, 2025). Adaptive explainability is another improvement in cognitive fit, as it adapts explanatory information to the user's expertise and context. However, highly personalised explanations raise concerns about scalability, standardisation, and epistemic consistency. Therefore, an explainability system's future should be characterised by responsiveness, usability, transparency, fairness, and interpretability across different user groups (e.g., Clark, 2013; BIS, 2024).

As more and more adaptive and personalised explainability systems become available, there is an increasing need for ethical safeguards. While personalisation can help increase comprehension and cognitive alignment, it also has the potential to misrepresent information to users based on what it can infer about them. If not properly designed to counteract bias, adaptive systems can inadvertently exacerbate cognitive biases, can give unequal information to equal users making the same decision, or can be more transparent to one user – but not another – making the same decision.

There are, therefore, several governance protection aspects that need to be put in place. Adaptation mechanisms should be auditable, for example, to check and explain changes to explanations. Second, users need to be able to choose alternative presentation formats and levels of detail, rather than being forced to accept the system's chosen presentations. Third, adaptive systems should be periodically reviewed to ensure they are fair, consistent, and free of unintended behavioural consequences across diverse user populations. Lastly, transparency regarding personalisation should be kept, meaning that the users should be aware of how and why the explanatory information is being adapted. Personalisation is both important for understanding and protected by these safeguards, which prohibit its use to gain behavioural control over the individual or to exploit the mind (OECD, 2024; UNESCO, 2023; BIS, 2024).

There are new governance risks that arise with adaptive explainability beyond the traditional transparency issues. Personalised explanations can inadvertently reinforce existing misconceptions, be inequitable in terms of the information they provide, and enhance behavioural compliance rather than informed decision-making. Organisations should thus create auditing procedures that can assess behavioural influence, impacts on fairness, and consistency with user groups. Ethical considerations should recognise that adaptive explanations should be supportive rather than persuasive and respectful of meaningful human autonomy in decision-making (OECD, 2024; UNESCO, 2023).

Another novel area of XAI is the incorporation of neuro-symbolic methods, which unite the representational capabilities of neural networks with the logical clarity of symbolic reasoning. The goal of these hybrid models is to generate computationally robust and semantically meaningful explanations to overcome the long-standing criticisms of black-box opacity (Stammer et al., 2021; Arrieta et al., 2020). Neuro-symbolic systems achieve interpretability and increased trust by incorporating domain knowledge and causal structures into their learning architectures, thereby generating more human-like explanations. However, their practical use is limited by technical complexity and scalability, especially in areas characterised by high-dimensional, unstructured data.

Despite these advancements, there are still considerable ethical concerns, including manipulative explainability, persuasive interface design, cognitive exploitation, and the maintenance of user autonomy in increasingly adaptive AI environments. Explanations are not neutral descriptions but rather the stories they form, which can influence perception, judgment and behaviour.

The principles of cognitive ergonomics offer further insight into explanation design by highlighting the need to align the presentation of information with human perceptual and cognitive abilities. Research findings indicate that interface complexity, information volume, attention requirements, and memory capacity significantly affect the accuracy of interpretation a user can achieve when interpreting AI recommendations. Explainability systems that reduce human cognitive load while keeping the essential information needed to make decisions will thus more effectively enable human-AI collaboration and maintain good decision performance (Endsley, 2017).

Behavioural research has demonstrated such effects and shown that framing and the partisan presentation of information are systematic and that the information provided is correct. (Tversky & Kahneman, 1981). This raises ethical questions about

AI, as when AI generates an explanation that urges the user to do something they may not have wanted, it violates autonomy and informed consent (Miller, 2019).

Ethical issues of persuasive interface design go beyond typical "transparency" issues. A variety of cognitive biases, such as anchoring, default bias, framing effects, perceived urgency, and authority bias, can be exploited by structuring the interface. These design decisions can affect the interpretation and/or critical evaluation of the explanations and may influence user engagement with the system's recommendations. These design decisions can lead to different interpretations and/or critical evaluations of the explanation and affect users' engagement with the system's recommendations. Cognitive exploitation is the systematic use of explanation mechanisms to exploit human cognitive vulnerabilities and influence behaviour for the primary purpose of achieving system goals, rather than aiding the user.

Behavioural studies have shown that how a problem is presented, how the interface is designed, and the framing of judgments can make a significant difference in how people make judgments, even if the explanation given does not change (Tversky & Kahneman, 1981). These mechanisms give rise to ethical issues in AI systems, as the explanations might serve as both a means of behavioural manipulation and a source of information. This risk is especially pronounced when users do not know which persuasive design elements may be included in explanation interfaces. Ethical explainability must focus on maintaining the user's autonomy, supporting their consent, and not on design practices that exploit vulnerabilities in their cognitive abilities to steer their behaviour. Meanwhile, studies show that competing explanations can cause cognitive overload and anxiety and create decision overload when users have to assess ambiguous results in a time-constrained setting (Messner & Siering, 2025). These considerations underscore the need for an ethical commitment and for effective explanation.

This is especially important in an environment with an information imbalance, limited knowledge, or high cognitive load. The user feels in control, but there is a covert explanation design that guides towards desired results. The architecture, presentation, and behavioural impact of interfaces, as well as the effects of persuasive mechanisms, must therefore be constantly evaluated to prevent any loss of user autonomy, fairness, or the integrity of the decision (Tversky & Kahneman, 1981; Miller, 2019; Poursabzi-Sangdeh et al., 2021).

Research in recent years has revealed high cognitive demands of explainability, which may lead to higher cognitive load, anxiety, and decision fatigue, and limit its usefulness in practice (Messner & Siering, 2025). These pressures are exacerbated by the need to balance personalisation and scalability across all of this. Cognitive fit and inclusiveness can be achieved through adaptive explanations, which are more difficult to implement but can be achieved with more data, computational power, and governance oversight, while standardised explanations are easier to implement but do not work for the cognitive and cultural differences (Vattonen & Reunanen, 2024; Petrik & Varnum, 2024). At a more basic level, explainability continues to face conflicting pressures of transparency and predictive power, simplicity and fidelity, and trust and critical assessment. The trade-offs suggest that no simple explanation always works, further underscoring the need for a context-sensitive approach that provides usable explanations and is accurate and responsible in decision support (Rudin, 2019; Essen & Meske, 2023).

The discussion above indicates that explainability should be considered a process rather than a single technical solution, as its application to a problem depends on context. There is likely to be further progress in interdisciplinary work that brings together the fields of machine learning, cognitive science, behaviour, and organisational theory. This integration allows explainability systems to be flexible across various decision contexts while simultaneously meeting multiple demands for explainability, usability, accountability, and prediction performance. This stance creates a more international and relevant basis for the design and governance of human-centred AI systems.

IMPLICATIONS

Explainable Artificial Intelligence (XAI) is a cognitive process and a socio-organisational practice where reconceptualisation is gaining momentum and importance as a key factor. While transparency, interpretability, and justification after the fact have been emphasised, there is mounting evidence that explanations are effective in operating within the bounds of rational thought (Simon, 1955; Arrieta et al., 2020) and in how people perceive, interpret, and apply information. Cognitive explainability, in turn, introduces the notion of the explanatory process as dynamic, influenced by the mental models, prior beliefs, and predictive reasoning mechanisms people use (Clark, 2013; Albini et al., 2023). This view brings decision intelligence to the level of integrating algorithmic outcomes with uncertainty and human judgement in a relational outcome, namely explainability (Berndt & Vössing, 2025).

For application, the need for effective decision-supporting systems implies that explanation strategies must be adaptive to cope with different levels of expertise, cognitive styles, and constraints. Empirical studies have shown that explanations aligned with users' mental models enhance understanding, trust calibration, and decision quality, while simplification can sometimes lead to fidelity issues and affect informed judgement (Essen & Meske, 2023; Poursabzi-Sangdeh et al., 2021). Human-AI collaboration also requires the ability to reflect critically and to be "calibrated," or appropriately trusting, between human and AI explanations, as a cognitively pleasing explanation can serve to reinforce algorithmic bias when taken at face value (Kuhl & Bush, 2025).

These are governance and policy challenges as well. While the value of transparency, fairness, and auditability is well recognised in regulation, the focus tends to be on interpretive effectiveness, with cognitive usability overlooked (BIS, 2024; Rosen & O'Leary, 2024). Therefore, there is a need for interdisciplinary approaches that combine methods from behavioural science, cognitive psychology, organisational theory, and computer science to realise explainability governance. There is also cultural and contextual variability that needs to be considered, making standardisation extremely difficult and underscoring the importance of adaptive governance mechanisms sensitive to varied institutional contexts (Petrik & Varnum, 2024). In conclusion, the future of explainability is about human-centred design, interdisciplinary research and collaboration, and context-based regulation to facilitate responsible, effective, and cognitively meaningful human-AI decision-making.

FUTURE RESEARCH AGENDA

The research program for explainable artificial intelligence (XAI) should view explanations as dynamic cognitive artefacts whose effects vary over time and across situations. Available evidence remains largely based on cross-sectional studies, which do not provide insights into the impact of repeated explanatory exposure on mental models, trust calibration, learning, and decision-making competence (Essen & Meske, 2023). It is therefore essential to conduct longitudinal studies to determine the extent to which explanations can lead to lasting understanding or, conversely, to exacerbate automation bias and reliance on algorithmic thinking (Clark, 2013; Simon, 1955). Additionally, there is a need to conduct cross-cultural validation, as preferences for explanations, decision heuristics, and trust-building in the socio-institutional context differ. These disparities call into question the universal and universalising nature of the explanations and underscore the importance of culturally responsive explainability paradigms that can support meaningful interpretation across a range of populations (Petrik & Varnum, 2024; Vattonen & Reunanen, 2024).

The need for the future is the development of a standardised set of Cognitive Explainability Indicators (CEIs) that enable consistent assessment across studies and applications. These can range from comprehension of explanations to calibrated trust ratios to cognitive load measurements to indices of decision-quality improvement to measures of bias mitigation to mental-model consistency assessments. The creation of validated CEIs would make it easier to compare explainability interventions and accelerate the development of cognitive explainability research (Albini et al., 2023; Rosen & O'Leary, 2024). Comparative multinational research can therefore be important not only for creating design principles that are relevant to the specific context of the problem, but also for addressing biases that can occur in current data and assessment tools.

Future research on XAI should evaluate beyond self-reported trust, employing neurocognitive measures of attention, cognitive load, and affective response during the explanation process. While these techniques offer promising possibilities for better understanding how users interpret and internalise AI explanations, they also have several drawbacks, such as scalability, privacy, the interpretation of neural data, and ecological validity (Friston, 2005; Seth & Friston, 2016; Thagard, 2003). Meanwhile, in the case of multi-agent and autonomous systems, it is essential to have systemic explainability frameworks as decisions are made through interactions among multiple human and artificial agents (Sado & Takac, 2023). A unified research program needs to consider the fields of neuroscience, cognitive psychology, computer science and organisational studies to create scalable and ethically sound models for complex human–AI decision-making environments (Weick et al., 2005).

CONCLUSION

In this chapter, explainability is returned to human cognition and socio-organisational processes, rather than being merely a technical property of AI systems. Based on the concepts of explainable AI, cognitive science, and decision intelligence, it suggests that good explanations arise from a dynamic interplay among algorithmic results, human thinking, prior beliefs, and conditions (Simon, 1955; Clark, 2013; Weick et al., 2005). The analysis shows that explanations, which are consistent with the mental models of users, have a positive impact on the understanding, calibration of trust and quality of decisions, while too complex explanations or explanations that are not suitable for the context increase the cognitive effort that the user must expend, can lead to the acceptance of the explanations uncritically, and can lead to misunderstandings (Essen & Meske, 2023; Messner & Siering, 2025). Moreover, the design of explanations has ethical implications, as it may either reduce or reinforce cognitive biases through framing, information selection, and interpretation. As a result, there are several ways to evaluate explainability beyond technical fidelity, including support for informed decision-making, context sensitivity, and responsible decision-making (Tversky & Kahneman, 1981; Miller, 2019).

The analysis also emphasises the ethical issues of the very process of explanation design. Explainability systems have become more adaptive, interactive and personalised, thus gaining more ability to influence users' perceptions and decision-making processes. This underscores the need to avoid manipulative explainability practices, safeguard user autonomy, mitigate cognitive exploitation, and ensure governance oversight of persuasive design mechanisms. Future studies of the role of explanations in understanding should also consider whether the explanation makes sense, is non-exploitative, fair, and decidable.

The chapter takes the concept of explainability beyond the narrower scope of decision explanation, viewing explanations as tools that enable interaction between human judgement and algorithmic reasoning (Berndt & Vössing, 2025). From this viewpoint, it is evident that the systems supporting decision-making must be cognitively compatible and must take into account user diversity, contextual variability, and evolving knowledge. On the other hand, the search for simpler and more faithful solutions, more transparent and more performative versions of explainability, and more trustful and critical versions of explainability illustrates the need for a flexible, rather than a one-size-fits-all, approach to explainability (Rudin, 2019; Poursabzi-Sangdeh et al., 2021). Existing governance models remain largely oriented towards transparency and accountability, with little attention to cognitive efficiency and cultural diversity (BIS, 2024; Rosen & O'Leary, 2024). To achieve this, a human-centred approach to algorithm design, multidisciplinary governance, and user-centric regulatory frameworks that facilitate informed, equitable, and responsible decision-making are needed to advance the cause of explainability. In conclusion, the benefit of explainable AI is not necessarily to make the algorithms' processes visible, but to enhance human understanding and judgement in situations of uncertainty (Petrik & Varnum, 2024).

REFERENCES

Abdelwanis, Y., Gafni, D., & Nissen, M. (2024). Cognitive processes while using Artificial Intelligence at work. *Formare, 45*(1), 23–41.

Albini, E., Arcangeli, M., & Barbu, E. (2023). Achieving descriptive accuracy in explanations via cognitive explainability. *PMC, 10117939.*

Arrieta, A. B., Díaz-Rodríguez, N., Del Ser, J., Bennetot, A., Tabik, S., Barbado, A., García, S., Gil-López, S., Molina, D., Benjamins, R., Chatila, R., & Herrera, F. (2020). Explainable Artificial Intelligence (XAI): Concepts, taxonomies, opportunities and challenges toward responsible AI. *Information Fusion, 58*, 82–115. DOI: 10.1016/j.inffus.2019.12.012

Berndt, J., & Vössing, M. (2025). *Decision intelligence: Merging human judgment and AI.* Harvard Business Review Analytics.

BIS. (2024). *Managing explanations: how regulators can address AI explainability* (FSI Papers No. 24). Bank for International Settlements.

Clark, A. (2013). Whatever next? Predictive brains, situated agents, and the future of cognitive science. *Behavioral and Brain Sciences, 36*(3), 181–204. DOI: 10.1017/S0140525X12000477 PMID: 23663408

Corlett, P. R., & Friston, K. J. (2020). What do we think about when we think about predictive processing? *Frontiers in Systems Neuroscience, 14*, 7509909. PMID: 32757598

Costa, R. D., & Silva, P. (2025). Reinforcement learning applied to a situation awareness model with cognitive explainability. *Information Sciences, 689*, 121–140.

Endsley, M. R. (2017). From here to autonomy: Lessons learned from human–automation research. *Human Factors, 59*(1), 5–27. DOI: 10.1177/0018720816681350 PMID: 28146676

Essen, T. V., & Meske, C. (2023). Expl(AI)ned: The impact of explainable artificial intelligence on users' information processing. *Information Systems Research, 34*(2), 460–481.

European Banking Authority. (2021). *Guidelines on loan origination and monitoring.*

Friston, K. (2005). A theory of cortical responses. *Philosophical Transactions of the Royal Society of London. Series B, Biological Sciences, 360*(1456), 815–836. DOI: 10.1098/rstb.2005.1622 PMID: 15937014

Kaur, H., & Singh, M. (2024). *Cognitive decision systems: Intelligent AI for better decisions*. Cognitive Economy.

Kuhl, U., & Bush, A. (2025). When bias backfires: The modulatory role of counterfactual explanations on the adoption of algorithmic bias in XAI-supported human decision-making.

Lee, E., & Kim, J. (2025). A comparison between SHAP and clinician-friendly explanations reveals enhanced trust in surgical decision-making. *PMC, 12475050.*

Long, D., & Zou, J. (2024). Sensible AI: Re-imagining interpretability and explainability using sensemaking theory. *Proceedings of the 2022 ACM Conference on Fairness, Accountability, and Transparency*, 789–801.

Lundberg, S. M., & Lee, S.-I. (2017). A unified approach to interpreting model predictions. *Advances in Neural Information Processing Systems*, *30*, 4765–4774.

Messner, W., & Siering, M. (2025). The cognitive cost of AI: How AI anxiety and attitudes influence cognitive strain. *Journal of Business Research*, *189*, 114–132.

Miller, T. (2019). Explanation in artificial intelligence: Insights from the social sciences. *Artificial Intelligence*, *267*, 1–38. DOI: 10.1016/j.artint.2018.07.007

Moore, J. H., & Williamson, C. (2022). Artificial intelligence explainability in clinical decision support systems. *PMC, 9931364*.

NIST. (2022). *Towards a standard for identifying and managing bias in artificial intelligence* (NIST SP 1270). National Institute of Standards and Technology.

OECD. (2024). *OECD Framework for the Governance of Artificial Intelligence Systems*. Organisation for Economic Co-operation and Development.

Petrik, M., & Varnum, M. (2024). *Cultural bias in explainable AI research: A systematic analysis. Research Portal*. Utrecht University.

Poursabzi-Sangdeh, F., Goldstein, D. G., Hill, J. M., & Lacotte, S. (2021). Manipulating and measuring model interpretability. *Proceedings of the 2021 CHI Conference on Human Factors in Computing Systems*, 1–52.

Regulatory Affairs Journal. (2023). Explainable Artificial Intelligence (XAI) Framework for Transparent Clinical Decision Support. *International Journal of Modern Trends in Life Sciences, 6*(4), 1427.

Ribeiro, M. T., Singh, S., & Guestrin, C. (2018). "Why should I trust you?": Explaining the predictions of any classifier. *Communications of the ACM*, *61*(8), 112–123.

Rosen, C., & O'Leary, C. (2024). Explainability in AI policies: A critical review of communications, reports, regulations, and standards in the EU, US, and UK. *ACM Computing Surveys*, *56*(8), 1–36.

Rudin, C. (2019). Stop explaining black-box machine learning models for high-stakes decisions and use interpretable models instead. *Nature Machine Intelligence*, *1*(5), 206–215. DOI: 10.1038/s42256-019-0048-x PMID: 35603010

Sado, F., & Takac, B. (2023). Explainable goal-driven agents and robots. *ACM Transactions on Interactive Intelligent Systems*, *13*(2), 1–34.

Seth, A. K., & Friston, K. J. (2016). Active interoceptive inference and the emotional brain. *Philosophical Transactions of the Royal Society B: Biological Sciences, 371*(1708), 20160007.

Shmueli, E. (2010). To explain or to predict? *Statistical Science*, *25*(3), 289–310. DOI: 10.1214/10-STS330

Simon, H. A. (1955). A behavioural model of rational choice. *The Quarterly Journal of Economics*, *69*(1), 99–118. DOI: 10.2307/1884852

Stammer, W., Friedrich, L., & Baier, D. (2021). *Revising neuro-symbolic concepts by interacting with their explanations*. CVPR Neuro-Symbolic XIL.

Thagard, P. (2003). Why cognitive science needs philosophy and vice versa. *Topics in Cognitive Science*, *1*(2), 237–254. DOI: 10.1111/j.1756-8765.2009.01016.x PMID: 25164930

Tversky, A., & Kahneman, D. (1981). The framing of decisions and the psychology of choice. *science, 211*(4481), 453-458.

Vattonen, A., & Reunanen, M. (2024). Preferences for AI explanations based on cognitive style and socio-cultural factors. *Proceedings of the ACM on Human-Computer Interaction, 8*(CSCW1), 1–26.

Wassim, M. (2023). *Explaining by conversing: The argument for conversational XAI* [Harvard undergraduate thesis]. Harvard DASH. https://dash.harvard.edu/server/api/core/bitstreams/934a4899-db75-4d91-9ed4-a259ec234485/content

Weick, K. E. (1995). *Sensemaking in organisations*. Sage Publications.

Weick, K. E., Sutcliffe, K. M., & Obstfeld, D. (2005). Organising and the process of sensemaking. *Organization Science*, *16*(4), 409–421. DOI: 10.1287/orsc.1050.0133

Chapter 3
From Brain to Machine: Cognitive Mechanisms Driving AI-Enhanced Decision Intelligence

G. Boopathi Raja
http://orcid.org/0000-0003-3345-4503
Velalar College of Engineering and Technology, India

ABSTRACT

Decision Intelligence is a new approach to AI that utilizes the insights we can gain from how humans make decisions. This chapter focuses on key cognitive processes (human brain functions) such as perception, attention, memory, learning and decision-making, with the goal of providing inspiration for developing more sophisticated AI systems. By bringing together knowledge bases from a variety of fields (neuroscience, psychology & computational intelligence) this chapter outlines how these human brain-like models can be adapted to machine learning architectures through the use of attention mechanisms, memory-augmented systems, neurosymbolic reasoning and affective computing. Ethical issues like transparency, bias and the potential dangers of brain/computer interface integration are examined to encourage innovation with responsibility. Future trends in neuromorphic computing, hybrid intelligence and human/computing collaboration will be identified to establish decision intelligence as an essential building block for the future of AI.

1. FOUNDATIONS OF COGNITIVE MECHANISMS

Decision Intelligence provides an alternative, empowered perspective of how AI systems will develop from the cognitive design of humans. The human brain is as much an organ as it is a means of processing information. As an example, the cognitive functions performed by our brain include perception and attention, mem-

DOI: 10.4018/979-8-3373-8988-2.ch003

ory and learning, and ultimately leading to decision-making through coordinated behavioural expressions. This section will discuss the neural architecture and cognitive processes of human beings in order to create a foundation for AI systems to replicate, compliment and/or augment our ability to reason with other people. Often, this neural organization and cognitive functioning of humans is a result of utilizing basic building blocks of brain-inspired AI systems. By integrating knowledge from the disciplines of neuroscience, psychology, and computational intelligence, we are able to provide a conceptual foundation to build decision intelligence systems.

Figure 1. From brain to machine: A schematic pathway to AI-enhanced decision intelligence

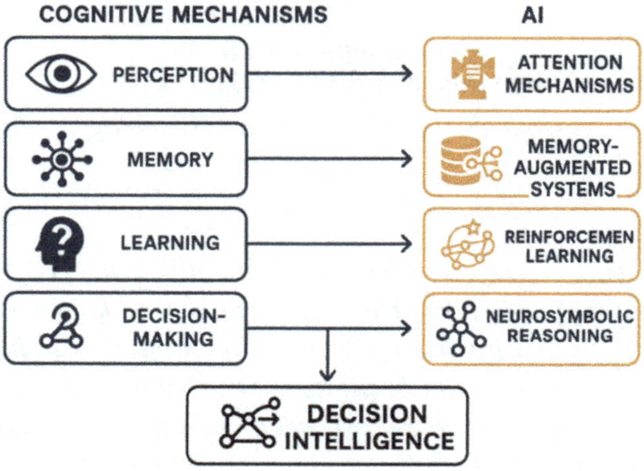

Figure 1 outlines how humans and their AI counterparts exhibit the same core cognitive functions—perception, memory, learning, and decision making—and portray them graphically. The human brain performs all four functions using specific neural processes: (1) attention to perceived inputs; (2) contextual information from memory; (3) reinforcement from memory; and (4) reasoning. The figure exhibits both an AI model and a high-level conclusion—by bridging human cognition and AI systems, the development of intelligent systems capable of independent and dynamic decision making across health care, finance, smart cities, and robotics—by establishing machine-level reasoning as the same as human cognition, which are biologically grounded and computationally enhanced as a synergistic whole(Boopathi Raja, 2021).

1.1 Neural Architecture and Information Processing in the Brain

There are estimated to be around 86 billion neurons in the human brain, all of which are connected through trillions of synapses that form a dynamic network that can perform parallel processing and distributed processing (Jiao et al., 2024). Neurons communicate with each other using both electrical impulses and chemical neurotransmitters. This communication provides for rapid signaling between neurons and allows for long-term modification of neuronal function. The architecture of neurons in the brain allows for the hierarchy of processing sensory information, which can be transformed into higher levels of meaning in our brain.

Figure 2. Neural architecture and information processing in the brain

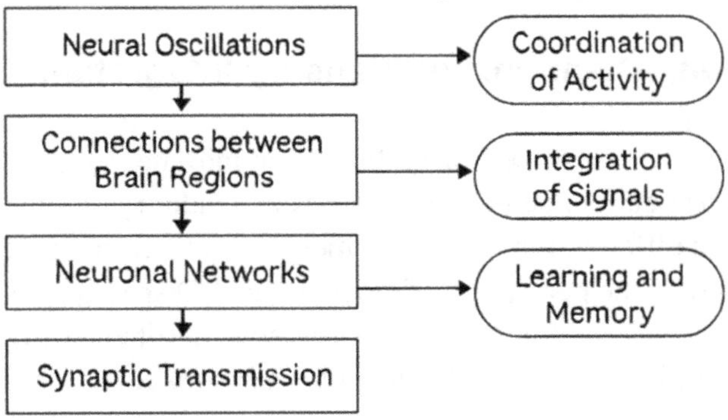

The architecture of brain is therefore similar to a very large multi-parallel processing system that has been optimized for massive amounts and constant change of information in the most energy efficient way. Additionally, the brain is able to integrate information over time and space through the mechanisms of neural oscillations, synaptic plasticity and recurrent connectivity. The emergence of emerging fields such as neuromorphic computing demonstrates the value of a hardware design and engineering processes that mimic the operation of spiking neurons to provide energy efficient computing (Wang et al., 2024; Zeng et al., 2023). Therefore, designing machine architectures based on the biological principles of the brain is a crucial part to ground the development of AI systems in cognitive neuroscience.

Figure 2 depicts the structure of a human brain's neuronal architecture, illustrating how synaptic transmission flows through a network of neurons into a connected brain region and then into a logged neuromodulatory signal. These four processes represent four distinct categories of cognitive functions—action coordination, signal integration, learning/memory—and demonstrate to what extent a brain can adaptively and energy-efficiently process information through parallel/distributed processing, synaptic plasticity, and recurrent connectivity; these processes establish the biological basis for brain-inspired AI systems. Table 1 provides the overview of mapping AI with brain.

Table 1. Overview of brain–AI mapping

Brain Region	AI Paradigm	Functional Parallel
Cortex	Deep Neural Networks	Hierarchical feature extraction
Thalamus	Modular Networks	Sensory routing and integration
Basal Ganglia	Reinforcement Learning	Action selection and reward processing
Cerebellum	Spiking Neural Networks	Timing, coordination, and adaptive control

1.2 Perception, Attention, and Sensory Integration

The process that converts sensory information into meaningful representations is referred to as perception. Sensory inputs (vision, auditory/cognitive, tactile, proprioception) from different modalities are merged together into perceptual units or representations (e.g., the perception of an object in 3-D space) by the brain. Inputs from different modalities are combined, processed, and filtered by cortical and subcortical regions of the brain (e.g., the thalamus, sensory cortices) to form the final perceptual representation through an integration process (Langdon et al., 2022).

Attention enables selective enhancement of relevant stimuli while suppressing distractions. Cognitive neuroscience research indicates there are two mechanisms for attention: bottom-up (or stimulus-driven) and top-down (or goal-directed). Examples of the brain networks that mediate these different types of attention processes reside in the parietal and prefrontal cortices (Harris, 2024). In contrast to cognitive neuroscience, the introduction of the concept of attention to AI has produced a major breakthrough in machine learning due to attention mechanisms built into transformer-based architectures which replay and update the weights attached to input features dynamically (Gonzalez, 2024). As a result of this emulation of human attention, machines can now concentrate their computational power on high-salience data.

An additional example of the brain's ability to do multimodal processing is the McGurk effect, where visual and auditory stimuli work together to create a perception. AI has also begun utilizing multimodal learning approaches to enhance perceptual encoding by integrating a combination of text, images and sensor information, which can be interpreted as evidence that the principles of cognitive neuroscience are being applied to the field of machine intelligence(Chen et al., 2022).

Figure 3. Mapping perception and attention: From sensory integration to AI models

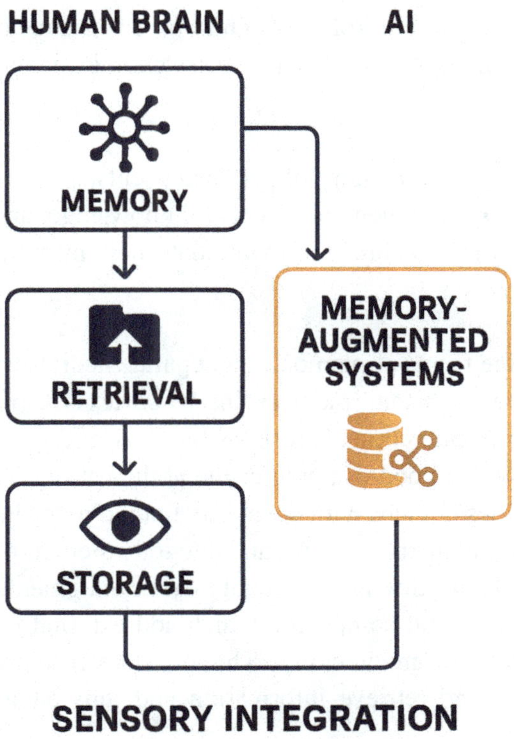

Figure 3 shows how sensory data - visual, auditory, tactile, and proprioceptive - from the human brain flows into perceptual representations, influenced by both bottom-up and top-down attention processes. Further, parallel processing related to perceptual attention can be seen in AI via transformer-type attention models and multimodal learning systems that utilize multiple forms of input, e.g., documents, images, and sensory data. The diagram illustrates the role that selective attention and the integration of visual and auditory stimuli play in an individual's ability to perform cognitive functions and decision making through the application of similar computational architectures in AI.

1.3 Memory Systems: Episodic, Semantic, and Working Memory

Memory plays an important role in cognition: It is the means of retaining and retrieving information over time. Cognitive psychology identifies three distinct forms of memory:

- ***episodic memory:*** the memory of personal events,
- ***semantic memory:*** the memory of general knowledge, and
- ***working memory:*** the ability to manipulate new information in short-term storage.

Each of these three forms of memory has separate neural structures: the hippocampus is responsible for episodic memory; the neocortex is responsible for semantic memory; and the prefrontal cortex is responsible for working memory.

Working memory is a good example of the ability of brain to actively process information. It provides humans with the ability to hold many items of information simultaneously, manipulate those items, and integrate them into a decision-making process. AI systems do this through the use of memory-augmented neural networks, such as differentiable neural computers, which add external memory modules to traditional neural network architectures. This provides machines with the ability to dynamically store and retrieve information and thus enhance reasoning and problem-solving.

Episodic and semantic memory serves as sources of inspiration for case-based reasoning and knowledge graphs within the realm of AI. By representing knowledge as a network of nodes, machines can simulate a semantic network and thus be able to make inferences and generalizations (Raikov & Pirani, 2022). The relationship between these three memory systems supports the importance of contextual and relational knowledge for both human and machine cognition.

1.4 Learning and Adaptation: Reinforcement and Hebbian Principles

Learning is the process through which experience leads to a change in cognition or behavior. Reinforcement learning and Hebbian learning are two principles that demonstrate how the brain adapts to the environment. Reward-based adaptation is at the core of reinforcement learning (RL). Dopaminergic pathways in the brain provide a mechanism for encoding prediction errors to provide input to the process of making decisions, and the mechanisms used to establish habits. RL algorithms within the AI realm replicate this process, allowing RL agents to improve upon their decision making through trial and error also known as learning. RL can be used in a diverse range of applications e.g. robotics, financial modeling, and therefore, demonstrates the cognitive-inspired nature of RL (Ren & Xia, 2024).

Hebbian learning can be summarized as cells that fire together wire together, and there is an emphasis on synaptic plasticity. When two neurons fire together, they grow closer together and are in a sense connected to one another, serving as a basis for associative learning. This idea has led to the development of unsupervised learning algorithms in AI, which rely on observations of correlations within sets of data to create representations of those data (Langdon et al., 2022). Hebbian-inspired neural models in AI include: self-organizing maps, spiking neural networks; thereby bridging the gap between human learning, and artificial learning through the use of Hebbian-based learning models.

RL and Hebbian-based principles provide further understanding of the dual nature of the brain's ability to learn in an associative and a goal-directed manner through combined methods and processes, which allow AI systems to utilize both approaches to develop hybrid models capable of balancing exploration/exploitation and pattern recognition (Shi et al., 2022).

1.5 Decision-Making Processes in the Prefrontal Cortex

Decision making represents both a culmination of all cognitive processes demonstrated through: perception, attention, memory, and learning through actions demonstrated in all cognitive processes. The prefrontal cortex plays a significant role in making decisions because its support of piecemeal executive functions associated with: planning, inhibition, and evaluation.

Studies from neuroeconomics show how the prefrontal cortex (PFC) helps to encode values, probabilities, and risks; which enables humans to make effective decisions in complex environments. Humans do not make completely rational decisions; different emotions, biases, and social contexts affect each person's perspective. For this reason, theories such as bounded rationality point to the fact that humans have limits on their ability to think, reason, solve problems, communicate, and make decisions based on the information available. In this regard, many AI systems mimic the cognitive processes used by humans and are built based on heuristics to make decisions under uncertain circumstances through methods of probabilistic reasoning, Bayesian inference, and decision-making, which attempt to find the right balance between accuracy versus computational efficiency (Pratt et al., 2023).

Thus far, research has been focused on the complementarily between human decision-making and machine-based decision-making. Gonzalez & Heidari state that human-like AI can augment human judgment by alleviating cognitive burden and improving dynamic decision-making capability(Gonzalez & Heidari, 2025).

Further, hybrid intelligence-based frameworks show how humans and machines can work together to achieve improved outcome (Ye et al., 2022).

Each of these cognitive mechanisms contributes to a coherent understanding of intelligence through its neural architecture, perception, attention, working memory, long-term memory, learning, and decision-making mechanisms; all together provide different insights into how the brain thinks, adapts to its surroundings, and exhibits behaviours. As these principles are translated into AI models, AI models will not only emulate human-like cognitive thinking but also be able to extend human-like cognitive thinking through the application of cognitive neuroscience principles to establish decision intelligence as a framework that connects biology and computation.

By creating integration between computational and human-based reasoning, the integration will keep machine reasoning with aligned to all aspects of human reason, including value systems, capabilities and limits of human capabilities. As discussed in the subsequent section, translating cognitive components into architectures to develop cognitive architectures lays basis for future neuromorphic computing systems, systems structured with both human memory and human intelligence, and hybrid intelligent systems (IIIS), therefore placing brain-mimicked intelligence at the leading edge of innovations for the future.

2. TRANSLATING BRAIN FUNCTIONS INTO AI MODELS

Translating function of human brain to architecture used for AI is one of the most phenomenal and disruptive approaches to creating computational intelligence. The human brain continues to be the most advanced and effective means to process information. However, with the recent advancements in the areas of machine learning, neuromorphic computing, and cognitive-inspired architectures, researchers have been able to identify and replicate many of the basic functional components of the effects of perception, attention, memory, emotion and reasoning on consciousness. Therefore, this section will highlight how cognitive-functional components are transformed into computer intelligence systems and provide information on the relationship between neuroscience and computational intelligence as it relates to cognitive-functional processing. In addition, the authors will reference other disciplines to provide the following: design principles to consider when designing systems; the various cerebral architectures that can be used to replicate the human cognitive functions; and different forms of machine intelligence that exemplify the translation of human cognitive functions into machine intelligence systems.

Figure 4. Translating brain functions into AI models: A cognitive-to-computational mapping

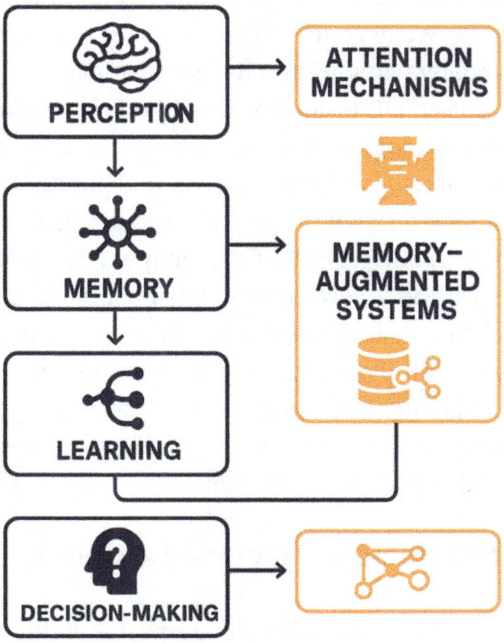

Figure 4 shows a conceptual translation of human cognitive processes—perception, memory, learning, and decision-making—into AI functions: attention mechanisms, memory augmentation, reinforcement learning, and neural representation of abstract concepts. Each of these cognitive functions has an analogue in computing, and the relationship of the analogues demonstrates how biology informs machine architecture. This illustrates that the mapping of the cognitive processes onto machines is what creates the basis of decision intelligence, empowering AI systems to mimic and augment human cognitive functions and provide dynamic context-sensitive solutions across multiple domains such as healthcare, financial services, and robotics.

2.1 Neural Networks and Neuromorphic Computing

Artificial neural networks (ANNs) are modeled after developing a system of artificial neurons that are similar in structure and function of biological neurons. Each artificial neuron utilizes a technique called weighted summation to process incoming signals, after which it generates an output by passing the summation of the weighted signals through a non-linear activation function that simulates the manner in which neurons synaptically integrate and fire their output signal to other neurons. ANNs prior to deep learning were simplistic representations of the human cognitive functions. ANNs since the emergence of deep learning architectures now have complicated multi-layer, hierarchical structures that function similarly to human cortex (Ren & Xia, 2024; Wang et al., 2024).

Neuromorphic computers go one step further than many of the other computing systems we have seen so far. Instead of simply running software that emulates the spiking behaviour of biological neurons, they are able to build hardware that replicates this behavior directly and thus creates a computing system similar to the human brain. For example, through the use of a spiking neural network (SNN), a neuromorphic computer can display all of the temporal dynamics required to process data in real-time with very low energy consumption (Zeng et al., 2023). As evidenced by the Intel Loihi neuromorphic chip's ability to perform event driven computations, there are multiple stages of power consumption reductions, while still enabling reliable adaptive behaviour. As such, they are paradigm of how to translate the neural principles of the brain into machine architectures and provide avenues for creating scalable, biologically plausible forms of AI.

2.2 Attention Mechanisms in Machine Learning

On the other hand, attention mechanisms underpin the ability as human beings to attend to and concentrate on one stimulus whilst filtering out all other distractions. Hundreds of applications have relied on this principle to transform how AI solves problems. For example, by employing a transformer architecture based on self-attention, long range dependencies across the entire set of input features are now dynamically weighted by the model as it captures contextual relationships from all locations in the input space(Ajitha et al., 2025).

This new way of processing information in a machine shares similar functions to the attention networks found in the human brain (e.g., the prefrontal and parietal cortices) by modulating incoming sensory data (Harris, 2024). The success of attention models in tasks such as machine translation, image captioning and multimodal learning has demonstrated that they can perform at or near the level of state-of-the-art AI systems (Gonzalez, 2024). Additionally, attention mechanisms in AI systems reduce the cognitive load required by the AI system, thereby behaving much like the highly efficient manner in which our brains manage very complex, large quantities of incoming information (Gkintoni et al., 2025).

2.3 Memory-Augmented AI Systems

Multiple types of memory systems combine to form the basis for human cognition; i.e., episodic, semantic, working memory are types of memory systems that work together as a complex unified system—the contributions of each play an important role in how humans understand their world through memories. A logical extension of these functions has been the development of memory-augmented architecture for AI. As an extension to traditional neural networks, external memory modules have been added to enhance the functioning of differentiable neural computers (DNC) and Neural Turing Machines, both of which enable dynamic allocation of memory and facilitate the storage and retrieval of information.

A working memory model can be used to emulate working memory by allowing the AI system to hold intermediate representations during the problem-solving process. Episode-based memory is the basis for case-based reasoning based on the use of prior episodic memory to make inferences about current problem solving based on past experiences (Raikov & Pirani, 2022). Similarly, knowledge graphs can provide their structure for information organization like semantic memory does—consolidated by nodes, and connected through one another to facilitate inference and generalization (Chen et al., 2022). By placing memory into AI, researchers and scientists are able to improve AI's ability to utilize reasoning and planning effectively, as well as increase AI's ability to adapt to changes in the environment, thereby eliminating the gap between statistical-based processes and symbolic-based processes of knowledge.

2.4 Emotion-Inspired and Affective Computing Models

Emotions are an essential part of human cognition and play an influential role in how we perceive our environment, how we remember things and how we make decisions. As supported by theories about the role of affective neuroscience in the way in which our behavior is formed from the interaction of the limbic system (emotion) and the neocortex (reason), machine emulation of these relationships has resulted in the development of affective computing for machines to detect, comprehend and show empathy towards emotion.

By integrating physiological signals, non-verbal (facial expression) and verbal (speech) characteristics to establish an affective state, a model based upon human emotions can be developed to help applications in healthcare monitoring, adaptive tutoring and improve human–computer interaction (Saw et al., 2025).

Integrating emotional intelligence with AI enhances the levels of empathy, personalization and contextual awareness as a component of the overall AI experience. There are, however, many ethical issues with implementing emotionally intelligent AI regarding privacy, bias, and anthropomorphism (Floridi & Nobre, 2024). To ensure responsible design of emotionally intelligent AI, the positive impact of technological innovation on society needs to be considered and balanced.

2.5 Neurosymbolic Integration for Reasoning and Decision-Making

Neurosymbolic integration creates the capability to use both statistical learning (e.g. neural networks) and symbolic logic (e.g. human-like learned reasoning capabilities) in the development of AI. In doing so the limitation of neural networks in respect to symbolic reasoning and/or interpretability is alleviated. Neurosymbolic integration effectively creates an implementation of the human brain's two processing systems - (i) associative networks that facilitate pattern recognition, and (ii) rule-based reasoning systems that facilitate abstract thinking (Sheth & Roy, 2024).

Neurosymbolic AI is distinguished by the ability to engage in logical deduction, causal reasoning, and the representation of knowledge whilst still benefiting from the neural/learned capability of neural networks. Utilize the power of both connectionist and symbolic paradigms to bring the application of neurosymbolic integration to the forefront of AI innovation and create cognitive-inspired AI systems.

2.6 Translating Brain Functions into AI Models

Translating brain function into AI models remains challenging despite significant progress because cogitation is complex, non-linear and contextual, thus directly emulating the brain is difficult; neuromorphic hardware has limited scalability, and ethical issues exist with respect to affective computing due to privacy and manipulation concerns. Neurosymbolic integration is promising, but advances are needed to create heterogeneous representations into a unified representation to be achieved via interdisciplinary collaboration including neuroscience, psychology, computer science and ethics.

Translating brain functions into AI models, as a result of the convergence of biology and computation is illustrated via the use of neural networks, attention mechanisms, memory-augmented systems, affective computing, and neurosymbolic integration; however, they not only emulate human cognition, also extend human cognition therefore allowing AI systems to achieve decision intelligence via complex environments.

By grounding AI in cognitive neuroscience, researchers can maintain alignment between machine reasoning and human capability, as well as with human values. Through this synergy, brain-inspired AI is positioned to become a seminal basis for future innovation, thus bridging the divide between natural and artificial forms of intelligence. Table 2 provides the comparison of core architectural principles of brain and AI.

Table 2. Brain vs. AI: Core architectural principles

Aspect	Human Brain	AI Systems
Basic Unit	Neurons (electrochemical signaling)	Artificial neurons (mathematical functions)
Connectivity	Dense, dynamic synaptic networks	Weighted connections in neural networks
Processing Style	Parallel, distributed, adaptive	Parallelized but often rigid (layered)
Energy Use	Extremely efficient (~20W for whole brain)	High computational cost, especially in training
Learning	Synaptic plasticity, reinforcement, Hebbian learning	Gradient descent, backpropagation, reinforcement learning
Specialization	Different brain regions recruited selectively	Mixture-of-experts, modular architectures (e.g., GPT-4)

2.7 Case Studies of Translational Models

a. Healthcare Diagnostics

Wearable EEG systems represent a new frontier of translating how the human brain functions into AI implementation. Shin et al. have developed brain/AI closed loop systems to enhance the autonomous decision-making capabilities of machines by utilizing neural signals to make machine decisions in real-time as a result of perceived sensory stimuli that require attention and/or decision-making(Shin et al., 2022).

b. Finance and Risk Modeling

Decision intelligence frameworks use attention-based and memory-augmented models in finance to improve fraud detection and portfolio management, based on models that mimic how humans make decisions under uncertainty and under dynamic conditions for robust model performance in dynamic environments (Pratt et al., 2023).

c. Smart Cities

The use of AI models to optimize traffic management, resource management, and energy management in smart cities, is based on cognition and senses integration, and attention mechanisms, as cognition is translated into machine architectures, thus able to make adaptive, effective decisions in urban systems (Luo et al., 2022).

3. COGNITIVE MECHANISMS DRIVING DECISION INTELLIGENCE

Decision intelligence represents the merger of cognitive science with AI. It is the systematic translation of knowledge and insight from human cognition into machine reasoning frameworks. The technical translation of brain function into AI models, this section will address the cognitive mechanisms themselves that serve to drive decision intelligence: how they create human-AI complementarity; how they reduce cognitive complexity; and, how they enable artificial agents to have a "theory of mind" (ToM) -- all of which serve to build hybrid intelligence. Ultimately, by understanding these mechanisms, we can identify the pathways through which cognitive-inspired AI will enhance decision making in dynamic and uncertain environments.

3.1 Human–AI Complementarity in Dynamic Decision-Making

Adaptability, context, and bounded rationality are characteristic of human cognition. However, due to cognitive biases, memory limitations, and the inability to process more than a limited amount of information, human beings often are unable to make optimal decisions. In contrast, AI systems have a distinct advantage over human beings; that is, they are capable of indexing, aggregating, processing, and analyzing the vast amount of data that is generated every day, performing probabilistic reasoning and optimization. Complementarity occurs when synergies between human judgement and machine computation reinforce one another.

According to Gonzalez & Heidari, dynamic decision making benefits from the application of this complementary relationship or symbiosis(Gonzalez & Heidari, 2025). Human beings interpret contextually, make ethical decisions, and imagine possibilities; at the same time, AI systems add computational precision and run predictive models. In healthcare diagnostics, for example, human doctors use patient narratives and additional contextual cues to interpret their exams, while AI systems examine patients by analyzing diagnostic imaging and genetics. AI and human diagnostics together will produce a higher level of accuracy than either can individually (Onciul et al., 2025).

Additionally, reciprocity can be applied to strategic areas such as finance and defense. Intuitive human judgement will assist in long-term planning and AI will assist in tactical execution. This is an example of the cognitive principle of distributed intelligence, which is when tasks are assigned based on capabilities, thus providing a level of resiliency and adaptability in environments with multiple levels of complexity.

3.2 Cognitive Load Theory and AI's Role in Reducing Complexity

The cognitive load theory (CLT) states that human working memory is limited in capacity and that providing too much information can negatively impact a person's ability to learn or make decisions. Humans experience this in high-risk areas like medical diagnostics or financial markets; too much cognitive load leads to errors, lost time and inefficient use of resources. AI systems, influenced by CLT concepts, are designed to decrease complexity through filtering, organizing, and providing a structured framework for information.

According to Gkintoni et al., AI has the capability to redefine learning effectiveness through providing adaptive interfaces and personalized feedback in order to manage an individual's cognitive load(Gkintoni et al., 2025). AI can also assist in smart city management by processing sensor data to help identify irregularities within traffic and energy consumption patterns—thereby alleviating some of the cognitive strain placed upon humans when attempting to manage these systems (Luo et al., 2022).

Through mimicking attentional control and the management of working memory, an AI system functions as a cognitive prosthetic, extending human capacity to manage complex systems. Hence, through utilizing this principle of CLT, decision intelligence systems will remain focused on being human-centered, enabling and enhancing the cognitive function of humans instead of overwhelming them.

3.3 Theory of Mind and Preference Learning in Artificial Agents

The theory of mind (ToM) has enabled humans to associate thoughts, emotions, and intentionality with either themselves, or with others. Thus, ToM plays an essential role in how one develops social cognition, empathy, and collaborative decision making. In order to apply ToM to AI agents, an AI agent must be capable of inferring user preferences through determining future actions and adjusting interactions to reflect users' preferences accordingly (Langley et al., 2022).

Preference learning, which is a computation-based form of modeling user choices and subsequently refining a recommendation made by the AI agent, will assist AI agents in making these types of decisions. Essentially, in educational settings, for instance, an AI tutor will provide tailored instruction based upon a student's expressed preferences or preferred learning style, resulting in improved experiences through more effective instruction (Gkintoni et al., 2025). In finance, deploying preference learning to optimize an investment portfolio will enable AI agents to mirror an individual's risk appetite while improving the chances of financial performance through alignment of individual investment preferences and investment vehicles (Pratt et al., 2023).

Gonzalez proposes cognitive algorithms that emulate human decision-making in dynamic environments, incorporating ToM principles to anticipate user needs(Gonzalez, 2024). This enhances human–AI collaboration, as machines become capable of interpreting not only explicit instructions but also implicit intentions. Such agents embody social cognition, enabling richer and more intuitive interactions.

3.4 Hybrid Intelligence: Parallel Cognition and Collaborative Systems

Hybrid intelligence integrates human and machine cognition into collaborative systems. Ye et al. introduce the concept of parallel cognition, where human and AI agents operate simultaneously, exchanging information and adapting strategies in real time(Ye et al., 2022). This framework mirrors distributed cognition in human groups, where collective intelligence emerges from interaction.

Hybrid systems are particularly valuable in domains requiring rapid adaptation. In robotics, cognitive control frameworks enable autonomous agents to collaborate with humans, balancing machine precision with human creativity (Yalawar et al., 2023). In smart cities, parallel cognition allows human planners and AI systems to co-manage resources, ensuring both efficiency and social responsiveness (Luo et al., 2022).

Ren et al. propose evolutionary roadmaps for human–machine collaborative decision-making, emphasizing iterative adaptation and co-learning(Ren et al., 2023). These systems embody the principle of complementarity, where human intuition and machine computation converge to achieve outcomes unattainable by either alone. Hybrid intelligence thus represents the future of decision intelligence, positioning AI as a partner rather than a replacement.

Figure 5. Hybrid intelligence: Parallel cognition and collaborative systems

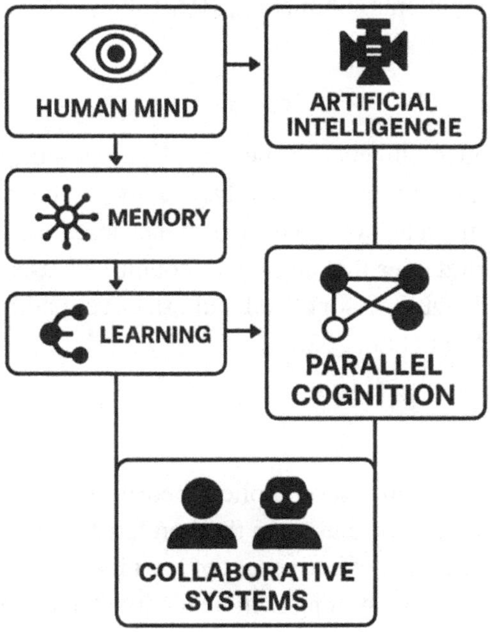

Figure 5 shows a framework for developing a unified hybrid intelligence based on human cognition and AI through a process of merging cognitive strengths into a Parallel Cognition system, where responsibilities are delegated according to each entity's respective capabilities. The final component will be through the establishment of Collaborative Systems, which can foster cooperative behavior between humans and machines in real time to increase efficiency, resilience, and ability to make better decisions in complex environments. This graphic illustrates one potential vision for upcoming AI technology and its use within the context of human intelligence—specifically as a partner to human intelligence rather than as a substitute for human cognitive function or thought process.

3.5 Cognitive Mechanisms in Practice

a. Healthcare

AI systems are utilizing cognitive load theory and preference learning to provide improved diagnostic information and personalized treatment. One way wearable electroencephalogram (EEG) devices can help accomplish this is through integrating various mechanisms of securing health-related (neurological) data through attention and perception systems to allow clinicians to make real-time therapeutic decisions that will most benefit patients (Shin et al., 2022).

b. Finance

The use of decision intelligence frameworks by combining the ability of humans and AIs to accomplish activities efficiently will also be beneficial for performing risk modeling and fraud detection. Using preference learning, these frameworks can develop investment strategies that align with cognitive biases and risk preferences to provide individuals with personalized and effective methods for making sound financial decisions (Pratt et al., 2023).

c. Education

AI-driven tutoring systems have applied theory of mind to assess and modify their interactions with students based on their preferences and cognitive load. This will significantly increase the effectiveness of learning as the content and pace provided to students using these systems will be identical to the principles of adaptive cognition (Gkintoni et al., 2025).

d. Robotics

The emergence of hybrid intelligence systems will allow autonomous agents (robots) to work collaboratively with humans in varying environments. Using cognitive control models will enhance the safety and effectiveness of robots in industrial and health care settings by allowing robots to adapt to the intentions of the human.

As the evolution of cognitive-based decision intelligence is driven by cognitive mechanisms (human/AI complementarily, cognitive load management, theory of mind, and hybrid intelligence), AI has improved its resiliency and adaptability and is able to provide tailored solutions by emulating and extending those same cognitive mechanisms found in humans. It is through the application of human values, capabilities, and limits that cognitive-based AI systems create a foundation for responsible innovation within the realm of decision intelligence.

4. APPLICATIONS OF BRAIN-INSPIRED DECISION INTELLIGENCE

Real-world implementations of decision intelligence offer the strongest validation of cognitive mechanisms in the creation of AI systems. With decision intelligence being created from architecture inspired by how the brain works (cognitive systems), it has moved beyond theoretical constructs and has been adapted to many practical applications, including healthcare, smart cities, robotics, finance, and education; each of these sectors contains complex decision-making environments that are characterized by uncertainty, dynamically changing data, and significant consequences. By the use of cognitive systems, AI models mimicking the processes used by the human brain to replicate how we perceive, pay attention to things, remember different types of information, learn, and make decisions, the integration of cognitive-based AI systems has improved the efficiency, accuracy, and adaptability of decision-making processes. In this section, we will discuss the many uses of decision intelligence or cognitive-based decision-making environments and provide examples of the potential for interdisciplinary convergence and ethical implications.

4.1 Healthcare: Diagnostics, Monitoring, and Personalized Treatment

Cognitive-inspired AI is revolutionizing the healthcare industry. The complexity of medical decision-making which requires integrating patient history, diagnostic imaging, genetic data, and clinical expertise parallels the principles of decision intelligence.

Figure 6 illustrates the AI-enabled healthcare pipeline. All sources of data used in healthcare (EEG, imaging, previous patient history, and wearables) ultimately feed into an attentional mechanism, an external memory, and/or an affective computational model. The ultimate result will be three different types of actionable data: 1) Diagnostics (discovery of a medical condition); 2) Real-time Monitoring (notify patients and/or caregivers of unusual activity); and/or 3) Personalized Treatment (develop adaptive treatment plans). The diagram clearly conveys how AI that uses cognitive principles will enhance the quality, timeliness, and customization of care within the healthcare system of the future.

4.1.1 Diagnostics

AI systems inspired by attention mechanisms and memory augmentation improve the accuracy of medical diagnoses. Saw et al.emphasized the importance of explainable AI in medical imaging, how using attention-based AI models draws attention to salient features present in radiographs or MRIs; these models also reduce diagnostic errors and increase the transparency of the diagnostic process because they emulate human attentional control(Saw et al., 2025). Recent study exemplifies how cognitive-influenced deep learning algorithms can be developed to identify breast cancers at an earlier stage and classify tumours based on genetic information (Raja, 2022b, 2025a).

Figure 6. Healthcare: Diagnostics, monitoring, and personalized treatment

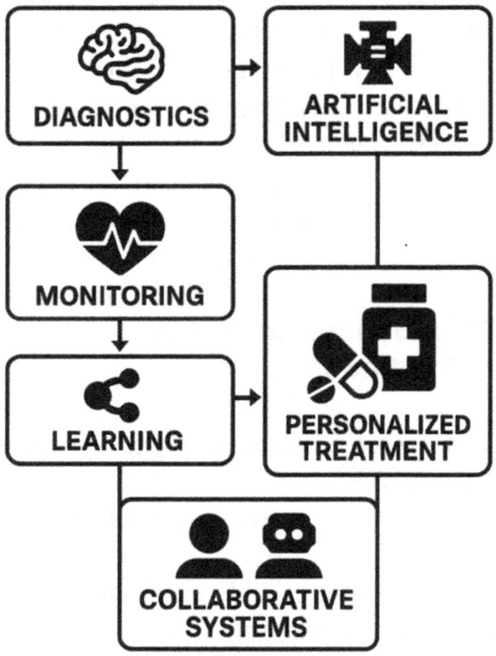

4.1.2 Monitoring

Wearable and implantable devices are examples of the integration between perception and decision-making processes. Recent research by Shin et al. has resulted in the development of wearable electroencephalogram (EEG) electronics that enable brain–AI closed-loop systems to monitor neurological states in real time(Shin et al., 2022). Such closed-loop systems incorporate sensory integration and attention control in order to support autonomous therapeutic applications. Moreover, Raja and Chakraborty have highlighted the ability of IoT-enabled wearable systems to facilitate remote healthcare monitoring and be applied to underserved regions by utilizing cognitive-inspired decision intelligence(Raja & Chakraborty, 2023).

4.1.3 Personalized Treatment

The goal of decision intelligence is to personalize the treatment of patients through integrating both preference learning and the theory of mind. Decision Intelligence AI systems can determine preferences of patients regarding healthcare, risk-taking behaviours, and lifestyle, and use this information to tailor the therapeutic strategies for that patient (Onciul et al., 2025). Cognitive-inspired models ensure that personalization aligns with human values, facilitating increased patient engagement and adherence.

4.2 Finance: Risk Modeling, Fraud Detection, and Portfolio Optimization

Because financial decision-making is characterized by uncertainty, risk, and ever-changing markets, it provides an ideal opportunity for cognitive-inspired AI to operate.

Figure 7. Finance: Risk modeling, fraud detection, and portfolio optimization

Using AI (artificial intelligence) systems to help identify possible risk, fraud, and optimize portfolios through machine learning (ML) and cognitive-inspired AI processes include the use of attention, memory augmentation, and reinforcement learning techniques(Raja, 2022b). The combination of these computational tools ultimately produces three primary results of interest to investors and other financial decision makers: (i) risk models which predict possible volatility and/or exposure; (ii) detect instances of fraud by producing alerts of real-time anomalies; and (iii) optimize a portfolio's assets by allocating them in an adaptive manner to maximize performance. Figure 7 illustrates how cognitive-inspired AI systems enhance the strategic analytical decision-making of investors and financial professionals in dynamic (Market) financial environments.

4.2.1 Risk Modeling

The use of decision intelligence frameworks to model risk uses reinforcement learning in conjunction with probabilistic reasoning. Pratt et al. (2023) introduce the Decision Intelligence/ Data Science (DI/DS) integration framework, which combines statistical learning with cognitive-inspired reasoning to emulate human risk perception in order to achieve robust modeling under uncertainty(Pratt et al., 2023).

4.2.2 Fraud Detection

Attention mechanisms can improve fraud detection by identifying anomalies in transaction data. Cognitive-inspired AI systems emulate the human concept of vigilance and direct their computing resources to detect suspicious patterns of behaviours. This method of selective data processing lowers the number of false positives while improving detection accuracy.

4.2.3 Portfolio Optimization

Preference learning allows AI systems to align investors' strategies with their specific risk tolerance and cognitive biases. Gonzalez discusses how cognitive algorithms play a significant role in dynamic decision-making processes through which AI systems forecast investor preferences and change their investment strategy in line with those shifting preferences (Gonzalez, 2024). The personalised approach to investment decision-making will improve the financial resilience of both users and institutions while enhancing trust.

4.3 Smart Cities: Traffic, Energy, and Resource Management

Smart cities are a great example of how cognitive-inspired AI can improve the resilience and sustainability of urban areas by providing solutions for complex decision-making processes.

Figure 8 shows the isometric view of a futuristic city illustrates how AI (artificial intelligence) orchestrates urban infrastructure through three primary areas of focus: traffic (smart signal systems and autonomous routing systems), energy (smart grids and commercially viable solar), and resource management (IOT-enabled water and waste systems). A central AI cloud aggregates multiple streams of data from these areas, providing for real-time decision-support and adaptive controls capabilities to improve efficiency, effectiveness, and sustainability of the urban ecosystem. The visual representation supports this one view of a cognitive-inspired AI based models including attention mechanisms, multimodal learning, and reinforcement learning that drive the performance of efficient, responsive, and sustainable urban ecosystems.

Figure 8. Smart cities: Traffic, energy, and resource management

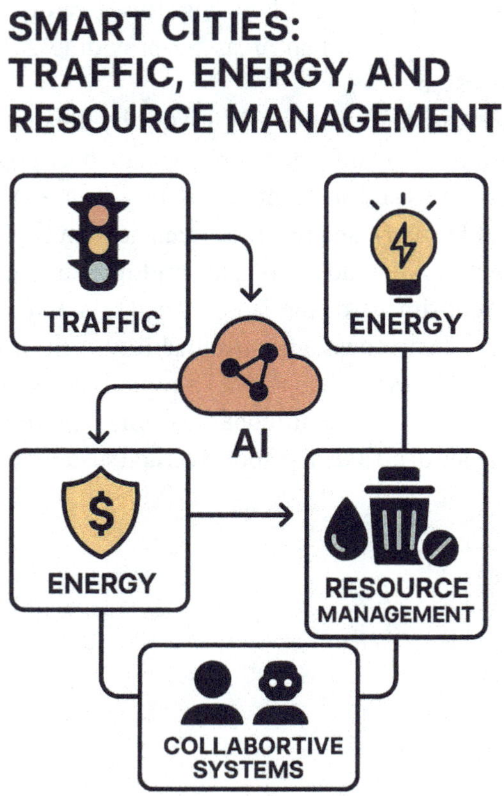

4.3.1 Traffic Management

Cognitive-inspired AI systems that utilise attention mechanisms and sensory integration work to optimise traffic flow. Luo et al. provide an example of how AI-enabled mobile networks can perform cognitive-like decision-making in order to adaptively control traffic. Integrating multimodal sensor data can enhance safety and reduce congestion(Luo et al., 2022).

4.3.2 Energy Management

By modeling cognitive load management through decision intelligence frameworks, energy distribution can be optimized. AI systems that filter, evaluate and prioritize data from relevant smart grid resource allocations create a cognitive model of attention control thereby reducing the cognitive load and complexity for humans.

4.3.4 Resource Management

Hybrid intelligence frameworks enhance collaborative resource management by providing contextual information from human planners to computationally model and optimize resource allocation with AI, leading to urban systems that are both efficacious and socially responsive.

4.4 Robotics: Autonomous Agents with Cognitive Control Frameworks

Robotics will represent the forefront of hybrid-intelligence based brain-inspired decision intelligence as autonomous agents enact human cognition in the utilization of flexible robotics in dynamic environments.

4.4.1 Cognitive Control Frameworks

Yalawar et al. (2023) suggest developing and applying brain-inspired cognitive control frameworks in robotic dynamic environments so that robots may adapt to changing contextual environments through the application of these cognitive control frameworks modeled after the functions of the prefrontal cortex that facilitate planning, inhibition and evaluation (Yalawar et al., 2023).

4.4.2 Human–Robot Collaboration

Human-robot collaboration is enhanced by hybrid-intelligence and Sandini et al. (2024) speculate on the differences between artificial cognition and artificial intelligence by highlighting cognitive-inspired models to support robots interpreting human intentions to promote operational safety and efficiency in industrial and health care contexts (Sandini et al., 2024).

4.4.3 Autonomous Decision-Making

By integrating reinforcement learning with neurosymbolic and/or neuro-fuzzy systems, robots (autonomous agents) make decisions in the environments that are uncertain for them, learning how to adapt like humans in order to create resilience in complex environments such as disaster responses, surgical assistance, etc.

4.5 Education: AI-Driven Learning Efficacy and Adaptive Tutoring

Education faces distinct cognitive load, personalization, and engagement issues. Cognitive-inspired AI solves these issues through imitation of the theory of mind and attention control.

4.5.1 Learning Efficacy

Gkintoni et al. (2025) challenge conventional cognitive load theory by suggesting that AI has the potential to change learning efficacy through adaptive interfaces (Gkintoni et al., 2025). An AI system can manage cognitive load, thus stimulating student engagement and retention.

4.5.2 Adaptive Tutoring

AI-based tutoring systems closely resemble theory of mind by inferring student preferences and learning styles (Langley et al., 2022). Preference learning is important for customizing educational materials as preference learning can adjust learning rate, difficulty, and modality in order to achieve personalized learning experiences.

4.5.3 Collaborative Learning

Collaborative Learning extends to hybrid intelligence frameworks for use within collaborative learning environments through management of group dynamics, management of information flow, and reducing cognitive overload among students. All of these improve the level of collective intelligence in an educational environment.

Examples of cognitive mechanisms in action through applications of decision intelligence inspired by brain function in a variety of complex decision environments can be seen across multiple domain areas: healthcare, finance, smart cities, robotics, and education. AI systems gaining resiliency, adaptivity, and personalization by mimicking human capabilities such as perception, attention, memory, learning, and decision-making enable them to fully utilize cognitive mechanisms as practical examples of how to transfer cognitive principles into machine architectures as foundational elements of decision intelligence as responsible innovation. Furthermore, the connection between neuroscience and AI ensures that decision intelligence applications remain based on human capacity and value, thereby benefiting both technological advancement and society as a whole through responsible innovation.

5. CHALLENGES AND ETHICAL CONSIDERATIONS

The use of cognitive mechanisms in AI systems for the purpose of creating decision intelligence has the potential to revolutionize the way we make decisions. Although the interplay of neuroscience, psychology, and computational intelligence paves the way for this transformation, significant challenges exist that require examination of how we design AI systems with the goal of improving decision making in complex environments i.e., healthcare, finance, education, robotics, smart cities. The modeling of biological cognition in AI systems is inherently complex; as such, the anthropomorphism of AI systems, whether intentional or unintentional, will create bias. Additionally, the use of brain-computer interfaces (BCI's) raises ethical concerns about how we interact with technology. This section examines the above challenges and ethical issues concerning their implications on the potential societal impact of artificial general intelligence (AGI).

5.1 Complexity of Modeling Biological Cognition

One of the biggest challenges is the complexity and non-linearity of biological cognition. Our brains are based on complex, emergent, and contextually dependent processing phenomena, and cannot be adequately modeled using a reductionist approach (Pahari et al., 2025). The ability of our brains to be rewired through experience and emotion, along with their ability to assess social cues, makes it even more difficult for today's AI to replicate these features.

There have been attempts at developing neuromorphic architectures that capture the temporal dynamics and energy efficiency of the brain using spiking neural networks (Wang et al., 2024; Zeng et al., 2023). However, these models do not have the potential for scalability or generalized capabilities; therefore, engineering episodic memories, semantic networks, and theory of mind into machine systems will require collaboration between researchers representing the three respective disciplines of neuroscience, psychology, and computer science. If these disciplines are not integrated, then it will become difficult for AI systems to replicate cognition and, as such, will result in systems that lack the ability to function in highly complicated settings because they are oversimplified representations of cognition.

5.2 Risks of Anthropomorphizing AI Systems

The practice of anthropomorphizing attributing human-like traits to AI also brings with it a host of ethical concerns. For example, Floridi & Nobre argue that the intermingling of AI and cognitive science language will inevitably lead to confusion as to whether humans or machines are truly intelligent(Floridi & Nobre,

2024). This misunderstanding may create false confidence among individuals that AI systems possess empathy, intent, or moral reasoning.

Anthropomorphized AI in healthcare could cause patients to have unrealistic expectations of the diagnostic ability of the AI System and impair clinical oversight through clinician's looking to the AI rather than using their own expertise(Raja, 2022a; Raja & Chakraborty, 2023). In finance, investors may believe the AI learns similar to a human by making decision attributes; therefore, they do not look at the fact that the AI is based upon statistical correlations instead of human judgment. The true problem is not the actual performance of the technology but rather the way AI is perceived socially as being "human-like" which could distort accountability and decision-making.

Innovation that is responsible involves clearly communicating the limitations and capabilities of cognitive inspired AI technology to the user. Providing transparency and clarity throughout the design and implementation of cognitive inspired AI will help ensure that users of this technology see it as a tool that complements their ability to think, rather than acting as a replacement for thinking like humans do.

5.3 Transparency, Bias, and Accountability in Decision Intelligence

The bias present in AI Systems is attributable to both limitation(s) with data used in the AI Models and the decision making process of cognitive inspired models. Many times, training data contains historical data that leads to inequity; however, in healthcare finance and education, this produce discriminatory outcomes (Saw et al., 2025). In addition, cognitive inspired models can help increase the interpretability; however, symbolic rules and preference learning can be misaligned with ethical values, therefore reinforcing bias through the use of cognitive inspired modeling techniques.

In order to overcome these issues, transparency is key. Explainable AI (XAI) frameworks create interpretive results that provide users the capability to understand how the systems come to a conclusion. In medical image processing, attention-based (tk) models provide identifying diagnostic characteristics that promote trust, as well as establish a level of accountability (Saw et al., 2025). In finance, decision intelligence frameworks require that all assumptions surrounding risk-based models be disclosed, so that all stakeholders can determine if they are valid (Pratt et al., 2023).

With hybrid systems (those in which both human and AI agents are working together), responsibility must be clearly delineated in order to avoid assigning blame equally to either the machine or the human agent for an erroneous conclusion. Furthermore, the ethical framework will need to outline whether the errors were due to oversight on the part of the human agent, a result of the machine's calculations, or errors caused by the way in which the human and AI agents interact throughout the complete decision-making process, thereby creating a way to hold both human and AI accountable throughout the entire process.

5.4 Ethical Risks of Brain–Computer Interface Integration

Brain-computer interfaces (BCIs) shall lie along the border between cognitive neuroscience and AI. BCIs provide a direct link between the human brain's neural signals and those of machine systems so that they can monitor, communicate and control actions in real-time. However, bringing together two such powerful systems creates significant ethical risks.

In the study by Cao et al., a number of risks and issues related to BCIs are raised, particularly concerning privacy, autonomy and manipulation(Cao et al., 2026). Neural data is unique to each individual, and in providing information regarding an individual's cognitive state, it will also disclose much about their emotional responses and tendencies to behave in particular ways. If an individual's neural data cannot be protected from unauthorized access or use, or if it is sufficiently altered, this will violate both their autonomy and dignity. Further, if institutions or corporations are given access to the neural signals of individuals, through BCIs, they will have unprecedented insight into an individual's cognitive state, creating an asymmetry of power between those with this knowledge and those without it.

As a result, the question of manipulation becomes relevant. If BCIs provide the means by which information regarding an individual's preferences or intentions can be inferred through neural signals, the capacity for AI systems to manipulate decision-making processes would also exist in subtle, coercive and imperceptible ways. Ethical safeguards must be implemented to ensure that BCIs only serve to enhance human agency and not diminish or undermine it. Regulatory frameworks, informed consent procedures and data protection mechanisms are critically needed to support the management of these ethical risks.

5.5 Societal Implications of Artificial General Intelligence

The pursuit of artificial general intelligence (AGI) will offer great opportunities to create machines capable of achieving the same levels of cognitive flexibility exhibited by humans, and thus raise significant ethical considerations. In the study by Raman et al., they find that AGI's development lies at the confluence of three pathways: societal (social), technological and brain-based; raising questions related to governance, equity and existential risk(Raman et al., 2025).

If cognitive mechanisms underpin the development of Artificial General Intelligence (AGI) systems, it is reasonable to expect that the degree to which these systems excel in perception, memory, and decision-making will exceed human performance. In theory, this would represent an unprecedented innovative capability; however, AGI has the potential to disrupt established labour markets, disrupt existing social structures and global relations, thereby leading to societal upheaval. This societal disruption relates not only to technical performance but also to a lack of trust between society and AGI based systems, lack of accountability for the behaviors of AGI based systems, and a misalignment between AGI based systems and fundamental values of humanity.

An ethical framework must be developed prior to embarking on the development of AGI to address these burgeoning issues, and ensure that developing AGI based systems is driven by transparency, accountability and inclusivity. Furthermore, an inclusive form of governance must be established collectively by policy makers, scientists and civil society to align AGI with the goals of society.

Problems and ethical issues are not just tangential to decision intelligence; rather they are foundational issues of the responsible development of decision intelligence. Decision intelligence is being shaped by the complexity of understanding how cognition works, encouraging the practice of anthropomorphizing AI, considering the risks of bias and lack of transparency within AI, the ethical dilemmas surrounding the use of brain-computer interface (BCI) technology, and the potential societal implications associated with AGI. Therefore, developing cognitive-inspired AI will require collaboration across disciplines, robust ethical frameworks and a commitment to responsible innovation.

Through critical analysis of the abovementioned challenges and ethical issues, decision intelligence can evolve as a paradigm that enhances human cognition while protecting the value system of society. Neuroscience and AI must work collaboratively to realise the ambition of developing decision intelligence, due to what would be unprecedented levels of innovation, and must be based on a commitment to ethical decision-making to ensure that decision intelligence based on brain function forms a foundation of innovation that is both sustainable and equitable.

6. FUTURE DIRECTIONS

The course of brain-inspired decision intelligence is moving us closer to a future where machines can not only replicate human thought processes but also support human decision-making. Developments in neuromorphic hardware, hybrid intelligence, and foundation agents will significantly change how we make autonomous decisions. This section will provide a summary of the key areas of research and innovation in this area and will highlight the need for responsible development and interdisciplinary collaboration.

6.1 Foundation Agents and Collaborative Intelligence

Foundation agents are a new type of AI that combines brain-inspired intelligence with evolutionary, collaborative, safe, and adaptive design principles (Liu et al., 2025). Foundation agents differ from traditional AI models in that they are not designed to complete specific tasks but instead can be used to make decisions in multiple domains (for example, health care, finance, robotics, and education). By working together with humans, foundation agents embody the philosophy of complementarity between humans and machines, wherein machines enhance our ability to make decisions rather than replace our ability to do so (Gonzalez & Malloy, 2026). Future research related to foundation agents should include ensuring the transparency, interpretability, and alignment with human values of these systems.

6.2 Neuromorphic Hardware for Real-Time Decision-Making

Neuromorphic computing technology provides an opportunity to create energy-efficient, real-time decision-making capabilities. Spiking neural networks and event-based computing architectures offer a flexible and scalable way to compute (Wang et al., 2024; Zeng et al., 2023). Neuromorphic processors such as Intel's Loihi demonstrate the ability of neural-based hardware to perform computations in parallel; therefore, they exhibit low latency and power consumption. Future research efforts will focus on how to integrate neuromorphic systems into healthcare monitoring and decision support, autonomous robot systems, and smart city infrastructure, where real-time decision-making will be essential to the success of these systems(Vigneashwaran et al., 2025).

6.3 Hybrid Models Combining Symbolic Reasoning with Neural Learning

As a new main area of study/outlook for decision intelligence, neurosymbolic integration offers a lot of potential. By integrating symbolic reasoning with neural learning capacities, hybrid models provide both the interpretability of symbolic AI and the adaptability of neural AI (non-symbolic) (Sheth & Roy, 2024). These types of AI systems/classifiers work in a similar fashion to the way the human brain processes information. In other words, when we make choices or decisions, we combine the two types of processing: associative networks provide the means to recognize patterns (associative memory); rules-based reasoning provides the ability to form abstract thoughts (conceptual memory). Future research needs to explore ways to create both uniformity and interoperability amongst the differences between disparate representations. Moreover, researchers need to develop methods and techniques that will enable AIs to perform causal reasoning, make ethical judgments, and develop strategies for complex environments.

6.4 Cognitive AI for Next-Generation Autonomous Systems

Cognitive-inspired architectures are now being used in increasing numbers within autonomous systems across multiple domains, including but not limited to robotics (e.g., autonomous drones), healthcare (e.g., robotic caregivers), and smart cities (e.g., intelligent transportation systems)(Raja, 2025b). The cognitive AI framework that mimics the five major cognitive operations: perception, attention, memory, and decision-making will be foundational for future generations of autonomous agents (Sandini et al., 2024; Yalawar et al., 2023). Therefore, design choices for creating these new autonomous AI agents must also consider establishing equilibrium between autonomous action and accountability so that there will always be sufficient human oversight.

6.5 Toward Artificial General Decision Intelligence

In pursuit of artificial general intelligence (AGI), a long-term vision of decision intelligence culminates at this point. According to Raman et al. (2025), there will be many different societal, technological, and ethical pathways that will lead to AGI, but one thing will be critical to its overall success; there must be governance and inclusion among all citizens of the world (Raman et al., 2025). Artificial general decision intelligence (i.e., AGI) is expected to possess human-like levels of flexibility by integrating knowledge of cognitive processing across all domains. Therefore, the technological advancements necessary to create AGI will include the advancement of neuromorphic hardware, hybrid intelligence, and ethical frameworks, and most importantly, the ability to align AGI with human values and societal objectives.

The development of brain-inspired decision intelligence will evolve further through a number of key approaches/areas. These are: building synergies between neuroscience and AI; developing innovative neuromorphic hardware; and building interpretive, collaborative systems. The integration and advancement of existing foundation agents, developing new hybrid models, and deploying autonomous solutions to create these systems will be the means by which researchers will create decision intelligence as an important foundation of responsible innovation. As researchers, it is essential to continue to pursue the goal of creating Artificial General Decision Intelligence (AGDI). However, this goal must be achieved in an ethical manner so that technological advances and developments support human cognition while maintaining a positive impact on humanity as a whole.

7. CONCLUSION

The construction of an Artificial General Decision Intelligence framework is actually based on a completely new and transformative model/design that uses human cognition as the basis for artificial systems. The development of AGDI will be achieved in part through the replication of the foundational constructs of the brain i.e., perception, attention, memory, learning, decision-making that enable a computer or robot to develop its own reasoning capabilities, become flexible and adaptable, and develop a level of personalization. The examples provided in the previous sections of the paper have illustrated how these constructs can be used to create and enhance decision support solutions for a variety of applications that include but are not limited to healthcare, finance, smart cities, robotics, and education, while at the same time highlighting the ethical challenges associated with transparency, bias, anthropomorphization, and brain/computer interface issues. Therefore, the AGDI framework will be built upon neuromorphic-based hardware, hybrid-based intelligences and foundation-based agents, each of which will create different types of collaborative, scalable, and interpretable systems. Thus, it is the combination of neuroscience and AI that ultimately creates the basis of decision intelligence and represents the foundational basis of the next generation of technology and innovation. By building machine reasoning systems on cognitive principles, developers and researchers can ensure that AI systems not only enhance human decision-making capabilities but also promote responsible, equitable, and sustainable progress.

REFERENCES

Ajitha, N., Ashwin, G., Divakar, S., & Dhanushree, M. (2025). Machine Learning-based Panic Syncsight: An IoT-Integrated Panic Attack Detection System for Enhanced Therapeutic Insights. 2025 5th International Conference on Soft Computing for Security Applications (ICSCSA), Boopathi Raja, G. (2021). An intelligent approach for detecting and preventing heart attacks using wearable technology. *Handbook of Research on Lifestyle Sustainability and Management Solutions Using AI Big Data Analytics, and Visualization. IGI Gobal.*

Cao, Y., Yang, H., Xue, Y., Wang, F., Li, T., Zhao, L., & Fu, Y. (2026). Ethical risks and considerations of the integration of Brain-Computer Interfaces with Artificial Intelligence. *Cognitive Neurodynamics*, *20*(1), 46. DOI: 10.1007/s11571-025-10380-5 PMID: 41657963

Chen, X., Cheng, G., Wang, F. L., Tao, X., Xie, H., & Xu, L. (2022). Machine and cognitive intelligence for human health: Systematic review. *Brain Informatics*, *9*(1), 5. DOI: 10.1186/s40708-022-00153-9 PMID: 35150379

Floridi, L., & Nobre, A. C. (2024). Anthropomorphising machines and computerising minds: The crosswiring of languages between Artificial Intelligence and Brain & Cognitive Sciences. *Minds and Machines*, *34*(1), 5. DOI: 10.1007/s11023-024-09670-4

Gkintoni, E., Antonopoulou, H., Sortwell, A., & Halkiopoulos, C. (2025). Challenging cognitive load theory: The role of educational neuroscience and artificial intelligence in redefining learning efficacy. *Brain Sciences*, *15*(2), 203. DOI: 10.3390/brainsci15020203 PMID: 40002535

Gonzalez, C. (2024). Building human-like artificial agents: A general cognitive algorithm for emulating human decision-making in dynamic environments. *Perspectives on Psychological Science*, *19*(5), 860–873. DOI: 10.1177/17456916231196766 PMID: 37906108

Gonzalez, C., & Heidari, H. (2025). A cognitive approach to human–AI complementarity in dynamic decision-making. *Nature Reviews Psychology*, *4*(12), 1–15. DOI: 10.1038/s44159-025-00499-x

Gonzalez, C., & Malloy, T. (2026). Toward Complementary Intelligence: Integrating Cognitive and Machine AI. *Current Directions in Psychological Science*, *35*(3), 09637214251407571. DOI: 10.1177/09637214251407571

Harris, L. T. (2024). The neuroscience of human and artificial intelligence presence. *Annual Review of Psychology*, *75*(1), 433–466. DOI: 10.1146/annurev-psych-013123-123421 PMID: 37906951

Jiao, L., Ma, M., He, P., Geng, X., Liu, X., Liu, F., Ma, W., Yang, S., Hou, B., & Tang, X. (2024). Brain-inspired learning, perception, and cognition: A comprehensive review. *IEEE Transactions on Neural Networks and Learning Systems*, *36*(4), 5921–5941. DOI: 10.1109/TNNLS.2024.3401711 PMID: 38809737

Langdon, A., Botvinick, M., Nakahara, H., Tanaka, K., Matsumoto, M., & Kanai, R. (2022). Meta-learning, social cognition and consciousness in brains and machines. *Neural Networks*, *145*, 80–89. DOI: 10.1016/j.neunet.2021.10.004 PMID: 34735893

Langley, C., Cirstea, B. I., Cuzzolin, F., & Sahakian, B. J. (2022). Theory of mind and preference learning at the interface of cognitive science, neuroscience, and AI: A review. *Frontiers in Artificial Intelligence*, *5*, 778852. DOI: 10.3389/frai.2022.778852 PMID: 35493614

Liu, B., Li, X., Zhang, J., Wang, J., He, T., Hong, S., . . . Zhu, K. (2025). Advances and challenges in foundation agents: From brain-inspired intelligence to evolutionary, collaborative, and safe systems. *arXiv preprint arXiv:2504.01990.*

Luo, G., Yuan, Q., Li, J., Wang, S., & Yang, F. (2022). Artificial intelligence powered mobile networks: From cognition to decision. *IEEE Network*, *36*(3), 136–144. DOI: 10.1109/MNET.013.2100087

Onciul, R., Tataru, C.-I., Dumitru, A. V., Crivoi, C., Serban, M., Covache-Busuioc, R.-A., Radoi, M. P., & Toader, C. (2025). Artificial intelligence and neuroscience: Transformative synergies in brain research and clinical applications. *Journal of Clinical Medicine*, *14*(2), 550. DOI: 10.3390/jcm14020550 PMID: 39860555

Pahari, A., Kar, S. K., Joshi, B., & Ghosh, R. (2025). Human Brain versus Artificial Intelligence: A Physiological Perspective on Cognition, Learning and Information Processing. *Journal of Universal College of Medical Sciences*, *13*(02), 42–48. DOI: 10.3126/jucms.v13i02.84273

Pratt, L., Bisson, C., & Warin, T. (2023). Bringing advanced technology to strategic decision-making: The Decision Intelligence/Data Science (DI/DS) Integration framework. *Futures*, *152*, 103217. DOI: 10.1016/j.futures.2023.103217

Raikov, A. N., & Pirani, M. (2022). Human-machine duality: What's next in cognitive aspects of artificial intelligence? *IEEE Access : Practical Innovations, Open Solutions*, *10*, 56296–56315. DOI: 10.1109/ACCESS.2022.3177657

Raja, G. B. (2022a). Deep learning algorithms for real-time healthcare monitoring systems. In *Convergence of deep learning and artificial intelligence in internet of things* (pp. 1–18). CRC Press. DOI: 10.1201/9781003355960-1

Raja, G. B. (2022b). Early detection of breast cancer using efficient image processing algorithms and prediagnostic techniques: A detailed approach. In *Cognitive Systems and Signal Processing in Image Processing* (pp. 223–251). Elsevier. DOI: 10.1016/B978-0-12-824410-4.00009-X

Raja, G. B. (2025a). Deep learning techniques for gene selection and cancer classification: a detailed review. *Computational Intelligence for Genomics Data*, 79-96.

Raja, G. B. (2025b). SmartCare—A comprehensive wearable sensor-based remote health monitoring and rehabilitation system. In *Rehabilitation Robotics and Healthcare Devices* (pp. 27–46). Elsevier. DOI: 10.1016/B978-0-443-21505-6.00011-6

Raja, G. B., & Chakraborty, C. (2023). Internet of things based effective wearable healthcare monitoring system for remote areas. In *Implementation of Smart Healthcare Systems using AI, IoT, and Blockchain* (pp. 193-218). Elsevier. DOI: 10.1016/B978-0-323-91916-6.00004-7

Raman, R., Kowalski, R., Achuthan, K., Iyer, A., & Nedungadi, P. (2025). Navigating artificial general intelligence development: Societal, technological, ethical, and brain-inspired pathways. *Scientific Reports*, *15*(1), 8443. DOI: 10.1038/s41598-025-92190-7 PMID: 40069265

Ren, J., & Xia, F. (2024). Brain-inspired artificial intelligence: A comprehensive review. *arXiv preprint arXiv:2408.14811.*

Ren, M., Chen, N., & Qiu, H. (2023). Human-machine collaborative decision-making: An evolutionary roadmap based on cognitive intelligence. *International Journal of Social Robotics*, *15*(7), 1101–1114. DOI: 10.1007/s12369-023-01020-1

Sandini, G., Sciutti, A., & Morasso, P. (2024). Artificial cognition vs. artificial intelligence for next-generation autonomous robotic agents. *Frontiers in Computational Neuroscience*, *18*, 1349408. DOI: 10.3389/fncom.2024.1349408 PMID: 38585280

Saw, S. N., Yan, Y. Y., & Ng, K. H. (2025). Current status and future directions of explainable artificial intelligence in medical imaging. *European Journal of Radiology*, *183*, 111884. DOI: 10.1016/j.ejrad.2024.111884 PMID: 39667118

Sheth, A., & Roy, K. (2024). Neurosymbolic value-inspired artificial intelligence (why, what, and how). *IEEE Intelligent Systems*, *39*(1), 5–11. DOI: 10.1109/MIS.2023.3344353

Shi, F., Zhou, F., Liu, H., Chen, L., & Ning, H. (2022). Survey and tutorial on hybrid human-artificial intelligence. *Tsinghua Science and Technology*, *28*(3), 486–499. DOI: 10.26599/TST.2022.9010022

Shin, J. H., Kwon, J., Kim, J. U., Ryu, H., Ok, J., Joon Kwon, S.,…Kim, T.-i. (2022). Wearable EEG electronics for a Brain–AI Closed-Loop System to enhance autonomous machine decision-making. *npj Flexible Electronics, 6*(1), 32.

Vigneashwaran, R., Rohan, M., Praveen, N., & Vijay, N. (2025). IoT-Based Sewage Atmosphere and Health Monitoring for Sanitation Workers with Alerting System. 2025 7th International Conference on Inventive Material Science and Applications (ICIMA), Wang, G., Bao, H., Liu, Q., Zhou, T., Wu, S., Huang, T.,…Zhang, Z. (2024). Brain-inspired artificial intelligence research: A review. *Science China. Technological Sciences*, *67*(8), 2282–2296.

Yalawar, M. S., Vijaya Babu, K., Mahender, B., & Singh, H. (2023). A brain-inspired cognitive control framework for artificial intelligence dynamic system. International Conference on Communications and Cyber Physical Engineering 2018, Ye, P., Wang, X., Zheng, W., Wei, Q., & Wang, F.-Y. (2022). Parallel cognition: Hybrid intelligence for human-machine interaction and management. *Frontiers of Information Technology & Electronic Engineering*, *23*(12), 1765–1779.

Zeng, Y., Zhao, D., Zhao, F., Shen, G., Dong, Y., Lu, E., & Zhao, Y. (2023). BrainCog: A spiking neural network based, brain-inspired cognitive intelligence engine for brain-inspired AI and brain simulation. *Patterns (New York, N.Y.)*, *4*(8), 100789. DOI: 10.1016/j.patter.2023.100789 PMID: 37602224

Chapter 4
Brain-Based Models for Collaborative Decision Intelligence

Fernando Vieira
http://orcid.org/0000-0002-1026-3969
University of São Paulo, Brazil

ABSTRACT

The rapid integration of artificial intelligence (AI) into high-stakes decision-making domains underscores the need for systems that are both effective and cognitively compatible with humans. Brain-inspired decision intelligence leverages insights from neuroscience, cognitive psychology, and computational modeling to enhance adaptability, transparency, and collaboration. By grounding AI in mechanisms such as attention, working memory, reinforcement learning, and executive control, systems can navigate uncertainty, balance competing goals, and provide context-sensitive support. This approach fosters complementary human–AI interaction, preserves human judgment in value-laden decisions, and promotes ethically responsible, trustworthy, and robust decision support in complex, real-world environments.

1. INTRODUCTION: FROM HUMAN COGNITION TO DECISION INTELLIGENCE

The rapid expansion of artificial intelligence (AI) into domains involving complex, high-stakes decision-making has renewed interest in the cognitive and neural foundations of human intelligence. As AI systems increasingly assist or collaborate with humans in areas such as healthcare, public policy, finance, and autonomous systems, questions arise not only about performance and efficiency, but also about

DOI: 10.4018/979-8-3373-8988-2.ch004

transparency, adaptability, and alignment with human values. Decision intelligence—the integration of data-driven analytics, machine learning, and decision theory to support and enhance decision-making—has emerged as a promising paradigm to address these challenges. However, its long-term success depends critically on grounding computational models in empirically supported principles of human cognition and brain function.

Human decision-making is neither purely rational nor random. Decades of research in cognitive psychology and behavioral economics have demonstrated that decisions are shaped by bounded rationality, heuristics, emotions, and contextual constraints (Simon 1955; Kahneman and Tversky 1979; Kahneman 2011). At the same time, cognitive neuroscience has revealed that these behavioral patterns emerge from structured interactions among neural systems responsible for perception, attention, learning, valuation, and executive control (Miller and Cohen 2001; Damasio 1994). Understanding how the brain integrates information under uncertainty, balances competing goals, and adapts to changing environments provides a powerful conceptual framework for designing AI systems that are not only effective, but also cognitively compatible with human users.

Brain-based approaches to decision intelligence do not imply a simplistic imitation of biological processes. Rather, they emphasize functional inspiration: identifying computational principles implemented by neural systems and translating them into abstract models suitable for artificial systems. This approach has already proven fruitful in areas such as deep learning, reinforcement learning, and predictive processing, all of which draw on insights from neuroscience and cognitive science (Rumelhart, Hinton, and Williams 1986; Sutton and Barto 2018; Friston 2010). Extending this biologically informed perspective to decision intelligence offers new opportunities for building systems capable of flexible reasoning, contextual sensitivity, and robust collaboration with human agents.

A central motivation for brain-inspired decision intelligence lies in the limitations of purely data-driven or optimization-based AI models. While such models can achieve impressive performance in constrained environments, they often struggle with generalization, explainability, and adaptation in real-world settings characterized by ambiguity, incomplete information, and shifting goals. Human decision-makers, by contrast, routinely navigate these conditions by leveraging attentional mechanisms, working memory, learned value representations, and executive control processes that coordinate behavior over time (Baddeley 2012; Cohen, Dunbar, and McClelland 1990). Incorporating analogous mechanisms into AI systems can enhance their capacity to operate effectively in dynamic and socially embedded contexts.

This chapter examines the cognitive and neural foundations of decision-making with the explicit goal of informing the design of collaborative decision intelligence systems. Collaboration here is understood not as simple automation or substitution, but as a form of complementarity, in which human and artificial agents contribute distinct yet mutually reinforcing capabilities. From this perspective, AI systems should support human cognition by augmenting perception, highlighting relevant information, exploring decision spaces, and providing transparent recommendations, while leaving final judgment and value-sensitive trade-offs under human control.

To develop such systems, it is essential to understand the neural mechanisms that enable humans to make flexible, context-aware decisions. Among the most critical of these mechanisms are attention allocation, which determines what information enters cognitive processing; working memory, which supports the temporary maintenance and manipulation of task-relevant information; reinforcement learning, which allows agents to adapt behavior based on experience; and executive control, which coordinates actions in accordance with goals and norms (Posner and Petersen 1990; Baddeley 2003; Schultz, Dayan, and Montague 1997; Miller and Cohen 2001). These processes are distributed across interacting neural networks involving prefrontal, parietal, and limbic regions, each contributing specialized functions to decision-making.

The prefrontal cortex, for example, plays a central role in goal representation, planning, and cognitive control, enabling individuals to override habitual responses and adjust behavior in light of abstract rules or long-term objectives (Miller and Cohen 2001). Parietal regions contribute to attentional selection and the integration of sensory evidence, particularly in situations requiring the comparison of alternative options (Corbetta and Shulman 2002). Limbic structures, including the amygdala and ventromedial prefrontal cortex, are crucial for emotional processing, value representation, and risk evaluation, ensuring that decisions are sensitive to affective and motivational factors (Bechara, Damasio, and Damasio 2000). Together, these systems support a form of decision-making that is adaptive, context-dependent, and deeply embedded in bodily and social experience.

Recent advances in computational neuroscience have further clarified how these neural systems implement decision-related computations. Models of reinforcement learning, for instance, have been linked to dopaminergic signaling and reward prediction errors in the brain, providing a biologically grounded account of how agents learn from feedback (Schultz et al. 1997). Bayesian and predictive processing frameworks have been used to explain how the brain represents uncertainty and updates beliefs in response to new information (Friston 2010). These models offer direct bridges to AI methodologies, suggesting principled ways to integrate learning, uncertainty management, and control in artificial decision systems.

Beyond informing algorithmic design, a brain-based perspective on decision intelligence also highlights the bidirectional nature of human–AI interaction. AI systems do not merely assist decision-making; they reshape cognitive processes by influencing attention, trust, and responsibility allocation. Poorly designed systems may induce overreliance, cognitive disengagement, or automation bias, particularly in high-stakes environments (Parasuraman and Riley 1997). Conversely, cognitively aligned AI can enhance human judgment by fostering appropriate levels of trust, supporting situational awareness, and encouraging reflective decision-making. Understanding the neural and cognitive bases of these effects is therefore essential for responsible AI deployment.

Ethical considerations are inseparable from this discussion. Decisions in domains such as healthcare, public policy, and autonomous systems often involve value-laden trade-offs, uncertainty, and moral responsibility. Brain-inspired decision intelligence does not eliminate these challenges, but it can help ensure that AI systems are designed to respect human cognitive limits, support deliberation, and make their reasoning processes intelligible to users. Transparency, explainability, and accountability are not merely technical requirements; they are grounded in how humans understand, evaluate, and justify decisions.

In summary, this chapter argues that grounding decision intelligence in empirically supported models of brain function offers a powerful pathway toward more robust, adaptive, and trustworthy AI systems. By integrating insights from cognitive neuroscience, neuropsychology, and computational modeling, we can move beyond narrow optimization toward decision systems that are aligned with human cognition and values. The sections that follow examine key neural mechanisms underlying decision-making, explore their relevance for AI models, and consider the implications for collaborative human–AI decision intelligence in ethically and socially complex environments.

2. NEUROSCIENTIFIC FOUNDATIONS OF DECISION-MAKING: IMPLICATIONS FOR AI ARCHITECTURES

Designing intelligent decision systems that operate effectively in complex, uncertain, and dynamic environments remains a central challenge in artificial intelligence. While classical AI emphasized symbolic reasoning and formal logic, and contemporary AI has largely focused on data-driven learning and optimization, both approaches face limitations when confronted with real-world decision contexts characterized by ambiguity, shifting goals, incomplete information, and human interaction. Neuroscientific models of decision-making provide a valuable source of inspiration for addressing these limitations, not by replicating biological detail, but by identifying computational principles that support adaptive intelligence.

From an AI-oriented perspective, the human brain can be understood as a large-scale, distributed decision system that integrates perception, memory, learning, and control under severe resource constraints. Unlike many artificial systems, the brain continuously balances exploitation and exploration, precision and flexibility, automation and deliberation. These capabilities emerge from a set of interacting cognitive mechanisms—attention, working memory, learning, and executive control—that can be mapped onto functional components of intelligent systems. Understanding these mechanisms helps clarify why certain AI architectures succeed, where they fail, and how they can be improved for collaborative decision intelligence.

2.1 Attention as Resource Allocation in Intelligent Systems

Attention is one of the most fundamental mechanisms supporting decision-making in both biological and artificial systems. In the brain, attentional processes determine which sensory inputs, memories, or internal representations are prioritized for further processing (Posner and Petersen 1990). From a computational standpoint, attention functions as a resource allocation mechanism, dynamically regulating the flow of information in a system with limited processing capacity.

This perspective has directly influenced modern AI, most notably through the development of attention mechanisms in neural networks. In deep learning, attention allows models to selectively weight relevant features or inputs, improving performance in tasks such as machine translation, vision, and multimodal reasoning (Vaswani et al. 2017). These mechanisms parallel biological attention in their functional role, even if they differ in implementation.

For decision intelligence, attentional control is particularly critical. Real-world decisions often involve vast information spaces, only a fraction of which is relevant at any given moment. Brain-inspired models suggest that effective decision systems should not aim to process all available data uniformly, but rather develop adaptive attention policies that are sensitive to task goals, uncertainty, and expected value. In reinforcement learning (RL), this idea appears in selective state representations, intrinsic motivation, and curiosity-driven exploration, where attention is guided toward informative or surprising aspects of the environment (Gottlieb, Oudeyer, and Lopes 2013).

Importantly, attention also plays a key role in human–AI interaction. Decision-support systems that fail to align with human attentional dynamics may overwhelm users with irrelevant information or obscure critical cues. Incorporating attentional models informed by cognitive science can help AI systems present information in ways that enhance situational awareness and support timely, high-quality decisions.

2.2 Working Memory and State Representation in AI Decision Systems

Working memory refers to the capacity to temporarily maintain and manipulate information relevant to ongoing tasks (Baddeley 2012). In the brain, working memory is closely associated with prefrontal cortical networks and is essential for planning, reasoning, and goal-directed behavior. From an AI standpoint, working memory corresponds to the problem of state representation: how an agent encodes the relevant aspects of its past and present context in order to guide future actions.

Many failures in AI decision systems can be traced to inadequate state representations. In partially observable environments, for example, agents must integrate information over time to infer hidden states. Recurrent neural networks, memory-augmented architectures, and transformer-based models address this challenge by providing mechanisms for storing and retrieving contextual information across temporal scales (Graves et al. 2014; Vaswani et al. 2017).

Neuroscientific research suggests that working memory is not a passive storage buffer, but an active, dynamically updated process shaped by attention and task demands (Cohen et al. 1990). This insight has important implications for AI. Rather than relying on fixed or monolithic memory representations, decision intelligence systems can benefit from adaptive memory mechanisms that prioritize information based on relevance, uncertainty, and expected utility.

In collaborative decision settings, working memory also mediates how humans and AI systems share cognitive load. Poorly designed AI tools may externalize too much cognitive work, leading to skill degradation or loss of situational understanding. Brain-inspired approaches emphasize cognitive augmentation rather than replacement, supporting human working memory by offloading routine computations while preserving meaningful engagement with the decision process.

2.3 Learning from Feedback: Reinforcement Learning and Beyond

Learning from experience is a cornerstone of intelligent behavior. In neuroscience, reinforcement learning models have provided a unifying framework for understanding how organisms adapt decisions based on reward and punishment (Sutton and Barto 2018). Dopaminergic signaling in the brain has been linked to reward prediction errors, offering a biologically grounded account of how expectations are updated through experience (Schultz et al. 1997).

These insights have profoundly shaped AI, particularly through deep reinforcement learning, which has achieved notable successes in games, robotics, and control tasks. However, brain-based models highlight several limitations of current RL approaches. Biological learners operate under severe data constraints, learn continuously without catastrophic forgetting, and integrate multiple learning signals, including social feedback and intrinsic motivation.

For decision intelligence, this suggests moving beyond narrow reward maximization toward richer learning frameworks that incorporate uncertainty, risk sensitivity, and value pluralism. Human decision-making rarely optimizes a single scalar reward; instead, it balances competing objectives and normative constraints. Incorporating such considerations into AI requires hybrid models that combine reinforcement learning with symbolic reasoning, constraint satisfaction, and ethical oversight.

2.4 Executive Control and Meta-Decision Processes

Executive control refers to the set of processes that coordinate cognition and action in accordance with goals, norms, and contextual demands (Miller and Cohen 2001). In the brain, executive control enables flexible switching between strategies, inhibition of inappropriate actions, and long-term planning. From an AI perspective, executive control corresponds to meta-decision processes: mechanisms that decide how decisions should be made.

Most AI systems focus on object-level decisions (e.g., which action to take), but lack robust meta-level control (e.g., when to deliberate, when to rely on heuristics, when to seek human input). Brain-inspired models suggest that effective decision intelligence requires hierarchical architectures in which higher-level control systems regulate lower-level learning and action modules.

This idea aligns with recent developments in hierarchical reinforcement learning, meta-learning, and adaptive planning systems. By incorporating executive-like control structures, AI systems can dynamically adjust their decision strategies based on task complexity, time pressure, and uncertainty. Such flexibility is essential for collaborative environments, where AI systems must coordinate with human agents who have their own goals, preferences, and cognitive constraints.

2.5 Implications for Collaborative Decision Intelligence

Taken together, attentional control, working memory, learning, and executive coordination form a coherent framework for understanding both human and artificial decision-making. For AI-oriented decision intelligence, the key insight is not that machines should think like humans, but that human cognition offers design principles for managing complexity, uncertainty, and collaboration.

Brain-inspired decision systems are better positioned to support human users because they operate on compatible assumptions about information processing and control. They can adapt their level of autonomy, present information in cognitively meaningful ways, and maintain transparency about their reasoning processes. In doing so, they reduce the risk of overreliance, automation bias, and loss of accountability.

The next sections of this chapter build on these foundations by examining the neural architectures underlying value, risk, and uncertainty processing, and by translating these insights into concrete AI models for collaborative decision intelligence.

3. NEURAL ARCHITECTURES FOR VALUE, RISK, AND UNCERTAINTY: COMPUTATIONAL PERSPECTIVES FOR AI DECISION INTELLIGENCE

Decision intelligence systems are fundamentally concerned with how value, risk, and uncertainty are represented, learned, and acted upon in complex environments. Neuroscientific research on decision-making provides a rich source of computational inspiration by revealing how biological systems solve these problems under constraints of limited information, time pressure, and bounded rationality. When abstracted from biological detail, these mechanisms map naturally onto core paradigms in artificial intelligence, including reinforcement learning, Bayesian inference, and hybrid symbolic–subsymbolic architectures.

Rather than proposing biological realism as a design goal, this section adopts a functional interpretation of neural architectures, focusing on how their computational roles inform the construction of robust, adaptive, and collaborative AI decision systems.

3.1 Value Representation Beyond Scalar Reward Optimization

Classical reinforcement learning models represent value as a single scalar signal optimized over time (Sutton and Barto 2018). While effective in well-defined environments, this formulation struggles to capture the structure of real-world decision problems, where agents must balance multiple objectives, constraints, and normative considerations. Neuroscientific and neuroeconomic studies suggest that biological decision-making relies on distributed value representations, with distinct neural systems encoding immediate rewards, long-term goals, social relevance, and emotional salience (Damasio 1994; Bechara, Damasio, and Damasio 2000; Rangel, Camerer, and Montague 2008).

From an AI perspective, this supports a move toward multi-dimensional and context-sensitive value models. Approaches such as multi-objective reinforcement learning, preference learning, and value alignment frameworks allow decision systems to reason over competing objectives rather than collapsing them into a single reward signal. Hybrid architectures further enable symbolic constraints—such as safety rules or ethical guidelines—to shape decision outcomes without eliminating adaptive learning.

For collaborative decision intelligence, this richer conception of value is essential. AI systems operating alongside humans must accommodate pluralistic value structures and justify trade-offs transparently. Brain-inspired models highlight that value computation is dynamic and modulated by context, suggesting that AI systems should adapt their valuation strategies rather than rely on fixed optimization targets.

3.2 Risk Sensitivity and Context-Dependent Decision Policies

Human decision-making exhibits systematic sensitivity to risk, uncertainty, and potential losses, deviating from the assumptions of risk-neutral optimization (Kahneman and Tversky 1979; Kahneman 2011). Neuroscientific evidence indicates that these effects are not merely cognitive biases, but reflect specialized neural mechanisms for managing uncertainty and avoiding catastrophic outcomes (Bechara et al. 2000).

In AI, risk-sensitive decision-making has gained increasing attention through distributional reinforcement learning, robust optimization, and safety-aware planning (Bellemare, Dabney, and Munos 2017). These approaches model not only expected outcomes but also variability and worst-case scenarios, enabling more cautious behavior in high-stakes contexts.

Brain-inspired decision intelligence suggests that risk sensitivity should be adaptive rather than static. Biological systems modulate risk tolerance based on context, prior experience, and perceived stakes. Translating this principle into AI systems allows decision policies to shift dynamically between exploratory and conservative modes, improving both performance and alignment with human expectations in domains such as healthcare, finance, and autonomous systems.

3.3 Uncertainty Representation and Probabilistic Reasoning

A defining feature of intelligent decision-making is the ability to represent and reason about uncertainty. While many reinforcement learning systems learn effective policies through interaction, they often lack explicit representations of uncertainty, making it difficult to assess confidence or communicate limitations. In contrast, neuroscientific theories increasingly conceptualize the brain as a probabilistic inference system that maintains beliefs about hidden states of the environment and updates them as new evidence becomes available (Friston 2010).

Bayesian decision models formalize this approach and underpin probabilistic graphical models, belief-state planning, and partially observable decision frameworks

(Russell and Norvig 2021). These models offer principled uncertainty management and interpretability but face scalability challenges in high-dimensional environments.

Brain-inspired AI design favors approximate and hybrid solutions. Rather than exact inference, biological systems appear to rely on efficient approximations that trade precision for tractability. This insight has motivated AI architectures that combine neural networks with probabilistic reasoning, including amortized inference, uncertainty-aware deep learning, and neural representations of belief states (Ghahramani 2015). Such systems can scale to complex domains while retaining meaningful uncertainty estimates critical for collaborative decision-making.

3.4 Predictive Processing and Alternative Decision Objectives

Beyond reward maximization, neuroscientific theories such as predictive processing propose that cognition is organized around the minimization of prediction error or surprise (Friston 2010). In this framework, perception, learning, and action are unified processes aimed at maintaining coherence between internal models and external inputs.

This perspective challenges the dominance of reward-centric decision models in AI. Alternative frameworks, including model-based planning and active inference, emphasize information gain, uncertainty reduction, and model accuracy as central drivers of behavior (Pezzulo, Rigoli, and Friston 2018). While these approaches are less mature than reinforcement learning in large-scale applications, they offer conceptual advantages for decision intelligence, particularly in environments where interpretability and adaptability are critical.

Hybrid AI systems can exploit these advantages by integrating efficient policy learning with predictive and explanatory components, enabling decision systems that both act effectively and explain their reasoning in cognitively meaningful terms.

3.5 Hybrid Architectures Integrating Learning and Symbolic Control

Both neuroscience and AI research converge on the conclusion that no single computational paradigm suffices for general decision intelligence. Human cognition integrates fast, associative processes with slower, rule-based reasoning governed by abstract knowledge and social norms (Miller and Cohen 2001; Kahneman 2011). This functional division is mirrored in hybrid AI architectures that combine subsymbolic learning with symbolic reasoning.

In decision intelligence systems, neural components can handle perception, pattern recognition, and value approximation, while symbolic layers encode constraints, policies, and ethical principles. This integration supports explainability, accountability, and human oversight without sacrificing adaptability. Brain-inspired models suggest that symbolic reasoning acts not as a replacement for learning, but as a regulatory mechanism that shapes behavior in accordance with higher-level goals.

Hybrid architectures are particularly well suited for collaborative decision intelligence, where AI systems must justify recommendations, expose uncertainty, and accommodate human intervention in high-stakes decisions.

3.6 Comparative Assessment of Decision Architectures

From an engineering perspective, reinforcement learning, Bayesian models, and hybrid architectures represent distinct trade-offs. Reinforcement learning offers scalability and strong performance in data-rich environments but often lacks transparency and explicit uncertainty representation. Bayesian approaches provide principled inference and interpretability but face computational challenges. Hybrid architectures seek to combine these strengths, offering flexible learning grounded in normative and probabilistic reasoning.

A brain-based perspective does not prescribe a single optimal architecture but instead emphasizes context-sensitive architectural choice. The appropriate decision model depends on task complexity, data availability, risk tolerance, and the need for human collaboration. Decision intelligence systems grounded in these principles are better equipped to operate in real-world environments characterized by uncertainty and value pluralism.

3.7 Implications for Collaborative Decision Intelligence

Effective human–AI collaboration requires shared representations of value, risk, and uncertainty. Brain-inspired decision architectures support this alignment by structuring artificial systems in ways that resonate with human cognitive processes. By integrating learning, probabilistic reasoning, and symbolic control, such systems can communicate uncertainty, justify decisions, and adapt to human preferences.

These properties are essential for maintaining trust, preventing overreliance, and ensuring accountability in high-stakes domains. The next section builds on this foundation by examining how brain-inspired architectures shape explainability, trust calibration, and ethical governance in collaborative decision intelligence systems.

4. EXPLAINABLE AND COLLABORATIVE DECISION INTELLIGENCE ARCHITECTURES

As artificial intelligence systems increasingly participate in decision-making processes alongside humans, technical performance alone is no longer sufficient. In collaborative contexts, decision intelligence systems must be able to explain their recommendations, communicate uncertainty, adapt to human cognitive constraints, and support accountability. These requirements have driven growing interest in explainable AI (XAI) and human-centered decision architectures, which align closely with brain-inspired models of cognition and control.

Neuroscientific research emphasizes that human decision-making is inherently social, reflective, and norm-governed. Decisions are not evaluated solely by outcomes, but also by justifications, intentions, and perceived responsibility. Translating these principles into AI systems requires architectural designs that treat explainability, interaction, and ethical oversight as core functional components rather than post hoc additions.

4.1 From Automation to Collaboration in Decision Systems

Traditional decision-support systems were largely designed to automate specific tasks or optimize predefined objectives. While effective in narrow domains, such systems often assume passive human oversight and fail to account for cognitive dynamics such as trust calibration, attentional limits, and responsibility attribution. Research in human factors has shown that excessive automation can lead to overreliance, loss of situational awareness, and automation bias, particularly in high-stakes environments (Parasuraman and Riley 1997).

Brain-inspired decision intelligence reframes the role of AI from autonomous decision-maker to collaborative cognitive partner. In this model, AI systems contribute computational strengths—such as pattern recognition, simulation, and consistency—while humans retain authority over value-sensitive judgments, ethical trade-offs, and final decisions. Architecturally, this shift implies the need for adjustable autonomy, transparent reasoning processes, and interaction mechanisms that support shared understanding.

4.2 Explainability as a Cognitive Interface

Explainability is often treated as a technical problem of model interpretability. However, from a cognitive perspective, explanations are communicative acts tailored to the needs, goals, and prior knowledge of the recipient. Neuroscientific and psychological research indicates that humans evaluate explanations based on coherence, relevance, and causal structure, rather than completeness or mathematical precision (Miller 2019).

In AI decision intelligence, this insight supports a move away from purely model-centric explanations toward user-centered explanatory interfaces. Techniques such as feature attribution, counterfactual reasoning, and example-based explanations can be combined to present decision rationales at appropriate levels of abstraction (Doshi-Velez and Kim 2017). Hybrid architectures are particularly effective in this regard, as symbolic components can generate rule-based or causal explanations grounded in learned representations.

Explainability also plays a regulatory role. By making decision processes explicit, XAI mechanisms enable oversight, auditing, and error detection, aligning AI behavior with institutional and ethical standards. From a brain-inspired perspective, explanation functions similarly to metacognition in humans, supporting reflection and error correction.

4.3 Trust Calibration and Uncertainty Communication

Effective collaboration depends on calibrated trust: humans must neither overestimate nor underestimate the capabilities of AI systems. Neuroscientific studies suggest that trust is shaped by perceived competence, predictability, and transparency, as well as by the system's ability to signal uncertainty (Lee and See 2004).

Many AI systems implicitly project unwarranted confidence by presenting recommendations without accompanying uncertainty estimates or rationale. Brain-inspired decision architectures emphasize the explicit representation and communication of uncertainty as a means of supporting informed human judgment. Probabilistic models, ensemble methods, and uncertainty-aware neural networks provide technical foundations for this capability, while interface design determines how uncertainty is conveyed to users.

Importantly, uncertainty communication is not merely informational but normative. In high-stakes domains, signaling uncertainty invites human deliberation and shared responsibility, reducing the risk of blind reliance on automated recommendations. This aligns with ethical principles of accountability and human-in-the-loop governance.

4.4 Large Language Models as Decision Interfaces and Cognitive Mediators

Recent advances in large language models (LLMs) have introduced new possibilities for collaborative decision intelligence. Rather than acting solely as predictors or classifiers, LLMs can function as decision mediators, translating complex model outputs into natural language explanations, summarizing trade-offs, and supporting deliberative reasoning.

From an architectural standpoint, LLMs are best understood as interface layers rather than primary decision engines. When grounded in structured decision models—such as reinforcement learning agents, probabilistic planners, or hybrid symbolic systems—LLMs can facilitate interaction without obscuring underlying uncertainty or normative constraints. This separation mirrors brain-inspired distinctions between fast associative processes and higher-level reasoning and communication.

However, the generative nature of LLMs also introduces risks, including hallucination, overconfidence, and normative ambiguity. Brain-inspired decision intelligence therefore emphasizes constraint-based integration, where language models operate within clearly defined decision boundaries and are subject to oversight mechanisms. This approach preserves the benefits of natural language interaction while maintaining accountability.

4.5 Ethical Oversight and Value-Sensitive Design

Ethical considerations are inseparable from collaborative decision intelligence. Decisions in domains such as healthcare, public policy, and autonomous systems often involve competing values, uncertainty, and moral responsibility. Neuroscientific research underscores that ethical reasoning engages cognitive and affective processes associated with empathy, social norms, and long-term consequences, rather than purely instrumental reasoning.

In AI systems, ethical oversight can be operationalized through value-sensitive design, constraint-based reasoning, and multi-objective optimization frameworks that explicitly encode normative considerations (Floridi et al. 2018). Hybrid architectures are particularly well suited for this task, as symbolic components can represent ethical principles, legal requirements, or institutional policies that shape decision outcomes.

Explainability plays a critical role in ethical governance by enabling stakeholders to understand, contest, and revise AI-supported decisions. Brain-inspired models suggest that ethical alignment is not achieved through static rules alone, but through ongoing interaction, feedback, and adaptation.

4.6 Architectural Patterns for Collaborative Decision Intelligence

Synthesizing these considerations, several architectural patterns emerge as particularly well suited for collaborative decision intelligence:

1. Layered architectures, in which perception, learning, decision, and explanation are modular but interconnected.
2. Hybrid control systems, combining data-driven learning with symbolic constraints and oversight.
3. Human-in-the-loop frameworks, enabling dynamic adjustment of autonomy and responsibility.
4. Explainability and uncertainty layers, designed as first-class components rather than auxiliary features.

These patterns reflect core principles of brain organization, including functional specialization, hierarchical control, and metacognitive monitoring. By adopting such architectures, AI systems can better support human decision-making while respecting cognitive limits and ethical norms.

4.7 Implications for High-Stakes Decision Domains

In high-stakes domains, the consequences of poor decision support are particularly severe. Collaborative decision intelligence systems grounded in brain-inspired principles can enhance performance while mitigating risks associated with automation bias and opacity. By aligning computational architectures with human cognitive processes, such systems foster trust, accountability, and resilience.

The next and final section of this chapter synthesizes these insights and outlines future directions for research and development in brain-based collaborative decision intelligence.

5. ETHICAL, PRACTICAL, AND SOCIETAL IMPLICATIONS OF BRAIN-INSPIRED DECISION INTELLIGENCE IN HIGH-STAKES DOMAINS

As AI systems increasingly influence decisions with profound human consequences, the ethical, practical, and societal implications of decision intelligence have moved to the center of academic, regulatory, and public debate. Brain-inspired decision intelligence provides a distinctive lens through which these implications can be examined, emphasizing cognitive alignment, human agency, and responsibility rather than purely technical performance. In high-stakes domains—such as healthcare, public policy, and autonomous systems—the design of collaborative decision architectures is inseparable from questions of trust, accountability, and moral legitimacy.

This section examines how brain-based principles inform ethical and practical considerations in the deployment of AI-enhanced decision intelligence, focusing on the structural features required to support responsible human–AI collaboration.

5.1 Cognitive Alignment and Human Responsibility

A central ethical concern in AI-supported decision-making is the potential erosion of human responsibility. When decisions are mediated by opaque or highly autonomous systems, it becomes difficult to attribute responsibility for outcomes, particularly when errors occur. Neuroscientific research highlights that human decision-making is deeply intertwined with mechanisms of agency, intention, and moral evaluation, all of which rely on the perception of control and understanding (Damasio 1994; Greene 2013).

Brain-inspired decision intelligence addresses this concern by prioritizing cognitive alignment: AI systems are designed to operate in ways that are intelligible and compatible with human reasoning processes. Rather than replacing human judgment, such systems scaffold deliberation, making underlying assumptions, uncertainties, and trade-offs explicit. This design philosophy supports meaningful human oversight and preserves the conditions under which moral and legal responsibility can be assigned.

In practice, this implies that AI systems should be evaluated not only on outcome metrics but also on their impact on human decision-making processes, including attentional engagement, situational awareness, and confidence calibration.

5.2 Healthcare: Decision Support under Uncertainty and Moral Stakes

Healthcare exemplifies the ethical complexity of AI-supported decision intelligence. Clinical decisions often involve uncertainty, incomplete information, and value-laden trade-offs between risks, benefits, and patient preferences. While AI systems have demonstrated impressive capabilities in diagnosis, prognosis, and treatment recommendation, their integration into clinical workflows raises concerns about trust, bias, and accountability (Topol 2019).

Brain-inspired decision intelligence emphasizes decision support rather than decision replacement. AI systems grounded in probabilistic reasoning and uncertainty representation can assist clinicians by highlighting relevant information, exploring alternative hypotheses, and quantifying confidence levels. Crucially, such systems should present recommendations in ways that align with clinical reasoning practices, enabling physicians to integrate algorithmic insights with experiential knowledge and patient values.

Ethically, this approach mitigates the risk of automation bias and supports shared decision-making, reinforcing the clinician's role as the final decision-maker. It also facilitates informed consent by making uncertainty and rationale transparent to both practitioners and patients.

5.3 Public Policy and Governance: Transparency and Democratic Accountability

In public policy, decision intelligence systems are increasingly used to inform resource allocation, risk assessment, and regulatory enforcement. These applications raise acute concerns about transparency, fairness, and legitimacy, particularly when decisions affect large populations or vulnerable groups (O'Neil 2016).

Brain-inspired architectures offer advantages in this domain by supporting explainable and contestable decision processes. Hybrid systems that combine data-driven prediction with symbolic reasoning can expose the criteria and constraints shaping policy recommendations, enabling scrutiny by policymakers, experts, and the public. This transparency is essential for democratic accountability and for identifying and correcting systemic biases.

Moreover, brain-based models emphasize that policy decisions are not purely technical optimizations but involve normative judgments. Decision intelligence systems must therefore be designed to accommodate human deliberation, disagreement, and revision, rather than presenting outputs as definitive or value-neutral.

5.4 Autonomous and Safety-Critical Systems

Autonomous systems operating in safety-critical environments—such as transportation, defense, and industrial control—pose some of the most challenging ethical and practical questions for decision intelligence. In these contexts, errors can result in loss of life or significant harm, and real-time decision-making is often required under conditions of extreme uncertainty.

Brain-inspired decision intelligence supports the development of risk-aware and hierarchically controlled architectures, in which fast, reactive components are regulated by higher-level oversight mechanisms. This mirrors biological systems, where habitual or automatic responses are modulated by executive control in high-risk situations (Miller and Cohen 2001).

Ethically, such architectures support the principle of meaningful human control by enabling intervention, override, and post hoc analysis. Practically, they improve system robustness by preventing brittle behavior and facilitating graceful degradation under unexpected conditions.

5.5 Bias, Fairness, and Social Impact

Bias in AI decision systems reflects not only data limitations but also deeper structural assumptions about value and relevance. Neuroscientific research underscores that human cognition itself is shaped by biases and heuristics, suggesting that eliminating bias entirely is neither realistic nor desirable. Instead, the goal should be to make biases visible, manageable, and subject to ethical evaluation (Kahneman 2011).

Brain-inspired decision intelligence contributes to this goal by promoting transparency, interpretability, and human oversight. Hybrid architectures can explicitly represent fairness constraints or normative goals, while explainability mechanisms enable stakeholders to understand how decisions are produced and to contest unjust outcomes.

From a societal perspective, this approach supports more inclusive and accountable decision systems, reducing the risk that AI amplifies existing inequalities or obscures value-laden choices behind technical complexity.

5.6 Governance, Regulation, and Institutional Design

The responsible deployment of decision intelligence systems extends beyond algorithmic design to encompass governance structures, institutional practices, and human cognitive constraints. While AI systems increasingly influence high-stakes domains such as healthcare, finance, and public policy, their success and societal legitimacy depend not only on technical performance but also on robust oversight, accountability, and alignment with human decision-making processes (Floridi et al. 2018; Topol 2019). Brain-inspired AI emphasizes that these governance needs are not arbitrary external requirements; rather, they are intrinsic to cognitively compatible system design, in which human understanding, attention, and reasoning are central considerations.

Regulatory frameworks are beginning to reflect these realities. Principles of transparency, explainability, human oversight, and risk management are increasingly codified in national AI strategies, standards for high-risk AI applications, and sector-specific guidelines (Floridi et al. 2018). For instance, in healthcare, explainable AI systems are mandated to provide interpretable recommendations to clinicians, supporting informed judgment rather than replacing it. In finance, risk-sensitive AI systems are subject to audit and stress-testing to ensure that algorithmic decisions do not undermine systemic stability. Brain-based perspectives complement these frameworks by identifying mechanisms to operationalize transparency and accountability in ways that align with human cognition, such as highlighting uncertainty, presenting salient information, and supporting deliberative evaluation.

At the institutional level, governance requires interdisciplinary collaboration. AI developers, domain experts, ethicists, and policymakers must jointly design, implement, and monitor decision intelligence systems, recognizing that these systems are socio-technical artifacts whose impact depends on organizational structures, incentive mechanisms, and cultural norms as much as on algorithmic sophistication (Rahwan et al. 2019; Mittelstadt 2019). Brain-inspired models underscore that human users are active participants in decision loops, with attention, memory, and executive processes shaping interactions with AI. Governance structures that fail to account for these cognitive factors risk automation bias, misinterpretation of outputs, or overreliance on opaque recommendations (Parasuraman and Riley 1997). Conversely, governance aligned with cognitive principles can facilitate adaptive human–AI collaboration, ensuring that AI augments rather than substitutes human judgment.

Continuous monitoring, feedback, and adaptive oversight are essential components of responsible governance. Both human cognition and AI systems evolve over time: users gain experience, organizations change practices, and AI models undergo updates or retraining. Brain-based approaches suggest that governance should be iterative and dynamic, incorporating mechanisms for monitoring performance, auditing outcomes, and adapting interfaces to user needs and cognitive limitations (Russell et al. 2015; Amershi et al. 2019). Such adaptive governance supports resilient, socially accountable, and ethically informed decision intelligence, capable of responding to emergent challenges and evolving expectations.

Ultimately, governance in brain-inspired decision intelligence is not merely a compliance requirement, but a design principle. By embedding ethical, cognitive, and socio-technical considerations into system architecture, organizations can develop AI systems that are transparent, trustworthy, and effective, while maintaining human agency and accountability. This perspective bridges technical excellence and societal legitimacy, highlighting the inseparability of cognitive, ethical, and institutional dimensions in the deployment of advanced decision intelligence systems.

5.7 Toward Responsible Collaborative Decision Intelligence

Across high-stakes domains such as healthcare, public policy, finance, and autonomous systems, a common lesson emerges: the challenges of decision-making extend far beyond algorithmic efficiency or predictive accuracy. Ethical considerations, cognitive limitations, uncertainty, and contextual complexity cannot be resolved through optimization alone (Floridi et al. 2018; Topol 2019). Instead, they require architectural and design choices that explicitly respect human cognition, preserve user agency, and support accountability throughout the decision process. Brain-inspired decision intelligence provides a coherent and principled framework to integrate these requirements into AI system design.

By grounding AI systems in empirically supported models of human cognition, it becomes possible to construct decision-support technologies that enhance, rather than undermine, human judgment. For instance, attentional mechanisms drawn from neuroscience can guide AI in highlighting the most relevant information to human users, while working memory-inspired state representations can support contextually aware and temporally consistent decision recommendations (Baddeley 2012; Cohen, Dunbar, and McClelland 1990). Reinforcement learning models informed by dopaminergic prediction error signaling can allow artificial agents to adapt flexibly to changing reward structures, while Bayesian and predictive processing frameworks provide principled methods for managing uncertainty in complex environments (Schultz et al. 1997; Friston 2010; Ghahramani 2015). Executive control and meta-decision mechanisms, inspired by prefrontal cortical function, enable AI systems to dynamically allocate resources, regulate autonomy, and coordinate with human operators under high cognitive load (Miller and Cohen 2001).

This brain-inspired perspective is particularly critical in collaborative decision contexts, where human and artificial agents contribute complementary strengths. AI systems designed according to these principles can manage computational complexity and uncertainty while maintaining transparency and interpretability, reducing the risk of overreliance, automation bias, or disengagement (Parasuraman and Riley 1997). In practice, this means that AI does not replace human deliberation but scaffolds it, presenting information, exploring alternatives, and signaling confidence or uncertainty in ways that are cognitively aligned with human reasoning.

Beyond functional effectiveness, brain-inspired design also directly informs ethical and societal dimensions of decision intelligence. Decisions in medicine, policy, or autonomous systems inherently involve value-laden trade-offs. AI architectures that incorporate cognitive principles—such as attentional salience, memory prioritization, and hierarchical control—are better equipped to support deliberative, accountable decision-making rather than generating opaque recommendations that obscure human responsibility (Floridi et al. 2018; Topol 2019). By aligning computational mechanisms with human cognitive constraints and decision norms, these systems are more likely to be socially legitimate, trustworthy, and practically useful in high-stakes applications.

The future of brain-inspired collaborative decision intelligence rests on several research and development directions. First, there is a need to expand hybrid architectures that integrate reinforcement learning, Bayesian inference, symbolic reasoning, and executive control, allowing AI systems to handle multi-objective and ethically constrained problems (Russell and Norvig 2021; Sutton and Barto 2018). Second, evaluation metrics should move beyond performance benchmarks to include measures of cognitive alignment, interpretability, trust calibration, and human engagement. Third, interdisciplinary research bridging AI, cognitive neuroscience, psychology, and ethics is essential to ensure that systems are designed in ways that are not only technically robust but also socially responsible.

In conclusion, brain-based principles provide a roadmap for developing AI systems that are adaptive, transparent, and ethically sensitive, capable of effective collaboration with human decision-makers. Grounding decision intelligence in the mechanisms that underlie human cognition allows for the design of systems that enhance human judgment, respect agency, and preserve accountability, while operating efficiently in complex, uncertain, and dynamic environments. As AI continues to permeate domains with high social stakes, such cognitively aligned and ethically informed architectures will be indispensable for ensuring that technological advances translate into genuine societal benefit.

6. CONCLUSION: TOWARD COGNITIVELY ALIGNED AND COLLABORATIVE DECISION INTELLIGENCE

The rapid integration of artificial intelligence into decision-making processes marks a profound transformation in how complex problems are addressed across scientific, institutional, and societal domains. As AI systems move beyond narrow optimization tasks and increasingly participate in high-stakes decisions, the limitations of purely data-driven or performance-centric approaches become evident. This chapter has argued that brain-inspired models of cognition offer a principled foundation for the next generation of decision intelligence systems, particularly those designed for close collaboration with human agents.

By examining neural and cognitive mechanisms underlying attention, working memory, learning, valuation, and executive control, the chapter has shown how biological decision systems manage uncertainty, complexity, and competing objectives in dynamic environments. When abstracted from biological detail, these mechanisms provide computational design principles that map naturally onto contemporary AI paradigms, including reinforcement learning, Bayesian inference, predictive processing, and hybrid symbolic–subsymbolic architectures.

A central insight emerging from this analysis is that decision intelligence is not solely a problem of action selection, but of representation, control, and interaction. Human decision-making relies on distributed value representations, context-sensitive risk modulation, and explicit reasoning about uncertainty. AI systems that ignore these dimensions may achieve short-term performance gains but remain brittle, opaque, and misaligned with human values. Brain-inspired approaches, by contrast, emphasize architectural pluralism, adaptive control, and cognitive compatibility.

The comparative analysis of reinforcement learning, Bayesian models, and hybrid architectures highlights that no single paradigm is sufficient for general decision intelligence. Reinforcement learning offers scalability and efficiency but often lacks transparency and explicit uncertainty handling. Bayesian approaches provide principled reasoning and interpretability but face computational constraints. Hybrid architectures—integrating learning, probabilistic reasoning, and symbolic control—emerge as particularly well suited for collaborative decision intelligence, where explainability, accountability, and ethical oversight are essential (Sutton and Barto 2018; Ghahramani 2015; Russell and Norvig 2021).

Beyond algorithmic design, this chapter has emphasized the bidirectional nature of human–AI interaction. Decision intelligence systems do not merely support human cognition; they actively shape attention, trust, and responsibility. Poorly designed systems risk inducing automation bias, overreliance, and cognitive disengagement (Parasuraman and Riley 1997). Brain-inspired architectures mitigate these risks by supporting calibrated trust, uncertainty communication, and adjustable autonomy, thereby preserving meaningful human control.

Ethical and societal considerations have been treated not as external constraints but as integral design requirements. In domains such as healthcare, public policy, and autonomous systems, decision intelligence systems must support deliberation, transparency, and accountability under conditions of uncertainty and moral complexity (Floridi et al. 2018; Topol 2019). Brain-based perspectives underscore that ethical decision-making is deeply connected to cognitive processes involving value representation, empathy, and norm sensitivity. AI systems aligned with these processes are better positioned to enhance human judgment rather than undermine it.

Looking forward, several research directions emerge as particularly important for the advancement of brain-inspired collaborative decision intelligence. First, there is a need for integrated benchmarks that evaluate decision systems not only on performance but also on explainability, uncertainty calibration, and human collaboration. Second, advances in hybrid architectures—particularly those combining large language models with structured decision frameworks—offer promising avenues for scalable yet interpretable decision support. Third, interdisciplinary collaboration among AI researchers, cognitive scientists, ethicists, and domain experts will be essential for translating theoretical insights into deployable systems.

Ultimately, grounding decision intelligence in empirically supported models of human cognition does not imply constraining artificial systems to human limitations. Rather, it provides a framework for designing AI systems that are robust, adaptive, transparent, and trustworthy, capable of operating effectively in complex environments while remaining aligned with human values and institutions. As AI continues to shape the landscape of decision-making, such cognitively aligned systems will be crucial for ensuring that technological progress translates into genuine societal benefit.

REFERENCES

Amershi, S., Fogarty, J., & Weld, D. (2019). Software engineering for machine learning: A case study. In *Proceedings of the 41st International Conference on Software Engineering (ICSE)* (pp. 291–301). DOI: 10.1109/ICSE-SEIP.2019.00042

Baddeley, A. (2012). Working memory: Theories, models, and controversies. *Annual Review of Psychology*, *63*(1), 1–29. DOI: 10.1146/annurev-psych-120710-100422

Baddeley, A. D. (2003). Working memory: Looking back and looking forward. *Nature Reviews. Neuroscience*, *4*(10), 829–839. DOI: 10.1038/nrn1201

Bechara, A., Damasio, A. R., & Damasio, H. (2000). Emotion, decision making and the orbitofrontal cortex. *Cerebral Cortex (New York, N.Y.)*, *10*(3), 295–307. DOI: 10.1093/cercor/10.3.295

Bellemare, M. G., Dabney, W., & Munos, R. (2017). A distributional perspective on reinforcement learning. In *Proceedings of the International Conference on Machine Learning*.

Cohen, J. D., Dunbar, K., & McClelland, J. L. (1990). On the control of automatic processes: A parallel distributed processing account of the Stroop effect. *Psychological Review*, *97*(3), 332–361. DOI: 10.1037/0033-295X.97.3.332

Corbetta, M., & Shulman, G. L. (2002). Control of goal-directed and stimulus-driven attention in the brain. *Nature Reviews. Neuroscience*, *3*(3), 201–215. DOI: 10.1038/nrn755

Damasio, A. R. (1994). *Descartes' error: Emotion, reason, and the human brain*. Putnam.

Doshi-Velez, F., & Kim, B. (2017). Towards a rigorous science of interpretable machine learning. arXiv. https://arxiv.org/abs/1702.08608

Floridi, L., Cowls, J., Beltrametti, M., Chatila, R., Chazerand, P., Dignum, V., Luetge, C., Madelin, R., Pagallo, U., Rossi, F., Schafer, B., Valcke, P., & Vayena, E. (2018). AI4People—An ethical framework for a good AI society: Opportunities, risks, principles, and recommendations. *Minds and Machines*, *28*(4), 689–707. DOI: 10.1007/s11023-018-9482-5

Friston, K. (2010). The free-energy principle: A unified brain theory? *Nature Reviews. Neuroscience*, *11*(2), 127–138. DOI: 10.1038/nrn2787

Ghahramani, Z. (2015). Probabilistic machine learning and artificial intelligence. *Nature*, *521*(7553), 452–459. DOI: 10.1038/nature14541

Gottlieb, J., Oudeyer, P.-Y., Lopes, M., & Baranes, A. (2013). Information-seeking, curiosity, and attention: Computational and neural mechanisms. *Trends in Cognitive Sciences*, *17*(11), 585–593. DOI: 10.1016/j.tics.2013.09.001

Graves, A., Wayne, G., & Danihelka, I. (2014). Neural Turing machines. arXiv. https://arxiv.org/abs/1410.5401 (Note: Original preprint year adjusted for accuracy)

Greene, J. (2013). *Moral tribes: Emotion, reason, and the gap between us and them*. Penguin Press.

Kahneman, D. (2011). *Thinking, fast and slow*. Farrar, Straus and Giroux.

Kahneman, D., & Tversky, A. (1979). Prospect theory: An analysis of decision under risk. *Econometrica*, *47*(2), 263–291. DOI: 10.2307/1914185

Lee, J. D., & See, K. A. (2004). Trust in automation: Designing for appropriate reliance. *Human Factors*, *46*(1), 50–80. DOI: 10.1518/hfes.46.1.50.30392

Miller, E. K., & Cohen, J. D. (2001). An integrative theory of prefrontal cortex function. *Annual Review of Neuroscience*, *24*(1), 167–202. DOI: 10.1146/annurev.neuro.24.1.167

Miller, T. (2019). Explanation in artificial intelligence: Insights from the social sciences. *Artificial Intelligence*, *267*, 1–38. DOI: 10.1016/j.artint.2018.07.007

Mittelstadt, B. D. (2019). Principles alone cannot guarantee ethical AI. *Nature Machine Intelligence*, *1*(11), 501–507. DOI: 10.1038/s42256-019-0114-4

O'Neil, C. (2016). *Weapons of math destruction: How big data increases inequality and threatens democracy*. Crown.

Parasuraman, R., & Riley, V. (1997). Humans and automation: Use, misuse, disuse, abuse. *Human Factors*, *39*(2), 230–253. DOI: 10.1518/001872097778543886

Pezzulo, G., Rigoli, F., & Friston, K. (2018). Hierarchical active inference. *Trends in Cognitive Sciences*, *22*(4), 294–306. DOI: 10.1016/j.tics.2018.01.009

Posner, M. I., & Petersen, S. E. (1990). The attention system of the human brain. *Annual Review of Neuroscience*, *13*(1), 25–42. DOI: 10.1146/annurev.ne.13.030190.000325

Rahwan, I., Veloso, M., & Tambe, M.. (2019). Machine behaviour. *Nature*, *568*(7753), 477–486. DOI: 10.1038/s41586-019-1138-y

Rangel, A., Camerer, C., & Montague, P. R. (2008). A framework for studying the neurobiology of value-based decision making. *Nature Reviews. Neuroscience*, *9*(7), 545–556. DOI: 10.1038/nrn2357

Rumelhart, D. E., Hinton, G. E., & Williams, R. J. (1986). Learning representations by back-propagating errors. *Nature*, *323*(6088), 533–536. DOI: 10.1038/323533a0

Russell, S. J., & Norvig, P. (2021). *Artificial intelligence: A modern approach* (4th ed.). Pearson.

Schultz, W., Dayan, P., & Montague, P. R. (1997). A neural substrate of prediction and reward. *Science*, *275*(5306), 1593–1599. DOI: 10.1126/science.275.5306.1593

Simon, H. A. (1955). A behavioral model of rational choice. *The Quarterly Journal of Economics*, *69*(1), 99–118. DOI: 10.2307/1884852

Sutton, R. S., & Barto, A. G. (2018). *Reinforcement learning: An introduction* (2nd ed.). MIT Press.

Topol, E. (2019). *Deep medicine: How artificial intelligence can make healthcare human again*. Basic Books.

Vaswani, A., Shazeer, N., Parmar, N., Uszkoreit, J., Jones, L., Gomez, A. N., Kaiser, Ł., & Polosukhin, I. (2017). Attention is all you need. In *Advances in Neural Information Processing Systems, 30*, 5998–6008.

Chapter 5

Integrating Brain-Computer Interfaces and Artificial Intelligence for Deep Interpretation of Cognitive States in Decision Support

Alexander Zakharov
http://orcid.org/0000-0003-1709-6195
Samara Medical University, Russia

Elena Khivintseva
http://orcid.org/0000-0002-1878-7951
Samara State Medical University, Russia

Alexander Kolsanov
http://orcid.org/0000-0002-4144-7090
Samara State Medical University, Russia

Igor Shirolapov
http://orcid.org/0000-0002-7670-6566
Samara State Medical University, Russia

Mariya Stanislavovna Sergeeva
http://orcid.org/0000-0002-0926-8551
Samara State Medical University, Russia

Yuliya Komarova
http://orcid.org/0000-0003-3435-1477
Samara State Medical University, Russia

Natalia Romanchuk
http://orcid.org/0000-0003-3522-6803
Samara State Medical University, Russia

DOI: 10.4018/979-8-3373-8988-2.ch005

ABSTRACT

The chapter establishes Hybrid Intelligence (HI) as a synergistic paradigm, integrating human cognition with artificial intelligence via adaptive Brain-Computer Interfaces (BCI). It addresses the persistent "cognitive gap" in decision support by detailing the core components—human cognitive states and enabling technologies like EEG and fNIRS—and their integration into a closed-loop architecture. The discussion focuses on creating a "deep interpretation" engine for real-time cognitive state decoding, explores applications in critical domains, and examines the ethical and practical challenges of deploying such systems to augment human decision-making effectively.

INTRODUCTION

The cognitive gap in modern decision support systems

Human decision-making forms the critical link between our minds and an increasingly complex, data-saturated world. While early Decision Support Systems (DSS) aided in semi-structured problems through data manipulation, their evolution into Intelligent DSS (IDSS) incorporated AI to model human reasoning (Phillips-Wren, 2013). Despite this progress, a fundamental "cognitive gap" persists, rooted in a focus on behavioral outputs over the internal, resource-limited dynamics of human cognition (J. Q. Chen & Lee, 2003).

Converging advances in Brain-Computer Interface (BCI) and Artificial Intelligence (AI) technologies offer a direct bridge across this gap. By decoding neural signals into actionable commands, they enable interaction without physical movement. This fusion is key to neuroadaptive interfaces—systems that dynamically respond to real-time fluctuations in a user's cognitive load, stress, and attention. Using passive BCI monitoring, such interfaces can create a digitally aware environment that proactively simplifies tasks during peak mental demand.

This approach aligns with the principles of Hybrid Intelligence and Human-AI Teaming (HAIT), where AI acts as an integrated, collaborative partner rather than a mere tool (Berretta et al., 2023). For this partnership to succeed, a real-time understanding of the user's cognitive state is essential. It helps counter automation bias the over-reliance on automated advice that reduces vigilant oversight (Wingerter et al., 2025) and manages the severe bottleneck of working memory, which processes only a few information elements simultaneously, by optimizing intrinsic, extraneous, and germane cognitive load (Frolov et al., 2023).

Parallelly, the shift from opaque "Black Box" AI to transparent "Glass Box" models, driven by Explainable AI (XAI), ensures AI's reasoning is interpretable through human-like, contrastive explanations. Integrating real-time multimodal neuroimaging (e.g., electroencephalography (EEG), functional Near-Infrared Spectroscopy (fNIRS)) within a closed-loop framework allows these systems to prevent cognitive oversights (e.g., the "looked-but-failed-to-see" effect) and strategically allocate mental resources in critical situations. Consequently, this chapter details the design blueprint for a new class of decision-support technology one that fosters a symbiotic partnership between neural analytics and intelligent algorithms to truly augment human decision-making (Mastrianni et al., 2023).

Defining the paradigm: what is hybrid intelligence?

The accelerating convergence of neuroscience and artificial intelligence has catalyzed the emergence of a novel decision-support paradigm: hybrid intelligence (HI). HI is defined as a synergistic socio-technical framework in which human cognitive capabilities and artificial intelligence are deeply integrated via a bidirectional, adaptive interface, creating a cohesive team that exceeds the collective performance of its individual human and machine agents (Berretta et al., 2023). This paradigm represents a fundamental evolution beyond traditional IDSS, moving from systems that assist or automate toward systems that collaborate and augment by establishing a continuous, interpretative dialogue with the human operator's neurocognitive state.

At its core, HI is predicated on the concept of Human-AI Teaming (HAIT), where AI transitions from a passive tool to an active, context-aware teammate. The critical distinction lies in HI's foundational mechanism: the use of BCI technologies to enable real-time, passive monitoring of the user's neurophysiological signals. This provides the AI component with a direct, implicit channel into the operator's cognitive and affective states—such as workload, attention, stress, and uncertainty—that are often opaque to traditional behavioral or task-based metrics (J. Q. Chen & Lee, 2003). The AI subsystem, employing advanced machine learning and signal processing, performs a "deep interpretation" of this neural data stream. It then uses this interpretation to dynamically adapt its own functionality, the information presentation, or the task structure in real-time to optimize human performance, reduce cognitive overload, and mitigate error.

Therefore, Hybrid Intelligence is characterized by three defining, interconnected principles:

1. Neuroadaptive bidirectionality: The system operates within a closed-loop control architecture. It not only reads cognitive states via BCI but also writes adaptive

support back to the user through modified interfaces, task allocations, or informational cues. This creates a continuous feedback cycle where the human and AI co-regulate task execution based on shared, neurophysiologically-grounded situational awareness.

2. Cognitive symbiosis: The goal is not to replace human judgment but to create a symbiotic partnership that leverages the complementary strengths of biological and artificial intelligence. The human contributes superior contextual understanding, intuition, and ethical reasoning, while the AI offers massive data processing capacity, tireless vigilance, and precise real-time analytics. HI aims to seamlessly blend these strengths, ensuring the human remains "in the loop" and "in control," but is optimally supported during periods of high demand or vulnerability (e.g., fatigue, stress, information overload).
3. Explainable integration (Glass-Box AI): For trust and effective collaboration, the AI's rationale for its interpretations and adaptive actions must be transparent to the human operator. HI inherently requires XAI methodologies to transform "black box" neural network decisions into semantically intelligible justifications that align with human contrastive reasoning. This allows the operator to understand why the system is adjusting support, fostering calibrated trust and enabling effective human override when necessary (Mastrianni et al., 2023).

In essence, hybrid intelligence is the engineering discipline of creating integrated human-machine systems where the machine gains a form of machine empathy—a computationally derived understanding of human cognitive states—and uses this understanding to act as a proactive, responsive partner. This paradigm aims to bridge the persistent "cognitive gap" in decision support by making computational systems fundamentally more aligned with the biological realities and finite resources of the human mind, thereby augmenting human potential in complex, high-stakes decision environments.

Objective and scope: from signal to supportive action

The primary objective of this monograph is to provide a comprehensive, engineering-focused framework for the design and implementation of HI systems specifically tailored for critical decision support. It moves beyond theoretical discussion to outline a coherent pipeline—from the acquisition of raw neurophysiological signals to the delivery of context-aware, supportive actions—that is both scientifically grounded and practically actionable. The scope is deliberately bounded to systems where the enhancement of human cognitive performance in complex, dynamic environments is the central goal, thereby excluding purely assistive or restorative medical BCIs, though the foundational technologies remain relevant (Agapov et al., 2017).

This work is structured to guide the reader through the three core translational stages of building an effective HI system:

The first scope encompasses the reliable capture and decoding of decision-relevant cognitive states. This involves a detailed examination of available BCI modalities focusing on the practical trade-offs between non-invasive workhorses like EEG and fNIRS and emerging high-fidelity platforms—and the subsequent AI-driven transformation of this data into interpretable cognitive constructs. The objective here is to establish a robust "deep interpretation" engine capable of disambiguating states such as escalating cognitive load, waning vigilance, or rising uncertainty from noisy, multimodal physiological data in real-time. This stage relies heavily on advanced signal processing and machine learning techniques, including end-to-end deep learning models for spatio-temporal pattern recognition (Roy et al., 2019).

The second scope addresses the crucial translation of a diagnosed cognitive state into an appropriate support strategy. This monograph details the architectural principles for building closed-loop systems where the AI component functions as an adaptive controller. The objective is to map specific cognitive state signatures (e.g., high theta power indicating overload) to tailored interventions (e.g., simplifying a dashboard, triggering an alert, or reallocating a subtask). This involves fusion methodologies for integrating neural data with other contextual streams (e.g., eye-tracking, system logs) and the implementation of shared control paradigms that respect human autonomy while providing timely support, thereby mitigating risks like automation bias (Wingerter et al., 2025).

The final scope concerns the effective communication of the system's support and the maintenance of a productive human-AI partnership. This covers the design of neuroadaptive interfaces that deliver feedback in an intuitive, non-disruptive manner, and the imperative of XAI to make the system's rationale transparent. The objective is to ensure that supportive actions are not only effective but also trusted and learnable by the human operator, fostering true teaming rather than passive reliance. The scope includes evaluation methodologies for assessing HI system performance in ecologically valid settings, moving beyond simple classification accuracy to metrics that reflect improved joint decision-making outcomes.

The human component: decodable cognitive states for decision-making

Effective HI rests upon a fundamental premise: that key dimensions of human cognition and affect, which critically influence decision-making quality, are not merely internal and subjective but are manifest in neurophysiological signals that can be quantified and interpreted. This chapter establishes a taxonomy of these decision-relevant neurocognitive states, providing a conceptual and empirical foundation for their measurement. It moves beyond a simple list to articulate how these states interact within the cognitive architecture under load, defining the target phenomena for the "deep interpretation" engine of an HI system.

Cognitive workload refers to the total amount of mental resources being expended to meet task demands. It is not inherently negative but becomes detrimental when demand exceeds available capacity, leading to performance breakdown. In HI, distinguishing between intrinsic load (inherent to task complexity), extraneous load (caused by suboptimal presentation), and germane load (effort toward schema construction) is crucial for targeted support. Neurophysiologically, workload is robustly associated with spectral power changes in EEG. A key marker is the suppression of alpha-band (8-13 Hz) power, particularly in parietal and occipital regions, associated with the active processing of sensory information, and an increase in frontal midline theta power (4-8 Hz) (Kirasirova et al., 2021), linked to focused attention and working memory engagement (Rottenberg et al., 2003). fNIRS studies further show a correlated increase in prefrontal cortex oxygenation under high working memory load (Agapov et al., 2017). For HI, continuous estimation of workload allows for dynamic task management, such as deferring non-essential notifications during periods of peak load or simplifying information displays to reduce extraneous cognitive burden (Agapov et al., 2017). For HI, continuous estimation of workload allows for dynamic task management, such as deferring non-essential notifications during periods of peak load or simplifying information displays to reduce extraneous cognitive burden.

This domain encompasses the allocation of processing resources to specific stimuli or thought streams. Sustained attention or vigilance is the ability to maintain focus over time, typically declining monotonically, leading to the "vigilance decrement." Focused/selective attention involves filtering distractors, while divided attention strains capacity. EEG markers are highly sensitive to attentional shifts: the attenuation of the posterior alpha rhythm is indicative of visual attention engagement, while the P300 event-related potential component reflects the conscious allocation of attention to a task-relevant, infrequent stimulus (Bulanov et al., 2020). A decrease in the ratio of beta to theta activity is a noted correlate of vigilance decrement. For decision-making, lapses in attention can result in critical information being "looked at but not seen." An HI system can counteract this by using real-time attention monitoring to highlight overlooked data or prompt re-checking when neural signatures of mind-wandering or disengagement are detected.

Optimal decision-making often requires overriding automatic responses, resolving conflict between competing options, and monitoring for errors. This executive function domain is primarily governed by the anterior cingulate cortex (ACC) and dorsolateral prefrontal cortex (DLPFC). A quintessential neural signature is the Error-Related Negativity (ERN), a fronto-centrally negative EEG deflection peaking within 100 ms. after an error commission, reflecting the brain's internal performance monitoring system.

Similarly, the N200 component and feedback-related negativity (FRN) signal response conflict and negative feedback, respectively (Shirolapov et al., 2024). Decoding these signals allows an HI system to identify moments of high decision conflict or self-recognized errors before they are behaviorally manifested. Decoding these signals allows an HI system to identify moments of high decision conflict or self-recognized errors before they are behaviorally manifested. This enables interventions such as suggesting a decision review, providing additional clarifying data, or reinforcing correct outcomes to strengthen learning (Maslova et al., 2023).

Decision-making is not coldly rational but is profoundly influenced by affective states. Acute stress, often measured via increased amygdala activity and sympathetic nervous system arousal, can narrow attention and bias decisions toward habitual responses. Uncertainty and doubt are cognitive-affective states associated with activation in the ACC and insula. Conversely, decision confidence the subjective certainty in a choice has been linked to specific post-decision signals in the prefrontal and parietal cortices. While often measured with peripheral physiology (e.g., heart rate variability, galvanic skin response), neural correlates from EEG (asymmetric frontal alpha activity for approach/withdrawal) and fNIRS (prefrontal oxygenation patterns under stress) are increasingly decodable. For HI, interpreting affective state is vital for managing automation bias over-reliance on AI under stress or uncertainty and for tailoring the system's communication style, perhaps adopting a more reassuring or explicitly confirmatory tone when low confidence is detected.

The taxonomy of decision-relevant cognitive states provides a functional framework; however, the engineering of Hybrid Intelligence systems requires mapping these abstract constructs to specific, measurable neurophysiological phenomena. This section delineates the neural correlates the brain regions and network dynamics underpinning each state and translates them into the observable signals captured by modern non-invasive and BCI technologies. This translation from neural activity to quantifiable data streams is the critical bridge between cognitive theory and practical system implementation.

Cognitive workload is primarily orchestrated by a fronto-parietal network. The DLPFC is central to working memory maintenance and manipulation, while the ACC monitors effort and conflict.

fNIRS as a rise in oxygenated hemoglobin (HbO) concentration. Simultaneously, electrophysiological recordings via EEG reveal two key oscillatory signatures: an increase in frontal midline theta (4-8 Hz) power, sourced to the ACC and medial prefrontal cortex, and a decrease in posterior alpha (8-13 Hz) power, reflecting the disinhibition of parietal and occipital cortices for enhanced sensory processing (Ragni et al., 2020). The integrated interpretation of these hemodynamic and electrophysiological changes provides a robust, multi-modal index of mental effort for HI systems.

The neural substrates of attention are modality- and type-specific. Sustained visual vigilance engages a right-dominant fronto-parietal-thalamic network, with the intraparietal sulcus and frontal eye fields playing key roles. The desynchronization (power decrease) of the alpha rhythm over occipital-parietal electrodes is a primary EEG correlate of visual attention engagement, indicating cortical activation. For auditory or cross-modal vigilance, theta synchronization in frontal regions may be more prominent. The P300 event-related potential, a positive deflection maximal over parietal sites around 300 ms post-stimulus, is generated by a network involving the temporo-parietal junction and inferior parietal lobule, and its amplitude is a direct gauge of attentional resource allocation to a task-relevant event (Picton, 1992). fNIRS can complement this by tracking sustained prefrontal and parietal oxygenation, which correlates with the maintenance of top-down attentional control.

Affective states such as stress, uncertainty, and confidence involve a complex interplay of limbic and cortical regions. Acute stress triggers the amygdala and hypothalamus, with downstream effects measured via peripheral physiology. However, its cortical impact, particularly the disruption of prefrontal executive functions, is observable as reduced functional connectivity within fronto-parietal networks via EEG coherence analysis and as deoxygenation in the prefrontal cortex via fNIRS under severe stress. Decision uncertainty and low confidence are associated with increased activation of the ACC and the anterior insula, regions integral to interoceptive awareness and salience detection (Droutman et al., 2015). While challenging to isolate with EEG due to deep sources, fNIRS shows promise in detecting insular and ACC engagement. Conversely, high confidence in a decision correlates with post-decision synchronization in the beta/gamma bands over sensorimotor and prefrontal cortices.

Despite significant challenges—including EEG's volume-conducted, ill-posed source localization, fNIRS's limited cortical surface coverage and lagging hemodynamic response, and substantial inter-subject variability in signal morphology and topography—the most robust approach for HI is a multi-modal fusion strategy that leverages the millisecond precision of EEG to capture rapid ERPs and oscillatory shifts and the superior spatial fidelity of fNIRS to anchor these dynamics to specific cortical networks, thereby enabling reliable deep interpretation of cognitive states for effective neuroadaptive decision support systems.

The technological component: BCI and AI foundations

The efficacy of a HI system is fundamentally constrained by the quality and practicality of its neural data stream. Therefore, the selection of an appropriate BCI modality is a critical first step, dictated by a trade-off between signal fidelity, user safety, portability, and robustness to real-world environments. Modern BCI technologies span a continuum from fully non-invasive, wearable systems to surgically implanted devices, each with distinct advantages and limitations for deployment outside controlled laboratory settings (Abiri et al., 2019; H. Zhang et al., 2024).

Non-invasive BCIs, which record neural activity from the scalp, are paramount for scalable HI applications due to their safety and relative ease of use (Lotte et al., 2018).

EEG is the predominant non-invasive modality, favored for its high temporal resolution (milliseconds), portability, and relatively low cost. Modern dry-electrode and wireless systems significantly improve usability for real-world scenarios. However, EEG signals are severely attenuated and blurred by the skull and scalp, resulting in poor spatial resolution and high vulnerability to physiological (e.g., eye blinks, muscle activity) and environmental artifacts. Despite these challenges, its practicality makes EEG the workhorse for most contemporary HI research, including real-time cognitive state monitoring and neuroadaptive systems (B. Xu et al., 2021).

EEG records the summation of postsynaptic potentials from synchronously firing pyramidal neuron populations, captured as oscillatory voltage fluctuations on the microvolt scale at the scalp surface, a process detailed in the seminal review "The origin of extracellular fields and currents--EEG, ECoG, LFP and spikes" (Buzsáki et al., 2012). Its paramount strength for HI lies in its millisecond-level temporal resolution, which is critical for tracking the rapid dynamics of cognitive processes such as error detection (manifested as error-related negativity, ERN) or shifts in attention. Modern advancements have significantly enhanced its applicability: dry-contact and semi-dry electrode systems reduce preparation time and improve comfort, while wireless amplifier units and miniaturized hardware enable ambulatory use in dynamic environments. However, the EEG signal is fundamentally constrained by its poor spatial resolution (approximately 1-2 cm) due to volume conduction effects, where electrical currents diffuse and blur through cerebrospinal fluid, skull, and scalp. This physical limitation makes precise localization of neural sources challenging without sophisticated inverse modeling techniques. Furthermore, EEG is highly susceptible to both physiological artifacts (electromyographic activity from facial muscles, electrooculographic signals from eye movements) and environmental noise, necessitating robust preprocessing and artifact rejection pipelines as a prerequisite for reliable cognitive state interpretation.

fNIRS measures hemodynamic responses associated with neural activity. It offers a better spatial resolution than EEG and is remarkably robust to motion and electrical artifacts, making it suitable for mobile applications. Its major drawback is low temporal resolution (seconds), limiting its use for decoding rapid cognitive transitions. fNIRS is often used complementarily with EEG to provide a multi-modal picture of brain function (Khan & Hong, 2017). fNIRS) operates on a distinct hemodynamic principle, measuring changes in the concentration of oxygenated (HbO) and deoxygenated (HbR) hemoglobin in the cortical vasculature following neural activation, as inferred from the differential absorption of near-infrared light, a methodology comprehensively outlined in the review "A review on continuous wave functional near-infrared spectroscopy and imaging instrumentation and methodology" (Scholkmann et al., 2014). This methodology provides several key advantages for HI systems. First, it offers superior spatial resolution to EEG (on the order of 1-2 cm), allowing for more anatomically specific probing of prefrontal or motor cortices crucial for decision-making and workload. Second, and critically for real-world deployment, fNIRS signals are inherently immune to electromagnetic noise and exhibit far greater resilience to motion artifacts compared to EEG, making them suitable for less restricted, naturalistic settings where some user movement is expected. The primary trade-off is a significantly slower temporal response, governed by the hemodynamic lag, with changes unfolding over a timescale of seconds. This limits fNIRS's ability to decode rapidly transient cognitive events but makes it exceptionally well-suited for monitoring sustained cognitive states such as mental workload, fatigue, or focal engagement over extended tasks.

The synergistic integration of EEG and fNIRS into hybrid neuroimaging systems represents a powerful frontier for HI, effectively mitigating the inherent limitations of each standalone modality, as empirically demonstrated in the foundational study "Enhanced performance by a hybrid NIRS-EEG brain computer interface" (Fazli et al., 2012). By simultaneously combining fast EEG signatures and slow fNIRS hemodynamics, this multi-modal fusion enables more robust decoding of cognitive states, surpassing either modality alone and establishing EEG-fNIRS as a gold-standard non-invasive framework for real-world HI systems, while the next subsection details how AI transforms these raw physiological streams into actionable insights.

Magnetoencephalography (MEG) records the magnetic fields generated by neuronal currents, providing excellent spatial and temporal resolution. However, MEG requires bulky, cryogenic shielding and is exorbitantly expensive, confining it to static, clinical settings and rendering it impractical for most real-world HI applications.

For clinical populations where maximum signal fidelity is required, invasive techniques offer superior performance (Cisotto et al., 2024).

Electrocorticography (ECoG) involves placing a grid of electrodes directly on the surface of the cortex. It provides signals with a much higher spatial resolution and signal-to-noise ratio than EEG, without the attenuation from the skull. While it requires craniotomy, its lower risk profile compared to deeper implants has enabled its use in long-term clinical BCI studies for communication and motor restoration.

Although ECoG remains a gold-standard semi-invasive modality offering superior spatial resolution and signal amplitude over EEG by bypassing the skull, the future of invasive BCI is being reshaped by novel form factors designed to mitigate the long-term biological challenges of traditional implants—exemplified by Synchron's endovascular Stentrode, which eliminates open-brain surgery but has limited signal resolution due to its intravascular position, and by Neuralink's and Paradromics' high-channel-count cortical arrays, which aim for unprecedented decoding bandwidth at the cost of full craniotomy risks.

Intracortical microelectrode arrays penetrate the brain tissue to record from small populations of neurons, offering the highest resolution signals for decoding fine-grained motor intentions and potentially complex cognitive states. The long-term biocompatibility and stability of these implants remain active challenges, and their use is currently restricted to pioneering clinical research.

The choice of modality for an HI-driven decision-support system must be application-specific. Table 1 summarizes the key operational and technical trade-offs of neural modalities in HI systems.

Table 1. Key operational and technical trade-offs of neural modalities in HI systems

Modality	Invasiveness	Spatial resolution	Temporal resolution	Portability	Robustness to noise	Primary hi use case
EEG	Non-invasive	Low (~1-2 cm)	High (ms)	High	Low	Ambulatory cognitive state monitoring, neurofeedback
fNIRS	Non-invasive	Medium (~1 cm)	Low (1-10 s)	High	High	Monitoring sustained cognitive workload, affect
ECOG	Semi-invasive	High (mm)	High (ms)	Low (clinical)	High	Precision decoding for clinical assistive devices
Intracortical	Invasive	Very High (μm)	High (ms)	Very Low	High	Foundational motor and cognitive decoding research

For broad-based HI systems aimed at enhancing decision-making in operational environments (e.g., control rooms, surgical suites, training simulators), non-invasive, portable modalities like EEG and fNIRS are the most viable (Aghajani et al., 2017; J. Chen et al., 2024). Invasive modalities, while powerful, remain reserved for specialized clinical interventions. Their continuous, unobtrusive operation aligns with the HI goal of providing implicit, real-time support without disrupting the user's primary task. In contrast, invasive modalities, while powerful, are currently reserved for specialized clinical interventions where the benefits justify the surgical risks. The subsequent section will detail how AI methodologies are engineered to overcome the inherent limitations of these practical BCI modalities, transforming raw, noisy signals into interpretable cognitive features (Aricò et al., 2018; Renton et al., 2019).

Miniaturized and minimally-invasive microsensors represent a frontier aimed at bridging the gap between the high noise of surface EEG and the risks of full cortical implants. A pioneering development in this domain is the wearable microneedle-BCI sensor, which utilizes an array of conductive polymer microneedles to penetrate the upper skin layers between hair follicles, achieving stable, high-fidelity neural signal capture for over 12 hours (Kim et al., 2025).

The core integration: architecting hybrid intelligence systems

The architecture of a HI system constitutes a strategic composition designed to synergistically unite the computational power of Artificial Intelligence (AI) with the cognitive, contextual, and ethical capabilities of the human operator. It is not merely a sequential processing pipeline but a dynamic, adaptive ecosystem where data, decisions, and control circulate according to a closed-loop principle. The key architectural challenge lies in bridging the semantic and operational gap between neural network models operating in statistical feature spaces and human cognition, which functions with symbols, abstractions, and value-based categories (Zakharov et al., 2025).

The fundamental architectural blueprint of an HI system can be decomposed into three interconnected tiers governing data flow and interaction protocols

1. The Sensor-Integrative Tier (Data and Perception Layer). This tier is responsible for aggregating, performing primary processing, and contextually enriching heterogeneous data. Inputs include structured data from databases, unstructured streams from IoT sensors, results of preliminary analytics, as well as explicit and implicit human inputs (queries, actions, biometric signals). A critical function at this level is the Contextual Anchoring Module, which uses ontologies and metadata to tag raw data with semantic labels, making it interpretable for both downstream AI models and human perception (W. Xu et al., 2021). For instance, a video feed from a surveillance camera is enriched with data on location, time, weather conditions, and historical incidents in that sector.
2. The Cognitive-Analytical Tier (Fusion and Interpretation Layer). This is the core of the system, where Fusion Methodologies are implemented at various levels. This tier comprises an ensemble of coordinated AI models (deep neural networks, support vector machines, Bayesian networks, etc.), each solving a

specialized task: pattern recognition, forecasting, anomaly classification. Their interaction is governed by a Fusion Arbiter – a software module that integrates outputs from individual models based on confidence scores, meta-features, or ranking rules. The architecture incorporates a parallel Interpretation Channel, which transforms the "black-box" outputs of models (e.g., activation vectors or saliency maps) into human-comprehensible formats: natural language descriptions, visual heatmaps, inference graphs (Adadi & Berrada, 2018). This channel serves as a bridge to the human-in-the-loop.

3. The Meta-Cognitive Tier (Orchestration and Control Layer). This is the strategic command level where human-AI partnership is physically enacted. It hosts the Shared Control Interface and the Adaptive Feedback Engine. The interface presents the arbiter's integrated result alongside its interpretation from the Interpretation Channel, allowing the human operator to calibrate the system's output, override decisions, or provide new constraints. The operator's actions, along with performance metrics and environmental feedback, are continuously captured by the Adaptive Feedback Engine. This engine implements the closed-loop principle, translating human corrections and real-world outcomes (success/failure signals) into tailored retraining signals, parameter adjustments for the arbiter, or updates to the contextual knowledge base (Bredeweg & Kragten, 2022). This creates a learning cycle where both the AI components and the human's mental model of the system co-evolve.

Data flows not only forward but also backward across these tiers. Data moves bidirectionally and asynchronously between these tiers. A forward flow (from Tier 1 to 3) represents the traditional analytic pipeline: from raw data to a recommended decision. The critical backward flow (from Tier 3 to 1 and 2) carries calibration signals, corrective labels, and heuristic rules, enabling the system to adapt in context. For example, if an operator consistently rejects a certain type of AI-generated alert in a specific scenario, the Feedback Engine can attenuate similar future alerts in that context or trigger a focused retraining of the relevant classifier.

Interaction is mediated by a standardized meta-protocol that accompanies all data packets. This protocol contains not only the payload (data or decision) but also metadata on provenance, confidence intervals, the fusion path traversed, and the available explanation primitives. This ensures that every component, including the human, operates with a traceable and auditable information state, which is fundamental for achieving explainability and establishing calibrated trust.

The efficacy of a HI system is fundamentally governed by the methodologies employed to fuse information from disparate artificial and human sources. Fusion is not a monolithic process but a hierarchical strategy, defined by the stage at which integration occurs within the information processing pipeline. The selection of the appropriate level—data, feature, or decision—entails critical trade-offs between informational fidelity, computational overhead, robustness to noise, and the capacity for human interpretability and intervention.

Data-level fusion, or raw data fusion, is the most foundational approach, characterized by the direct combination of heterogeneous data streams prior to any significant feature extraction or analytical processing. This methodology operates on the principle that the greatest potential for discovering novel, synergistic patterns lies in the unprocessed or minimally cleaned data corpus. Techniques involve the temporal and spatial alignment, registration, and concatenation of multi-modal data—such as co-registering LiDAR point clouds with synchronized camera images or aligning physiological sensor readings (e.g., EEG, galvanic skin response (GSR)) with user interaction logs (Lahat et al., 2015). The primary advantage is the preservation of maximal information content, allowing subsequent AI models to learn complex, cross-modal correlations inherently. For instance, a convolutional neural network (CNN) trained on fused pixel-and-depth data can learn object features intrinsically tied to geometry. However, this approach is computationally intensive, highly susceptible to the "curse of dimensionality," and critically sensitive to noise, missing data, and synchronization errors in any constituent stream. It also creates a "fused black box," where the origin of a signal becomes opaque, posing significant challenges for subsequent interpretation and debugging.

Feature-level fusion represents a more abstract and manageable paradigm, where distinct feature vectors are extracted independently from each data source or modality before being integrated into a unified, higher-dimensional representation. This is the most prevalent and flexible methodology in HI systems. The process involves transforming raw data into domain-relevant features (e.g., spectral components from audio, SIFT/SURF descriptors from images, statistical summaries from time-series) and then combining them via concatenation, transformation, or selective integration. Techniques such as Principal Component Analysis (PCA), Canonical Correlation Analysis (CCA), or modern deep learning-based encoders (e.g., multimodal autoencoders) are used to project features into a joint embedding space where modalities can interact (Ramachandram & Taylor, 2017). This level offers a balance: it reduces dimensionality and noise compared to data-level fusion while retaining more discriminative power than decision-level fusion. It allows for asynchronous processing of modalities and facilitates human understanding, as features often retain semantic meaning (e.g., "edge density," "heart rate variability"). A key challenge is the "semantic gap" between feature spaces—aligning the statistical representation of an image feature with a textual descriptor requires sophisticated learning to establish a common manifold.

Decision-level fusion is the highest abstraction layer, where integration occurs after each source or model has produced its own independent decision, classification, or confidence score. In an HI context, these sources include various AI models and explicit human inputs (e.g., a binary approval/rejection, a priority ranking). The fusion mechanism acts as an arbiter or meta-classifier, combining these outputs using rules or learned models. Common techniques include weighted voting (based on model confidence), Bayesian inference, Dempster-Shafer theory (for handling uncertainty), and supervised meta-learners (e.g., stacking) (C. Zhang & Ma, 2012). This approach offers maximum modularity, robustness, and explainability. Individual models can be developed, trained, and validated independently. The failure or low confidence of one source does not corrupt the entire pipeline. Most importantly, the human decision can be treated as just another expert input within the fusion framework, allowing for transparent calibration of human-AI trust (e.g., dynamically adjusting the weight of a human operator's input based on their measured fatigue or domain expertise). The primary limitation is the loss of rich correlational information present in the raw data or features, as fusion is based on summarized outputs. It also introduces latency, as each model must complete its processing before the final fusion can occur.

The choice of fusion methodology is not exclusive and is often architecturally nested. A sophisticated HI system may employ hybrid fusion strategies: using feature-level fusion for tightly coupled modalities (e.g., video and audio for emotion recognition) and decision-level fusion to combine this result with an independent analysis from a knowledge graph and a human oversight decision.

The selection is dictated by system requirements (Tab. 2):

1. Data-Level Fusion is preferred when data modalities are highly synchronized, correlated, and resources permit learning from the raw composite.
2. Feature-Level Fusion is optimal for balancing performance and complexity, especially when modalities are related but not perfectly aligned, or when partial interpretations of features are needed.
3. Decision-Level Fusion is essential for robustness, modular system design, and in contexts where human judgment, explainability, and the graceful degradation of the system are paramount.

Table 2. Comparison of fusion methodologies in hybrid intelligence systems

Criterion	Data-level fusion	Feature-level fusion	Decision-level fusion
Abstraction level	Low (Raw Data)	Intermediate (Features)	High (Decisions/Scores)
Information loss	Minimal	Moderate	Significant
Robustness to noise	Low	Moderate	High
Computational load	High	Moderate to High	Low to Moderate
Modularity & flexibility	Low	Moderate	High
Explainability potential	Very Low	Moderate	High
Human-input integration point	Difficult	Possible (as a feature source)	Natural (as an expert input)

The selection of fusion level is therefore a design decision driven by the primary system requirements: data-level fusion maximizes information but sacrifices modularity; feature-level offers balance; decision-level prioritizes robustness and explainability.

The "deep interpretation" engine: advanced methodologies

The transition from a basic BCI to a true HI system necessitates a paradigm shift from simple signal classification to a continuous, contextual, and personalized interpretation of the user's cognitive and affective state. This chapter delineates the advanced methodologies that constitute a "Deep Interpretation" engine, moving beyond discrete commands to model the nuanced dynamics of the human mind in symbiosis with AI.

Traditional neuroadaptive systems rely on classifying brain signals into a finite set of predefined states (e.g., "high workload," "low engagement," "error detection"). This approach is fundamentally limited, as cognitive states are not discrete but exist on a continuum and evolve dynamically over time. The cutting edge lies in continuous latent state modeling, where high-dimensional neural data (e.g., EEG, fNIRS) is used to infer a low-dimensional, continuously evolving latent variable representing the user's unobservable cognitive trajectory. Techniques such as State-Space Models (SSMs), Kalman and Particle Filters, and modern recurrent neural networks (RNNs) with latent variable models (e.g., Variational Autoencoders for time series) are employed for this purpose (Kao et al., 2014). For instance, instead of classifying workload as "high" or "low," a latent model can estimate a scalar "cognitive resource utilization" value that fluctuates smoothly with task demands. This enables more granular and temporally precise adaptations, such as dynamically adjusting the complexity of information presented on a display in real-time.

Furthermore, these latent spaces can be structured to separate confounding factors (e.g., general arousal from task-specific engagement), providing a cleaner signal for system adaptation (Gershman & Blei, 2012).

A core challenge in neural interpretation is the profound variability in brain signals across different individuals (inter-subject) and within the same individual across sessions or contexts (intra-subject). A robust HI system cannot operate with a one-size-fits-all model. Personalization involves creating user-specific models, but collecting extensive calibration data for each user is impractical. Advanced methodologies thus focus on few-shot learning, transfer learning, and domain adaptation. Here, a base model is pre-trained on a large, diverse population dataset to learn general neural representations. This model is then rapidly fine-tuned for a new user with only a small amount of calibration data, leveraging techniques like model reprogramming or hypernetwork-based adaptation (Yair et al., 2019). Second, to address intra-subject variability (e.g., due to fatigue, circadian rhythms, or learning effects), systems incorporate continual online adaptation. This involves incrementally updating the user model in the background using newly acquired data, often stabilized by experience replay or elastic weight consolidation algorithms to prevent catastrophic forgetting of previously learned patterns (Parisi et al., 2019). The HI framework is crucial here, as the system can actively query the user for confirmation ("Were you distracted just now?") to obtain high-quality labels for this ongoing adaptation.

The most accurate deep learning model is useless in a critical HI application if its reasoning is opaque. When an AI recommends diverting pilot attention or altering a medical diagnosis based on neural data, the human partner must understand why. XAI for neural networks is therefore non-negotiable. Post-hoc explanation techniques are applied to elucidate which features of the brain signal drove a particular interpretation. Spatiotemporal attribution methods, such as Layer-Wise Relevance Propagation (LRP) or Integrated Gradients applied to EEG spectrograms or source-localized activity maps, can highlight the specific brain regions and frequency bands (e.g., increased frontal theta power) that contributed to a "high workload" classification. For latent state models, explanation involves tracing how the latent variables evolve in response to stimuli. Furthermore, counterfactual explanations can be generated: "The system inferred low engagement because the alpha power in the occipital lobe was high; had it been 20% lower, the state would have been classified as engaged." This form of explanation fosters calibrated trust, allowing the human operator to accept the AI's judgment when the neurophysiological rationale aligns with their subjective experience or task context, and to reject it when it does not.

Brain signals alone provide an incomplete picture. The "Deep Interpretation" engine achieves robustness and contextual fidelity by fusing neural data with complementary behavioral and physiological streams. Eye-tracking provides a direct window into visual attention and cognitive processing (e.g., pupil dilation as a measure of cognitive load, fixations indicating information search patterns). Peripheral physiology—heart rate variability (HRV), electrodermal activity (EDA), and electromyography (EMG)—offers reliable indicators of autonomic arousal, stress, and affective valence. Interaction logs (keystrokes, mouse movements, response times) provide objective performance metrics. The power of HI is in the principled fusion of these modalities. For example, a rise in frontal EEG theta power might indicate working memory engagement, but if accompanied by decreased pupillary constriction velocity and increased error rate, the system can more confidently infer cognitive overload versus focused problem-solving (McDuff et al., 2014). Deep learning architectures like multi-modal transformers are specifically designed to learn joint representations from these asynchronous, heterogeneous data streams. This multi-modal context is also vital for explanation, as the system can present a unified narrative: "The user is likely frustrated (inferred from increased EDA and corrugator EMG) and struggling to locate relevant information (inferred from chaotic gaze scanpath), leading to increased cognitive workload (inferred from EEG)." This rich, multi-faceted interpretation forms the bedrock for truly intelligent and context-aware system adaptation.

To synthesize and contrast these core methodological pillars, their respective challenges, and technological implementations, the following table 3 provides a consolidated overview.

Table 3. Summary of advanced methodologies for deep interpretation

Methodology	Core challenge	Key techniques	Outcome for HI
Latent state modeling	Discrete states poorly model cognition	State-Space Models, RNN-VAEs, Kalman Filters	Continuous, smooth estimate of cognitive dynamics (e.g., engagement gradient)
Personalization	High inter-/intra-subject variability	Transfer Learning, Few-Shot Learning, Online Adaptation	A system that adapts to the individual user over time, minimizing calibration
Explainable ai (XAI)	Black-box models erode trust	LRP, Saliency Maps, Counterfactual Explanations	Transparent, auditable interpretations that enable calibrated human trust
Multi-modal fusion	Single modalities are ambiguous	Multi-modal Transformers, Late Fusion Architectures	Robust, context-aware interpretation disambiguating states like stress vs. focus

Applications in critical decision support domains

The true test and ultimate validation of HI systems lie in their deployment within high-stakes environments where decisions have significant consequences for human health, safety, and performance. These domains are characterized by complexity, uncertainty, time pressure, and severe penalties for error, creating an imperative for systems that augment rather than replace human expertise. The application of HI across these diverse fields demonstrates a unifying principle: the synergistic partnership between human judgment and artificial computation creates a new, emergent capability greater than the sum of its parts.

In medicine, HI systems are transcending the role of passive tools to become active collaborative partners. In surgical assistance, robotics integrated with HI principles provide not just enhanced precision but context-aware guidance. Systems like the da Vinci Surgical System can be augmented with real-time AI analysis of endoscopic video, overlaying critical anatomical structures, highlighting potential vessels, or warning of deviation from the planned surgical pathway. This creates a shared control paradigm where the surgeon maintains ultimate authority while the AI handles stability filters, tremor reduction, and context-specific visual enhancements (Lee et al., 2024). In neurological rehabilitation, HI drives adaptive neuroprosthetics and exoskeletons (Zakharov et al., 2020). By decoding user intent via EEG or EMG signals and integrating it with real-time kinematic and environmental data, these systems provide assistive forces only as needed, promoting neural plasticity and user agency. For diagnostic support, HI moves beyond mere pattern recognition. Systems integrate multi-modal patient data (imaging, genomics, proteomics, electronic health records) to present a differential diagnosis ranked by probability, accompanied by evidence (e.g., highlighted regions in an MRI, relevant literature excerpts). The clinician's role evolves to evaluating the AI's reasoning, incorporating nuanced patient history, and making the final ethical and contextual judgment call, forming a diagnostic dyad (Topol, 2019).

These domains demand continuous situational awareness and rapid response to rare, high-consequence events. In aviation, HI systems are evolving from autopilots to cognitive co-pilots. They integrate real-time sensor data (engine performance, weather radar, traffic) with flight plan databases and pilot state monitoring (e.g., gaze tracking, vocal stress) to predict and manage threats. The system might not just alert to a potential conflict but propose optimized resolution advisories, manage lower-priority subsystems, and adapt its interface to reduce cognitive load during high-workload phases (Hagemann et al., 2023). In air traffic control (ATC), HI assists controllers by predicting potential conflicts beyond a human-manageable horizon, simulating resolution strategies, and automating routine communications. The human controller oversees the strategic airspace picture, handles exceptional cases, and maintains the final authority for separation assurance. For power grid management, modern smart grids' scale and volatility exceed unaided human monitoring capacity. HI systems continuously analyze grid stability, predict load fluctuations from diverse sources (including renewables), and suggest optimal dispatch and switching actions. The human operator interprets these suggestions within a broader context of maintenance schedules, regulatory constraints, and long-term reliability goals, ensuring strategic oversight of the AI's tactical optimization.

HI is revolutionizing skill acquisition by enabling personalized, adaptive training environments that respond to the learner's cognitive and psychomotor state. Adaptive simulators in fields like medicine, aviation, and military operations use HI to dynamically adjust scenario difficulty, introduce complications, and provide targeted feedback. By fusing performance metrics (e.g., surgical instrument path efficiency, flight control inputs) with physiological indicators of stress or workload (e.g., heart rate variability, pupil dilation), the system can infer a trainee's cognitive readiness and challenge them at the optimal edge of their ability. Cognitive skill tutors extend this beyond procedural tasks to domains like critical thinking and command decision-making. These systems model the learner's knowledge state, identify misconceptions through dialog and problem-solving actions, and generate tailored hints or counter-examples. The HI paradigm ensures the tutor is not a rigid script but a collaborative coach that explains its reasoning, adapts to learning styles, and knows when to intervene versus when to let the learner struggle productively.

HI is fundamentally reshaping the human-machine interface by creating closed-loop systems that adapt to the user's cognitive and affective state. Neuroadaptive interfaces use real-time brain activity (typically via non-invasive EEG or fNIRS) to infer cognitive workload, attention, or error-related potentials. This allows the system to dynamically reallocate tasks—for instance, simplifying a display during high workload or taking over routine tasks when it detects attention is focused on a critical primary task. In assistive technologies, HI enables profoundly new levels of autonomy. A BCI for a paralyzed individual becomes far more effective when integrated with an HI framework. The user's neural commands are combined with contextual AI: computer vision for object recognition, natural language processing for communication prediction, and smart environment control. The user provides high-level intent (e.g., "drink"), and the HI system translates this into a safe, coordinated sequence of actions for a robotic arm, managing low-level trajectory planning while continuously seeking user confirmation via a secondary BCI channel or gaze tracking. This shared autonomy reduces user fatigue and error rates, making the technology genuinely usable (Edelman et al., 2019).

Across all these domains, the common thread is the shift from automation to augmentation. The HI system does not seek to remove the human from the loop but to strategically position them within it, leveraging human strengths—contextual understanding, ethical reasoning, and commonsense generalization—while mitigating human limitations in processing speed, attention consistency, and computational endurance. The resulting partnership enhances performance, safety, and resilience, setting a new paradigm for critical decision support in the 21st century.

Ethical, societal, and practical challenges

The development and deployment of HI systems, particularly those interfacing directly with the human brain, transcend technical hurdles and venture into profound ethical, societal, and practical territory. While the potential benefits for human augmentation are significant, they are accompanied by a complex array of risks that must be proactively identified, analyzed, and mitigated. This chapter examines the core challenges that must guide the responsible innovation of neuroadaptive HI technologies.

At the forefront of ethical concerns lies the unique nature of neural data as the purported "final frontier" of personal privacy. Unlike passwords or location history, brain signals can reveal unconscious preferences, emotional states, predispositions to certain health conditions, and covert cognitive processes. The collection, storage, and analysis of this neurodata raise critical questions of mental privacy and the potential for unauthorized access or coercive "brain-reading" (Zander & Kothe, 2011). Closely linked is the threat to personal autonomy and cognitive liberty—the right to self-determination over one's own mental processes. An HI system designed to optimize performance could, through subliminal neuroadaptation or overly persuasive conditioning, subtly nudge a user's decisions, eroding their sense of agency. This creates a paradox: a tool intended to augment the will could potentially undermine it. Establishing robust ethical and legal frameworks for neurodata sovereignty—defining who owns neural data and under what conditions it can be used—is therefore a foundational prerequisite for the field.

HI systems inherit and can amplify the biases present in their two core components: the human neurophysiological data they are trained on and the AI algorithms that process it. Neurophysiological data is not universal; it varies with age, sex, cultural background, and neurodiversity (e.g., ADHD, autism spectrum). Training datasets that lack demographic diversity will produce models that perform poorly for underrepresented groups, leading to discriminatory outcomes, for instance, in adaptive educational or employment systems (Liu et al., 2021). Furthermore, the AI algorithms themselves can introduce bias through their architecture or objective functions. A system designed to maximize engagement by adapting to a user's state might learn to exploit neurological vulnerabilities, prioritizing addictive interaction patterns over user well-being. Ensuring fairness requires active curation of diverse, representative datasets, the application of algorithmic fairness constraints during model training, and continuous auditing for disparate impacts across different user populations.

Beyond grand ethical principles, the real-world adoption of HI systems faces significant practical barriers. Usability remains a primary challenge. Current non-invasive neural interfaces (EEG, fNIRS) often require cumbersome setup, are sensitive to motion artifacts, and suffer from low signal-to-noise ratios in uncontrolled environments. For HI to move beyond the laboratory, seamless, ergonomic, and socially acceptable form factors are essential. Robustness is equally critical. A system that fails unpredictably—misinterpreting fatigue as disinterest, or being disrupted by environmental electromagnetic noise—is not just useless but potentially dangerous in critical applications. HI architectures must be designed for graceful degradation, with clear fallback protocols and real-time confidence estimation. Finally, scalability presents a dual challenge: scaling technically to handle multi-user, cloud-connected scenarios while maintaining data privacy, and scaling economically to ensure equitable access and avoid exacerbating the "neuro-divide" between those who can afford augmentation and those who cannot.

A profound societal risk associated with advanced automation is de-skilling—the attrition of human expertise as tasks are increasingly offloaded to AI. In an HI system for air traffic control or surgery, over-reliance on automated diagnosis and recommendation can lead to the erosion of the human operator's foundational knowledge and situation assessment skills, leaving them ill-prepared to take over during system failures (Jiang et al., 2019). Mitigating this requires HI design philosophies that emphasize human oversight and continuous learning, rather than full automation. Closely related is the problem of accountability. When a decision made by a human-AI partnership leads to harm (e.g., a medical error or a financial loss), attributing liability becomes complex. Was it a flaw in the algorithm, a misinterpretation by the human, a failure in their calibration, or a latent bug in the fusion logic? This "responsibility gap" challenges existing legal frameworks. Developing clear standards for system certification, audit trails for decision provenance (explainable AI logs), and defining the legal concept of a "duty of care" for HI operators and developers are urgent societal tasks.

These challenges are not prohibitive but constitutive; they define the responsible development pathway for HI. Addressing them requires a deeply interdisciplinary approach, integrating not only neuroscience and computer science, but also ethics, law, psychology, and social science. A proactive governance model—involving researchers, industry, policymakers, and the public—is needed to establish guidelines for ethical design, bias mitigation, and accountability before these technologies become widespread. The goal is not to stifle innovation, but to steer it toward a future where HI truly enhances human flourishing, agency, and equity.

CONCLUSION AND FUTURE HORIZONS

The exploration of HI systems charts a course toward a new paradigm of human-machine symbiosis, where artificial intelligence becomes a contextual, adaptive partner rather than a static tool. This monograph has established a framework for such systems, built upon a closed-loop architecture that integrates real-time neural interpretation, multi-modal fusion, and explainable AI to enable seamless, synergistic collaboration in critical domains.

The transition from concept to functional HI system requires a methodical, inter-disciplinary approach. It begins with a human-centric problem definition, followed by the architectural integration of neural data acquisition, a "Deep Interpretation" engine for continuous state modeling, and adaptive feedback mechanisms. Success is measured not by automation efficiency alone, but by enhanced human performance, elevated situation awareness, and calibrated trust, validating the system in complex, real-world scenarios as demonstrated in surgical suites, cockpits, and training simulators.

The future of HI is being shaped by several transformative frontiers. Passive BCIs will shift interaction from explicit command to implicit collaboration by continuously decoding cognitive states without conscious user effort. Collaborative HI will expand the dyad to human teams, with AI acting as a mediator to optimize group dynamics and shared awareness. The most speculative yet profound frontier involves brain-to-brain networks, where AI could facilitate direct information exchange or synthesize collective intelligence, demanding parallel advancements in neuroethics and security.

To navigate this future responsibly, stakeholders must adopt guiding principles. Researchers must prioritize ecological validity and the creation of diverse, open neural datasets to combat bias. Engineers and designers must embed explainability, robustness, and privacy-by-design into core architectures, ensuring human oversight remains paramount. Policymakers and ethicists must proactively develop governance frameworks for neural data sovereignty, algorithmic accountability, and equitable access to prevent exacerbating societal divides.

Ultimately, the goal of HI is not cognitive replacement but cognitive augmentation. By steadfastly centering human autonomy, dignity, and agency within the technological loop, we can steer this powerful convergence toward a future that amplifies our uniquely human capacities for judgment, creativity, and ethical reasoning.

REFERENCES

Abiri, R., Borhani, S., Sellers, E. W., Jiang, Y., & Zhao, X. (2019). A comprehensive review of EEG-based brain–computer interface paradigms. *Journal of Neural Engineering*, *16*(1), 011001. DOI: 10.1088/1741-2552/aaf12e PMID: 30523919

Adadi, A., & Berrada, M. (2018). Peeking Inside the Black-Box: A Survey on Explainable Artificial Intelligence (XAI). *IEEE Access : Practical Innovations, Open Solutions*, *6*, 52138–52160. DOI: 10.1109/ACCESS.2018.2870052

Agapov, S. N., Bulanov, V. A., Zakharov, A. V., & Sergeeva, M. S. (2017). Comparison of classifiers in the tasks of the single-trial vep classification. *Zhurnal Vysshei Nervnoi Deyatelnosti Imeni IP Pavlova, 67*(4), Article 4.

Aghajani, H., Garbey, M., & Omurtag, A. (2017). Measuring Mental Workload with EEG+fNIRS. *Frontiers in Human Neuroscience*, *11*, 359. DOI: 10.3389/fnhum.2017.00359 PMID: 28769775

Aricò, P., Borghini, G., Di Flumeri, G., Sciaraffa, N., & Babiloni, F. (2018). Passive BCI beyond the lab: Current trends and future directions. *Physiological Measurement*, *39*(8), 08TR02. Advance online publication. DOI: 10.1088/1361-6579/aad57e PMID: 30039806

Berretta, S., Tausch, A., Ontrup, G., Gilles, B., Peifer, C., & Kluge, A. (2023). Defining human-AI teaming the human-centered way: A scoping review and network analysis. *Frontiers in Artificial Intelligence*, *6*, 1250725. DOI: 10.3389/frai.2023.1250725 PMID: 37841234

Bredeweg, B., & Kragten, M. (2022). Requirements and challenges for hybrid intelligence: A case-study in education. *Frontiers in Artificial Intelligence*, *5*, 891630. DOI: 10.3389/frai.2022.891630 PMID: 36003066

Bulanov, V. A., Zakharov, A. V., & Khivintseva, E. V. (2020). Wavelet transform for the identification of P300. *IOP Conference Series. Materials Science and Engineering*, *862*(5), 052049. DOI: 10.1088/1757-899X/862/5/052049

Buzsáki, G., Anastassiou, C. A., & Koch, C. (2012). The origin of extracellular fields and currents—EEG, ECoG, LFP and spikes. *Nature Reviews. Neuroscience*, *13*(6), 407–420. DOI: 10.1038/nrn3241 PMID: 22595786

Chen, J., Yu, K., Bi, Y., Ji, X., & Zhang, D. (2024). Strategic Integration: A Cross-Disciplinary Review of the fNIRS-EEG Dual-Modality Imaging System for Delivering Multimodal Neuroimaging to Applications. *Brain Sciences*, *14*(10), 1022. DOI: 10.3390/brainsci14101022 PMID: 39452034

Chen, J. Q., & Lee, S. M. (2003). An exploratory cognitive DSS for strategic decision making. *Decision Support Systems*, *36*(2), 147–160. DOI: 10.1016/S0167-9236(02)00139-2

Cisotto, G., Zancanaro, A., Zoppis, I. F., & Manzoni, S. L. (2024). hvEEGNet: A novel deep learning model for high-fidelity EEG reconstruction. *Frontiers in Neuroinformatics*, *18*, 1459970. DOI: 10.3389/fninf.2024.1459970 PMID: 39759760

Droutman, V., Bechara, A., & Read, S. J. (2015). Roles of the Different Sub-Regions of the Insular Cortex in Various Phases of the Decision-Making Process. *Frontiers in Behavioral Neuroscience*, *9*. Advance online publication. DOI: 10.3389/fnbeh.2015.00309 PMID: 26635559

Edelman, B. J., Meng, J., Suma, D., Zurn, C., Nagarajan, E., Baxter, B. S., Cline, C. C., & He, B. (2019). Noninvasive neuroimaging enhances continuous neural tracking for robotic device control. *Science Robotics*, *4*(31), eaaw6844. Advance online publication. DOI: 10.1126/scirobotics.aaw6844 PMID: 31656937

Fazli, S., Mehnert, J., Steinbrink, J., Curio, G., Villringer, A., Müller, K.-R., & Blankertz, B. (2012). Enhanced performance by a hybrid NIRS–EEG brain computer interface. *NeuroImage*, *59*(1), 519–529. DOI: 10.1016/j.neuroimage.2011.07.084 PMID: 21840399

Frolov, N., Pitsik, E., Grubov, V., Badarin, A., Maksimenko, V., Zakharov, A., Kurkin, S., & Hramov, A. (2023). Perceptual Integration Compensates for Attention Deficit in Elderly during Repetitive Auditory-Based Sensorimotor Task. *Sensors, 23*(14). *Sensors (Basel)*, *23*(14), 6420. Advance online publication. DOI: 10.3390/s23146420 PMID: 37514714

Gershman, S. J., & Blei, D. M. (2012). A tutorial on Bayesian nonparametric models. *Journal of Mathematical Psychology*, *56*(1), 1–12. DOI: 10.1016/j.jmp.2011.08.004

Hagemann, V., Rieth, M., Suresh, A., & Kirchner, F. (2023). Human-AI teams—Challenges for a team-centered AI at work. *Frontiers in Artificial Intelligence*, *6*, 1252897. DOI: 10.3389/frai.2023.1252897 PMID: 37829660

Jiang, L., Stocco, A., Losey, D. M., Abernethy, J. A., Prat, C. S., & Rao, R. P. N. (2019). BrainNet: A Multi-Person Brain-to-Brain Interface for Direct Collaboration Between Brains. *Scientific Reports*, *9*(1), 6115. DOI: 10.1038/s41598-019-41895-7 PMID: 30992474

Kao, J. C., Stavisky, S. D., Sussillo, D., Nuyujukian, P., & Shenoy, K. V. (2014). Information Systems Opportunities in Brain–Machine Interface Decoders. *Proceedings of the IEEE, 102*(5), 666–682. DOI: 10.1109/JPROC.2014.2307357

Khan, M. J., & Hong, K.-S. (2017). Hybrid EEG–fNIRS-Based Eight-Command Decoding for BCI: Application to Quadcopter Control. *Frontiers in Neurorobotics, 11*. Advance online publication. DOI: 10.3389/fnbot.2017.00006 PMID: 28261084

Kim, H., Kim, J. H., Lee, Y. J., Lee, J., Han, H., Yi, H., Kim, H., Kim, H., Kang, T. W., Chung, S., Ban, S., Lee, B., Lee, H., Im, C.-H., Cho, S. J., Sohn, J. W., Yu, K. J., Kang, T. J., & Yeo, W.-H. (2025). Motion artifact–controlled micro–brain sensors between hair follicles for persistent augmented reality brain–computer interfaces. *Proceedings of the National Academy of Sciences of the United States of America, 122*(15), e2419304122. DOI: 10.1073/pnas.2419304122 PMID: 40193612

Kirasirova, L. A., Zakharov, A. V., Morozova, M. V., Kaplan, A. Y., & Pyatin, V. P. (2021). ERP Correlates of Emotional Face Processing in Virtual Reality. *Opera Medica et Physiologica, 8*(3), 3.

Lahat, D., Adali, T., & Jutten, C. (2015). Multimodal Data Fusion: An Overview of Methods, Challenges, and Prospects. *Proceedings of the IEEE, 103*(9), 1449–1477. DOI: 10.1109/JPROC.2015.2460697

Lee, A., Baker, T. S., Bederson, J. B., & Rapoport, B. I. (2024). Levels of autonomy in FDA-cleared surgical robots: A systematic review. *NPJ Digital Medicine, 7*(1), 103. DOI: 10.1038/s41746-024-01102-y PMID: 38671232

Liu, Y., Wang, T., Wang, K., & Zhang, Y. (2021). Collaborative Learning Quality Classification Through Physiological Synchrony Recorded by Wearable Biosensors. *Frontiers in Psychology, 12*, 674369. DOI: 10.3389/fpsyg.2021.674369 PMID: 33995232

Lotte, F., Bougrain, L., Cichocki, A., Clerc, M., Congedo, M., Rakotomamonjy, A., & Yger, F. (2018). A review of classification algorithms for EEG-based brain–computer interfaces: A 10 year update. *Journal of Neural Engineering, 15*(3), 031005. DOI: 10.1088/1741-2552/aab2f2 PMID: 29488902

Maslova, O., Komarova, Y., Shusharina, N., Kolsanov, A., Zakharov, A., Garina, E., & Pyatin, V. (2023). Non-invasive EEG-based BCI spellers from the beginning to today: A mini-review. *Frontiers in Human Neuroscience, 17*, 1216648. DOI: 10.3389/fnhum.2023.1216648 PMID: 37680264

Mastrianni, A., Sarcevic, A., Hu, A., Almengor, L., Tempel, P., Gao, S., & Burd, R. S. (2023). Transitioning Cognitive Aids into Decision Support Platforms: Requirements and Design Guidelines. *ACM Transactions on Computer-Human Interaction, 30*(3), 1–28. DOI: 10.1145/3582431 PMID: 37694216

McDuff, D., Gontarek, S., & Picard, R. (2014). Remote measurement of cognitive stress via heart rate variability. *2014 36th Annual International Conference of the IEEE Engineering in Medicine and Biology Society*, 2957–2960. DOI: 10.1109/EMBC.2014.6944243

Parisi, G. I., Kemker, R., Part, J. L., Kanan, C., & Wermter, S. (2019). Continual lifelong learning with neural networks: A review. *Neural Networks*, *113*, 54–71. DOI: 10.1016/j.neunet.2019.01.012 PMID: 30780045

Phillips-Wren, G. (2013). Intelligent Decision Support Systems. In Dcumpos, M., & Grigoroudis, E. (Eds.), *Multicriteria Decision Aid and Artificial Intelligence* (1st ed., pp. 25–44). Wiley., DOI: 10.1002/9781118522516.ch2

Picton, T. W. (1992). The P300 Wave of the Human Event-Related Potential. *Journal of Clinical Neurophysiology*, *9*(4), 456–479. DOI: 10.1097/00004691-199210000-00002 PMID: 1464675

Ragni, F., Tucciarelli, R., Andersson, P., & Lingnau, A. (2020). Decoding stimulus identity in occipital, parietal and inferotemporal cortices during visual mental imagery. *Cortex*, *127*, 371–387. DOI: 10.1016/j.cortex.2020.02.020 PMID: 32289581

Ramachandram, D., & Taylor, G. W. (2017). Deep Multimodal Learning: A Survey on Recent Advances and Trends. *IEEE Signal Processing Magazine*, *34*(6), 96–108. DOI: 10.1109/MSP.2017.2738401

Renton, A. I., Mattingley, J. B., & Painter, D. R. (2019). Optimising non-invasive brain-computer interface systems for free communication between naïve human participants. *Scientific Reports*, *9*(1), 18705. DOI: 10.1038/s41598-019-55166-y PMID: 31822715

Rottenberg, J., Wilhelm, F. H., Gross, J. J., & Gotlib, I. H. (2003). Vagal rebound during resolution of tearful crying among depressed and nondepressed individuals. *Psychophysiology*, *40*(1), 1–6. DOI: 10.1111/1469-8986.00001 PMID: 12751799

Roy, Y., Banville, H., Albuquerque, I., Gramfort, A., Falk, T. H., & Faubert, J. (2019). Deep learning-based electroencephalography analysis: A systematic review. *Journal of Neural Engineering*, *16*(5), 051001. DOI: 10.1088/1741-2552/ab260c PMID: 31151119

Scholkmann, F., Kleiser, S., Metz, A. J., Zimmermann, R., Mata Pavia, J., Wolf, U., & Wolf, M. (2014). A review on continuous wave functional near-infrared spectroscopy and imaging instrumentation and methodology. *NeuroImage*, *85*, 6–27. DOI: 10.1016/j.neuroimage.2013.05.004 PMID: 23684868

Shirolapov, I. V., Zakharov, A. V., Shishkina, A. A., Sergeeva, M. S., Komarova, Y. S., Romanchuk, N. P., Bannov, V. M., Kuznetsova, O. G., & Khivintseva, E. V. (2024). Efficiency of computerized cognitive training for prevention of cognitive impairments and stimulation of neuroplasticity. *Advances in Gerontology = Uspekhi Gerontologii, 37*(3), 221–229. *Scopus.* Advance online publication. DOI: 10.34922/AE.2024.37.3.007

Topol, E. J. (2019). High-performance medicine: The convergence of human and artificial intelligence. *Nature Medicine, 25*(1), 44–56. DOI: 10.1038/s41591-018-0300-7 PMID: 30617339

Wingerter, T. L., Straub, T., & Schweitzer, S. (2025). Mitigating Automation Bias in Generative AI Through Nudges: A Cognitive Reflection Test Study. *Procedia Computer Science, 270*, 2106–2114. DOI: 10.1016/j.procs.2025.09.331

Xu, B., Zhang, D., Wang, Y., Deng, L., Wang, X., Wu, C., & Song, A. (2021). Decoding Different Reach-and-Grasp Movements Using Noninvasive Electroencephalogram. *Frontiers in Neuroscience, 15*, 684547. DOI: 10.3389/fnins.2021.684547 PMID: 34650398

Xu, W., Dainoff, M. J., Ge, L., & Gao, Z. (2021). *Transitioning to human interaction with AI systems: New challenges and opportunities for HCI professionals to enable human-centered AI* (Version 4). arXiv. DOI: 10.48550/ARXIV.2105.05424

Yair, O., Ben-Chen, M., & Talmon, R. (2019). Parallel Transport on the Cone Manifold of SPD Matrices for Domain Adaptation. *IEEE Transactions on Signal Processing, 67*(7), 1797–1811. DOI: 10.1109/TSP.2019.2894801

Zakharov, A. V., Bratchenko, I. A., Bratchenko, L. A., Neupokoeva, A. V., Khivintseva, E. V., Shirolapov, I. V., & Zakharov, V. P. (2025). Deep learning of surface-enhanced Raman spectroscopy data for multiple sclerosis diagnostics. *The European Physical Journal. Special Topics, 234*(15), 4789–4797. Advance online publication. DOI: 10.1140/epjs/s11734-024-01449-z

Zakharov, A. V., Bulanov, V. A., Khivintseva, E. V., Kolsanov, A. V., Bushkova, Y. V., & Ivanova, G. E. (2020). Stroke Affected Lower Limbs Rehabilitation Combining Virtual Reality With Tactile Feedback. *Frontiers in Robotics and AI, 7*, 81. https://www.frontiersin.org/articles/10.3389/frobt.2020.00081. DOI: 10.3389/frobt.2020.00081 PMID: 33501248

Zander, T. O., & Kothe, C. (2011). Towards passive brain–computer interfaces: Applying brain–computer interface technology to human–machine systems in general. *Journal of Neural Engineering, 8*(2), 025005. DOI: 10.1088/1741-2560/8/2/025005 PMID: 21436512

Zhang, C., & Ma, Y. (Eds.). (2012). *Ensemble Machine Learning*. Springer. DOI: 10.1007/978-1-4419-9326-7

Zhang, H., Jiao, L., Yang, S., Li, H., Jiang, X., Feng, J., Zou, S., Xu, Q., Gu, J., Wang, X., & Wei, B. (2024). Brain–computer interfaces: The innovative key to unlocking neurological conditions. *International Journal of Surgery*, *110*(9), 5745–5762. DOI: 10.1097/JS9.0000000000002022 PMID: 39166947

Chapter 6

Brain-Inspired Artificial Intelligence Models for Intelligent Decision-Making

Annmol Damien
http://orcid.org/0009-0001-6800-6316
Christ University, Bengaluru, India

Jain Mathew
http://orcid.org/0000-0003-4019-5132
Christ University, Bengaluru, India

ABSTRACT

Brain-inspired artificial intelligence provides a promising approach for advancing intelligent decision-making in complex and uncertain environments. Integrating insights from neuroscience, cognitive science, and artificial intelligence, this chapter explores how human cognitive mechanisms such as learning, attention, memory, emotion, motivation, and bounded rationality can inform the design of adaptive and explainable decision systems. Cognitive architectures, neuro-symbolic models, and learning and attention-based approaches are examined to demonstrate how biologically inspired principles enhance flexibility, transparency, and robustness beyond traditional optimisation-driven AI. The discussion emphasises decision intelligence as a human-centred paradigm that supports rather than replaces human judgement, particularly in high-stakes contexts. Ethical considerations, explainability, and responsible design are highlighted as essential for trust and accountability, offering an interdisciplinary framework for developing cognitively aligned and context-sensitive AI systems.

DOI: 10.4018/979-8-3373-8988-2.ch006

1. INTRODUCTION

Artificial intelligence (AI) has revolutionized the world of decision-making at high speeds in various fields like finance, health care, business management, and intelligent systems. Modern AIs, especially those that are based on machine learning and deep neural networks, have proven to exhibit incredible pattern recognition, prediction, and optimisation abilities. Most of the current AI systems are, however, weak at their calculations when applied in complex, uncertain, and dynamic decision processes. These environments require accuracy and efficiency as well as adaptability, contextual awareness, transparency, and reference to human values and reasoning processes. These shortcomings have led to an increasing questioning of the strength, explainability, and reliability of AI-advanced decision systems.

Human decision making, on the contrary, though limited in terms of cognitive capacity and vulnerable to bias, is highly flexible and resilient in the face of uncertainty. In people, incomplete information is constantly being incorporated, competing goals are weighed, and decisions are changed with experience and feedback. It has been shown that these abilities are formed by the highly organized interaction of various neural and cognitive processes, among which are perception, learning, memory, attention, emotion, motivation, and valuation. The brain does not, therefore, engage in one specific computational process in decision-making, but a distributed and dynamic process that is influenced by both cognitive and affective influences.

Neuroscience and cognitive science lessons have continued to inspire the work on brain-inspired artificial intelligence, an interdisciplinary paradigm aimed at abstracting, generalizing, and applying biologically and cognitively inspired concepts to the design of AI systems. Instead of trying to recreate the brain on a biological basis, brain-inspired AI aims at simulating the functional processes of feedback-based learning, predicting rewards, selective attention, reasoning with memory, and adaptive control. Such mechanisms provide useful design guidelines on how AI systems can be designed to be more flexible and interpretable, and how they can be effectively deployed in real-life situations of decision making.

Decision intelligence has become a very important point of implementation within this paradigm. Decision intelligence are AI systems which are intended to help, assist, or co-operate with human decision-makers especially in an environment where there is uncertainty, risk and dynamic change (Cortés et al., 2000). Contrasting the conventional decision-support that is often based on fixed rules or more limited optimisation goals, brain-inspired decision intelligence aims to match the computational decision-making processes to human cognitive decision-making strategies. This kind of alignment is vital in enhancing transparency, trust, and usability especially in high stakes areas where human control and accountability are still invaluable.

The chapter discusses the brain-inspired AI work models of intelligent decision-making, with specific reference to their application in the fields of decision intelligence and human-centred artificial intelligence models. With a multi-level approach, the chapter incorporates neuroscientific basis of the decision making as well as the cognitive theories and computational abstractions. It explores how the mechanisms of learning, memory, attention, emotion and valuation in the human brain can be used to guide the design of adaptive and explainable AI designs. Particular focus is placed on hybrid cognitive-computational and neuro-symbolic models using the capabilities of neural learning with the potential of organized reasoning and knowledge representation.

Besides architectural concerns, the chapter also deals with the main challenges in using AI in decision-critical settings, such as decision-making under uncertainty, cognitive bias model and mitigation, and integration of both short-term and long-term goals. The importance of explainability and human-centred design is highlighted in all the above, as it is acknowledged that intelligent decision systems should not only be effective but also be able to explain their reasoning in cognitively effective manner. Brain-inspired designs can provide viable opportunities to create transparent, trustworthy and ethically responsible AI by basing AI decision processes on principles rooted in human cognition.

The chapter also contextualizes brain-inspired decision intelligence to real life application areas like financial decision-making, healthcare decision support and human-AI collaborative system. Such situations exemplify why AI systems should be used to supplement human judgement and not to substitute it especially in situations where judgement is uncertain, risky, and involves ethical implications. Lastly, the chapter considers the future directions of research, such as neuromorphic computing, scalable cognitive architecture and enhanced convergence between neuroscience and artificial intelligence.

As a whole, this chapter provides an interdisciplinary and integrative approach to brain-inspired artificial intelligence in decision-making. It provides concepts and design guidelines to the development of intelligent decision systems by synthesising knowledge in neuroscience, cognitive science, and AI to create adaptive, explainable, and human cognition-compatible intelligent decision systems. This chapter also aims to serve as a scholarly research source for researchers, a concept guide for system designers, and a reference base for higher students seeking the future of intelligent decision making.

2. NEUROSCIENCE OF DECISION-MAKING

Biological systems are complex, distributed, and dynamic in decision-making processes, which arise as a result of the coordinated action of numerous neural and cognitive subsystems (Kalaidjieva & Swanson, 2004). In contrast to classical models of computation, which presuppose the presence of a single function of optimisation or a linear chain of information processing operations, the human brain is based on interacting neural networks that can be performed in parallel and at several time scales. These networks combine both the input of the senses, inner physiological and emotional states, previous experience, and future expectations. This means that the process of human decision-making is necessarily adaptive, context-sensitive, and able to operate under uncertainty.

The neuroscientific studies have determined some of the fundamental brain parts and circuits, which complement each other during decision-making. Executive control, planning, maintenance of goals, and long-term consequences evaluation all center around the prefrontal cortex (Domenech & Koechlin, 2015). Various subregions within the prefrontal cortex are involved in various components of making a decision, including abstract reasoning, inhibitory control, and value comparison (Sakagami et al., 2006). Conflict monitoring, observing mistakes, and an indication of the necessity of cognitive regulation are functions of the anterior cingulate cortex that occur specifically in matters where alternative decisions are possible or where there is uncertainty.

The striatum and dopaminergic pathways are closely linked with reward processing and learning, encoding information regarding anticipated rewards and prediction errors (Keiflin & Janak, 2015). Through these mechanisms, people can change their behavior according to the response, and reinforce the behaviours that bring positive results and discourage those that cause losses. Limbic structures like the amygdala and insula aid emotional and affective aspects of the decision-making process and help in fear, anxiety, perception of risk, and interoceptive awareness (Zhang et al., 2024). Notably, these emotional cues do not simply intrude on thinking rationally, rather, they give vital data on value and importance, especially in doubtful or high-stakes situations. One of the most influential perspectives emerging from neuroscience is the recognition that decision-making involves the continuous interaction between cognitive and affective systems. Dual-process theories conceptualise this interaction as involving fast, intuitive processes and slower, deliberative processes. Intuitive processes enable rapid responses and pattern-based judgments, while deliberative processes support reflective reasoning and cognitive control. However, neuroscientific evidence suggests that these processes are deeply intertwined, with emotional and motivational signals shaping deliberative evaluation and executive decision-making (D. G. V. Mitchell, 2011). This insight challenges the simplistic dichotomies between emotion and reason and has important implications for computational models of intelligence.

Reinforcement learning provides a crucial bridge between neuroscience and computational decision-making. Neural mechanisms of reward prediction error allow the brain to compare expected and actual outcomes, updating value representations accordingly (Cohen, 2006). This learning process supports adaptation over time and enables decision-makers to refine strategies in dynamic environments. Attention and memory systems further modulate learning by determining which experiences are encoded and which cues are prioritised during future decisions. Collectively, these neuroscientific findings provide a rich foundation for brain-inspired artificial intelligence models that aim to capture the adaptive and context-sensitive nature of human decision-making.

3. COGNITIVE MECHANISMS BEHIND INTELLIGENT DECISION-MAKING

In the cognitive science view, the concept of limited rationality is the central determinant of human decision-making. Unlike idealised models of rational choice, which assume unlimited information, perfect foresight and unlimited computational capacity, the actual human decision-makers have to act under significant cognitive, informational and temporal constraints. Such constraints compel the application of simplifying strategies that enable people to make decisions effectively in complicated and uncertain conditions. Instead of searching through all viable options, humans tend to satisfice, in the sense that they find a solution that is good enough based on the available limitations. This view has far-reaching consequences on the formation of decisions and the systematic exclusion of the normative model of rationality.

Due to bounded rationality, heuristics (or mental shortcuts) are extensively used to lighten the cognitive load, in other words, to simplify complex decision problems. Heuristics allow quick judgment and act especially when it comes to cases of uncertainty, ambiguity, or time pressure. Heuristic-based decision-making in most real-life situations is very adaptive in such a way that a person is able to operate effectively without a thorough analysis ((Mahadeen et al., 2020). But the use of heuristics also has systematic patterns of bias so that the deviations in judgment and choice can be expected. These biases do not occur by mere chance but are due to the cognitive processes that make effective decision-making achievable.

One of the most important cognitive processes that determines which information enters conscious processing and which is ignored is "attention," making it a fundamental component of decision-making. Attention is a filtering mechanism that focuses on the priorities of particular stimuli, attributes, or options rather than others. Such selectivity is necessary because the amount of potentially relevant information inherent in most decision environments is way beyond the capacity of cognition. There are bottom-up and top-down processes that influence attentional control. Perceptual salience, which is novelty, intensity, or emotional relevance, enhances the bottom-up attention, whereas goals, expectations, and prior knowledge determine the top-down attention (Campos et al., 2019).

Attentional allocation is very crucial in the process of preference or decision-making. While focusing on specific attributes, e.g., possible gains, risks, or losses, the person is actually giving more weight to the attributes in their evaluations. As an example, the framing effects occur when various presentations of the same information cause attention to other facets of an option, resulting in a different selection of that option even when objective values are equal. Attention can therefore not only be seen as a passive filter, but also as an active determinant of the construction of subjective values and preferences.

Attention and working memory are closely associated such that it limits how much information a person can focus on and process at one time when making decisions. Working memory has a limited capacity in terms of both the number of items that can be stored at once and the duration or the time for which the information can be retained without rehearsal (Cowan, 2022). Consequently, people tend to make decisions based on simplified versions of decision problems, paying attention to some of their features or options instead of making thorough analyses. These limits promote the adoption of sequential, attribute-based, or rule-based decision methods instead of global optimisation.

The experiential basis of intelligent decision-making is given by memory systems in a broader sense. Epic memory helps one to remember certain past events, such as past decisions and results. This memory enables learning of success and failure towards adaptive adjustment of future behaviour. Episodic recall can specifically have a strong impact on uncertain decisions because the vivid or emotionally salient outcomes in the past can disproportionately affect expectations and perceptions of risks (Grossberg & Gutowski, 1987). Semantic memory, in its turn, contains the general knowledge, concepts, rules and schemas which organize the knowledge about the world. Semantic knowledge facilitates decision-making through the provision of abstract structures through which reasoning, categorisation and inferences can be made (Janjua et al., 2012). To illustrate, domain-specific expertise is largely based on properly developed semantic structures, which allow identifying patterns effectively and developing expectations. In commercial or financial decision-making, semantic memory is a part of the cognition of market rules, rules of strategies, and normative evaluation boundaries.

Besides episodic and semantic memory, emotional memory is an important element that influences decisions. Emotional memory is a form of memory that gives affective meaning to an experience, and what is stored in it is not just a depiction of what occurred, but also what it was like (Ochsner & Schacter, 2000). Such affective tags play a larger role in future preferences by having a biasing effect on attention, memory, and judgment in favor of the option with positive or negative emotional consequences. Emotional memory is, therefore, a fast valuation process through which one can utilize the experience of emotion without necessarily taking a lot of time in deliberation.

These memory systems together facilitate mental simulation, which is a mental process by which people imagine possible future events and assess how these events are likely to turn out. Mental simulation enables decision-makers to place themselves in the situation of hypothetical results, comparing alternatives, and estimating how they can emotionally react to the various options (Waring et al., 2010). The skills are especially necessary when it comes to complex decisions, which have long-term implications, are uncertain, or are linked to morality. Nevertheless, a mental simulation is limited by cognitive biases, as imagined futures tend to be influenced by salient memories, affective states and narrow perspective-taking.

Motivation and emotion are not the peripheral factors of making decisions, but rather central elements of thinking. Feelings act as quick and efficient evaluative cues, directing attention and development of preferences before deliberate decision-making (George & Dane, 2016). These affective cues may be adaptive, especially in cases where a quick response is needed or where a lot of analysis is not feasible. As an example, the negative effect can be an indicator of threat, whereas the positive effect can prompt exploration and interaction.

Motivational state also influences decision-making by employing effort allocation, persistence, and tolerance to uncertainty. Cognitive engagement may be enhanced by the high levels of intrinsic or extrinsic motivation, and more efforts may be directed by people towards the search and analysis of information. On the other hand, the lack of motivation can encourage default options or heuristic use. Motivation is also linked to risk preferences in which people with prized goals may be more willing to take the risk of uncertainty or in the short-term gain in favor of a long-term gain.

Notably, emotional and motivational forces are dynamic and situational. Attentional focus, memory retrieval, and risk sensitivity can change in response to stress, anxiety, fatigue, and social pressure, and cause the decision strategies to change. As an example, anxiety could reduce the attention range and induce risk-averse behaviour, whereas positive mood conditions could increase the range of attention and explore different options. Cognitive models that ignore these affective dynamics are in danger of simplifying decision processes and not being able to explain behaviour in the real world.

Knowledge of the interaction of attention, memory, emotion, and motivation is thus crucial to the creation of intelligent decision systems. Conventional artificial intelligence models, which often assume consistent preferences, complete information, and fully rational assessments, cannot be successfully applied in settings with a nature of ambiguity, volatility, and human interaction. Conversely, brain-inspired artificial intelligence tries to incorporate brain-like cognitive mechanisms that include the way in which humans actually process information and arrive at a decision.

Brain-inspired AI systems can be made more flexible and robust by modeling selective attention, memory-based reasoning, feedback-driven learning, and affective modulation. Such systems are more adapted to give priority to the relevant information, adapt to the current conditions, and give context-based recommendations. Furthermore, the inclusion of cognitive processes contributes to the increase of interpretability by allowing decision-making to be explained in terms of well-known human concepts like goals, attention, and learning from experience.

Finally, the process of intelligent decision-making cannot be comprehensively understood and imitated without references to cognitive limitations and processes that influence human behavior. Instead of considering bounded rationality and bias as constraints that need to be removed, brain-inspired methods consider them to be fundamental aspects of adaptive intelligence. Through these principles, the intelligent decision systems can go beyond the narrow-focus optimisation to more human-competent decision-making by helping to achieve quality, transparency, and context-sensitive decision-making in complex real-world environments.

4. BRAIN-INSPIRED ARTIFICIAL INTELLIGENCE MODELS

Brain-inspired artificial intelligence (BI-AI) is a broad emerging range of computational technique that conceptually and functionally simulate neural and cognitive ideas of the way human intellectual capacity is acquired and preserved. Rather than making an effort to model the biological brain in its analytical and physiological detail, brain-inspired methods aim to model significant processes that support intelligent behavior, which comprises learning by experience, selective attention, reasoning by memory, goal-directed action, and dynamic decision-making. The general goal of BI-AI is to develop systems capable of operating in intricate, unpredictable, and dynamic environments that are very close to the environment in which decisions must be made. Traditional systems of artificial intelligence have been very successful in tasks that can be considered to be narrowly focused, especially those that are related to pattern recognition, classification, and optimization. However, many of these systems fail in instances where they are required to make extrapolation beyond their training, to learn how to deal with new conditions, or even give a justification of why they have made a certain decision. The latter drawbacks are notably deplorable in the fields where decisions that most likely will influence the organization, finance, healthcare, policy, or governance are to be made, where the uncertainty, ambiguity, ethical limitations, as well as human control are put in focus. The concept underlying brain-inspired artificial intelligence addresses these challenges through the implementation of the concepts of human perception, learning, thinking, and decision-making.

The major characteristics of brain-inspired AI are its focus on functional mechanisms, but not biological fidelity. Other approaches are not based on neuroscience on a neural level, but instead run at a cognitive or computational level of abstraction, and they are informed by cognitive science, psychology, and decision theory. BI-AI can trade-off between biological plausibility, computational tractability, and practical applicability with the help of this multi-level orientation. The attained models are more interpretable and consistent, as well as more adaptive and robust with human decision-making processes.

Among the general framework of brain-based artificial intelligence, three types of models are especially applicable to intelligent decision-making. They are cognitive architectures, neuro-symbolic and hybrid models, and learning, attention, and memory-inspired models. All these methods represent different facets of human thinking and have unique benefits in the design of intelligent decision systems.

4.1 Cognitive Architectures

Cognitive architectures are one of the most explicit models of human-like intelligence in artificial systems. A cognitive architecture is a common computational system that defines the structure and organisation of the cognitive processes, such as perception, attention, memory, learning, reasoning, and decision-making. Cognitive architectures are not concerned with isolated functions, but are concerned rather with the way these processes interact with one another over time to create a global behavior.

Fundamentally, cognitive architectures are aimed at modeling the regularities of human thinking that are sustained by multiple tasks and areas. They usually contain modules or components that are associated with various cognitive functions, and communication and coordination mechanisms between these components. This is a modular but integrated design that allows cognitive structures to facilitate goal-oriented behaviour, planning, problem-solving, and adaptive control in challenging environments.

The main advantage of the cognitive architectures is that they are able to define explicit goals, constraints, and strategies. The decision-making in these architectures can be defined as the process of making decisions that address the existing objectives without violating the environmental factors and the internal resources. This framing has a strong resemblance to the way of human decision-making, in which the decisions are not made using only the immediate rewards, but also based on the long-term goals, norms, and context.

Learning through experience also is given emphasis on cognitive architectures. Learning mechanisms allow the systems to refresh knowledge representations, optimise decision strategies and adjust behaviour through feedback. Notably, learning is usually placed within a wider psychological environment, in interactions with memory, attention, and reasoning. This combination enables cognitive architectures to show cumulative learning and transfer knowledge across tasks, as opposed to task-specific optimisation.

Other benefits of cognitive architectures are transparency and interpretability. Due to the explicit representation of cognitive processes, the flow of information through the system and the way of decision making can be tracked. This traceability is especially useful in the field of decision intelligence, where the decision-makers need to know how and why to make such decisions. Explaining what decision was made and why it was made is of paramount importance in terms of trust, accountability, and adherence to ethics in areas like finance, healthcare, and governance.

Moreover, cognitive architectures suitably correspond to human AI cooperation. Such systems are able to express decisions, recommendations, and explanations in cognitively meaningful terms by following an operationally defined manner that is conceptually well grounded in human reasoning. Such an alignment allows interaction between human decision-makers and AI systems, enabling AI to be used as a cognitively useful partner instead of an unintelligent decision-maker.

Cognitive architectures have their weaknesses, especially in the area of scalability and computation efficiency. Detailed modeling of complex cognitive processes may be computationally intensive, and combining perceptual learning based on large-scale data is a continuing problem. However, brain-inspired decision intelligence has a potent base in cognitive architectures, especially in scenarios where interpretability, adaptability, and human concordance are the sole priority.

4.2 Neuro-Symbolic and Hybrid Models.

Neuro-symbolic and hybrid models are a significant development in brain-inspired artificial intelligence to combine the benefits of the neural and the symbolic approaches and address their limitations. The classical neural network frameworks are good at recognizing patterns, learning statistics, and adapting to big data, yet they are not very transparent, do not have structured reasoning abilities, and fail to disclose knowledge. Comparatively, symbolic artificial intelligence systems are founded on logical reasoning, rule-based inference, and the representation of knowledge and are less powerful in their capacity to learn noisy data and evolve with time.

Neuro-symbolic AI seeks to integrate these complementary paradigms with each other to create integrated systems that can be adaptive and interpretable. Perceptual processing, feature extraction, and data learning are usually done by neural components, whereas reasoning, planning, constraint satisfaction, and explanation are done by the symbolic ones. This combination captures the main features of human thinking, in which the unconscious, associative mind plays against the conscious, rule-oriented mind.

Neuro-symbolic models have a number of benefits in a decision-making context. To begin with, they facilitate systems to learn by doing and following explicit rules and constraints. This is especially necessary in the areas that are controlled by regulations, moral standards, or safety considerations, where any unregulated optimisation can result in negative or destructive results. Neuro-symbolic models ensure that the decisions do not go outside acceptable limits by instilling symbolic rules inside the learning systems.

Second, neuro-symbolic systems facilitate causal reasoning and explanation, which are critical in making high-stakes decisions. Symbolic representations enable the system to define the logical framework of the decisions, to distinguish the causal relationships, and explain the choices in a manner comprehensible to humans. This capability handles one of the most major complaints of pure neural decision systems, which is their opaqueness and unintelligibility.

Third, hybrid models are more robust and general. Neural learning adapts to new data and dynamic environments possible and symbolic knowledge makes it stable and structured. The interaction between this combination and neuro-symbolic systems allows generalisation to data beyond the training data to be more efficient than data-driven models. Such strength is important in a questionable or swiftly evolving choice atmosphere.

Neuro-symbolic approaches are also similar to the principles of responsible and ethical AI. They are aided by these systems to provide transparency, accountability, and human oversight through making reasoning processes transparent and through normative constraints. The decision-makers can study, criticize, and streamline the symbolic elements to ensure that the behaviour of AI does not go against human ideals and corporate goals.

Regardless of this, neuro-symbolic models have technical difficulties, such as integration of heterogeneous representations, as well as coordination of learning and reasoning. Nonetheless, it is in the process of continuous research that architectures and ways of ensuring a seamless integration continue to be refined. Therefore, neuro-symbolic AI is one of the foundations of brain-inspired decision intelligence, the interface between learning by data and reasoning by design.

4.3 Learning, Attention, and Memory-Inspired Models

Other significant brain-inspired artificial intelligence models are learning, attention, and memory-inspired models, which make direct reference to neural and cognitive operation principles. These models concentrate on the mechanisms that facilitate adaptability, selective information processing, and experience-related reasoning abilities, which are important in making intelligent decisions in changing conditions.

Reinforcement learning (RL) is one of the most outstanding learning paradigms based on neural reward mechanisms. In reinforcement learning, agents gain the ability to make decisions that maximise cumulative reward by interacting with the environment. This is a biological learning process in that the behavior is molded by errors in reward prediction, which indicates that there is a difference between the expected and actual behavior. (Schultz, 2024). Reinforcement learning enables the agents to develop their strategies over time and balance between exploration and exploitation of the existing rewards.

Scheduling sequential decision-making in uncertainty is best adapted to reinforcement learning. Through feedback, RL-based systems will be able to adapt to the evolving conditions and increase the refinement of the policies through experience. However, the traditional reinforcement learning models are often subject to heavy assumptions with respect to the reward functions, and may be confronted with scarce feedback, complex environments, or time dependencies. By providing more cognitive mechanisms, attempts have been made to overcome these limitations by using hierarchical and memory-augmented reinforcement learning and brain-inspired extensions.

Attention mechanisms are essential in coping with information overload through selective attention to information that is relevant. Based on human attention mechanisms, computational attention mechanisms enable models to dynamically attenuate inputs depending on their relevance, context, or task requirements (Diederen & Fletcher, 2020). Attention improves efficiency in the decision-making system by focusing on important information and eliminating noise. It is also useful in terms of interpretability, as it is used to help understand what factors contributed to making a specific decision.

Mechanisms of attention are extremely useful in high dimensional or multi-modal environments where it is not possible to exhaustively process data. Attention-based models facilitate context-sensitive and flexible decision-making by causing computational resources to be directed towards salient features. Such selective processing is very close to human cognition, wherein attention is guided by both perceptual salience and goal-directed control.

Memory, augmented models also improve the intelligence of decisions by providing the possibility of storing and recalling previous experiences. These models are also inspired by the human memory systems, enabling the AI systems to store information over time as part of long-term learning and contextual reasoning. The historical patterns and memory processes make the agents use past decisions and act according to previous experiences.

Learning, attention, and memory processes are used to achieve dynamic adaptation, situational awareness, and uncertainty resilience when combined. These models facilitate the process of updating decisions made by AI systems according to experience, prioritizing the information, and incorporating the historical background into the current assessments. These abilities are critical in making decisions in a real-life context where conditions are never fixed, and information is usually partly complete or uncertain.

Notably, these mechanisms assist in humanizing AI systems. Through their functioning that is similar to human learning and attention, brain-inspired models are likely to yield less human-understandable decisions and more intuitive decisions as well. This conformity promotes confidence and successful co-operation between people and artificial intelligence systems in making decisions.

4.4 Integrative Perspectives on Brain-Inspired Models

Although cognitive architectures, neuro-symbolic systems, and learning-attention-memory models can be discussed individually, they are better in their integration. Human intelligence is the result of the interaction between various cognitive processes, and so are intelligent artificial systems, which can be made to use complementary methods. Models inspired by the brain are trying to combine structured reasoning and adaptive learning, selective attention and memory in coherent structures.

This kind of integration allows AI systems to be applied on both the high and the low levels of abstraction in perception and high levels of reasoning and strategic decision-making. It is also scalable, robust, and explainable in character, which are desirable for decision intelligence in a complex setting. A future direction of making adaptive and trustworthy intelligent systems more generative is through integrative brain-inspired models, which apply neuroscientific, cognitive science, and artificial intelligence.

5. BRAIN-INSPIRED DECISION INTELLIGENCE

Decision intelligence has become a form of paradigm shift in artificial intelligence, where a system that tends to automate pre-determined processes is replaced by one that tends to assist, augment, and supplement the human decision-making process. Conventional decision support systems are mostly based on fixed rules, optimisation procedures, or predictive analytics. Although successful in highly organised settings, these systems tend to perform poorly in real-life settings where there are uncertainty, incomplete information, dynamic change, and conflicting agendas. Brain-inspired decision intelligence is an extension of the view of decision intelligence, which is based on the alignment of computational decision-making processes to the cognitive strategies used by human decision-makers (O'Callaghan, 2023).

Fundamentally, brain-inspired decision intelligence appreciates that human decisions are not often the result of pure optimisation, rather, they are the result of the interplay between learning, memory, attention, emotion, and reasoning processes that use limited rationality. Brain-inspired decision systems abstract these processes, thus better suited to operate in environments where the goals keep changing, information is uncertain, and outcomes are undetermined. These systems are especially useful in areas where decisions are to be constantly updated in accordance with feedback and evolving circumstances.

The ability to describe uncertainty explicitly is one of the largest differentiating characteristics of a brain-inspired decision system. These systems do not regard uncertainty as a noise that should be removed, as in the case with other systems that model uncertainty as a fundamental aspect of the decision environment. Brain-inspired systems can use probabilistic representations, or estimates of confidence, or uncertainty-aware learning processes to make risk, ambiguity, and incomplete knowledge reasoning in a way that is similar to human decision-makers. This helps them to come up with robust and adaptive recommendations as opposed to brittle ones.

Integration of long-run and short-run goals is another major characteristic. Human choices tend to have trade-offs between short-term and long-term rewards. Brain-inspired decision intelligence embodies such a time aspect by including learning and planning processes that balance the exploitation of strategies that are known, and also the exploration of alternative strategies. This balance reflects on the behavior of human decision-making and is essential in areas of life like finance, healthcare, and organisational management, where short-term profits can take advantage of the long-term viability.

Notably, models of cognitive bias are also brought into brain-inspired decision intelligence. Instead of perfect rational users, such systems understand how people can be predicted to behave in ways that follow predictable patterns, e.g., loss aversion, overconfidence, and framing effects. The prospect of bias in artificial intelligence systems makes them amenable to the idea of countering or balancing these impacts, such as by reframing data, reminding users of risks that they otherwise ignore, or prompting individuals to be more balanced in their judgments. By doing so, the brain-inspired decision intelligence is not just a replica of human cognition, but rather aims at complementing and enhancing their performance.

The most important aspect of this paradigm is the emphasis on human AI cooperation. The decision systems inspired by the brain are not supposed to replace human judgment in high-stakes or ethical situations. Instead, they act as the thinking allies that help the people who are making the decisions by providing systematic information, alternative opinions, and adaptive suggestions. The given orientation of cooperation does not deprive people of human autonomy and responsibility but leads to improved quality of the decisions.

6. EXPLAINABILITY AND HUMAN-CENTRED DESIGN

Explainability is among the most burning issues in modern artificial intelligence, especially when it comes to systems that contribute to consequential decision-making. A large number of developed AI models, particularly those that are founded on deep learning, are opaque in nature, which is a black box, hence hard to comprehend how it arrives at decisions. This deficiency in transparency diminishes trust, restricts adoption, and creates ethical and legal issues in particular areas, including finance, healthcare, and governance of the people.

Brain-inspired artificial intelligence has promising avenues of explainable decision intelligence, based on cognitively meaningful representations. The internal processes of brain-inspired models can also be described using more intuitive terms to human users because the models tend to be based on such principles as goals, attention, learning, and memory. This conformity to the way human beings think gives an innate basis of explanation. Cognitive architectures and hybrid models justify explainability using explicit reasoning processes. Goal activation, rule evaluation, memory retrieval, and learning updates can be followed in their sequences to make decisions. This traceability allows the systems not only to explain the outcomes but also to give process-based explanations that discuss how and why a specific decision was arrived at. This would be especially useful when accountability and justification are demanded. Attention-based mechanisms also contribute to better explainability as they can show what information was used to make a decision. Attention mechanisms enable systems to indicate the factors that were prioritised during the evaluation process by establishing salient features or inputs. This selective attention reflects the human attentional mechanisms and helps in the explanation that is succinct, relevant and cognitively tractable. Human based design principles are critical towards making sure that the explanations are technical as well as usable and meaningful. Explanations should also be made user-specific regarding cognitive abilities, skills, and requirements of users.

Brain-inspired decision intelligence focuses on the strategies of adaptive explanation that vary in depth and detail, depending on the situation and user attributes. Brain-inspired decision systems can be used to build trust and acceptance by having transparency, reliability, and alignment with human values as their primary goals. Explainability is not a technical property but a relational property, and it determines the way users interact with intelligent systems, perceive them, and depend on them.

7. APPLICATIONS

The models of brain-inspired decision-making have proven to be broadly applicable in areas that are characterised by complexity, uncertainty, and high-stakes outcomes. Brain-inspired systems, unlike traditional decision-support systems, focus on adaptive learning, representation of uncertainty, and match the human cognitive processes. This renders them especially appropriate in the real-world decision contexts where information is lacking, preferences are changing, and decisions are contingent on changing circumstances. Brain-inspired decision systems have great benefits in the field of financial decision-making, such as the analysis of investments, risk assessment, and portfolio management. Financial markets are risky in character and not only affected by evolving quantitative indicators, but also by psychological and behavioural variables such as risk perception, overconfidence, and loss aversion. Brain-inspired models have the ability to incorporate these cognitive and affective elements into traditional financial data to make more realistic depictions of investor behaviour. Such systems can change their strategies over time, react to regime changes, and change their exposure to risk on the fly by perpetual learning and feedback, and market results. In addition, through the explicit modelling of uncertainty and cognitive bias, brain-inspired systems will be able to help decision-makers to be aware of and reduce suboptimal investment behaviours and thus encourage more balanced and informed financial decisions. Brain-inspired decision intelligence has great potential in the healthcare sector with regard to assistance in making complex clinical decisions.

In clinical settings, it is necessary to combine heterogeneous sources of information, such as patient history, diagnostic tests, clinical guidelines, and experience to enhance the management of clinical environments. The brain-inspired systems are quite appropriate in this environment because they are able to synthesise various types of data with transparency and interpretability of their reasoning processes. These systems can be used by clinicians to think in terms of probabilistic uncertainty and learn in a way that is based on past cases to help them in their diagnostic reasoning, treatment planning, and risk stratification. Notably, brain-inspired solutions are consistent with the collaborative aspect of healthcare decision-making, which should be provided as a cognitive support and not a decision-making system by AI. This anthropocentric orientation advocates clinical judgment, improves trust, and assists in safe, critical environments with responsible adoption.

Brain-inspired decision systems have a significant role to play in improving joint decision-making in human AI collaborative settings to address the limitations in human cognitive capabilities and overcome the limitations that have been identified. These systems have the ability to be adjusted to the preferences of the individual user, the way an individual cognits and the expertise levels of the person so that decision support becomes personalised. Brain-inspired models enhance transparency and common ground between humans and AI through the explained generation of explanations, feedback-based learning, and the sense of context. The abilities are especially useful in organisational decision making, education, and public policy, where decision making processes may concern a variety of stakeholders, conflicting goals and ethical concerns. The brain-inspired systems with assistance in reflection, deliberation, and learning make the collaborative efforts of decision-making more inclusive and effective.

8. ETHICAL AND RESPONSIBLE BRAIN-INSPIRED AI

The increased influence of artificial intelligence systems on making decisions with social, economic, and personal implications that can be far reaching makes ethical issues the main concern when designing and deploying them. Certain ethical concerns, specifically, high level of similarity with the processes of the human brain, including its cognition, influence and decision making are brought up by the transparency of artificial intelligence. Although such cognitive congruence also makes decision support more intuitive, adaptive and efficient, there are some essential concerns with regard to manipulation, independence, accountability and responsible utilization of human-like reasoning processes.

One of the ethical issues is connected to human freedom and agency. The brain-inspired decision systems are intended to control the decisions based on attention, learning, emotion, and bias modeling. Otherwise, these systems can covertly influence users to make specific decisions in a manner that compromises independent decision-making or takes advantage of cognitive biases. Conscientious brain-inspired decision intelligence thus puts human agency at the centre to ensure that AI has been deployed as assistive systems and not as persuasive or coercive agents. Human-in-the-loop and human-on-the-loop frameworks should be incorporated, especially in system types with high stakes like finance, healthcare, and public policy, where a human decision-maker should be held to ultimate account.

Additional ethical challenges are accountability and responsibility. With brain-inspired systems evolving and learning throughout their lifetime, it is possible to get confused with tracing how a particular recommendation was obtained. There is thus a need to have clear governance structures defining responsibility among system designers, deployers and end users. Ethical system design should involve ways of auditing decisions, recording of model behaviour and the maintenance of transparent lines of responsibility, particularly in cases where AI-supported decisions have unfavourable consequences.

The other ethical requirements in brain-inspired decision intelligence are transparency and explainability. Since they affect human judgment, users should be in a position to know how the system recommendations were reached, whether by the influence of uncertainty, previous experience, or assumptions of cognitive modeling. Explainable mechanisms not only facilitate trust and user confidence but also permit useful contestability, where people can challenge, override, or disregard AI-generated advice. This is critical especially where the decision made has legal, medical or financial ramifications.

The issues of fairness and bias are of particular interest to brain-inspired AI, because the models of human cognition can inadvertently recreate or enhance the existing social biases. Uncritical political tendencies like risk aversion, stereotyping, or loss sensitivity may support inequality or disadvantage certain groups. The methods to be used as a responsible approach thus involve systematic bias detection, constant assessment and intentional inclusion of fairness constraints. The brain-inspired systems developed to emulate the ethical human brain should not just attempt to reconstruct the human way of thinking but also address the established shortcomings and ensure that more fair decisions are reached.

Finally, the design philosophy of responsible and ethical brain inspired AI must consider the values aligned design approach where ethical concerns are entrenched through the system life-cycle during problem formulation and model development through deployment and continued monitoring. Technologists, cognitive scientists, ethicists, and domain experts should work collaboratively on interdisciplinary work in order to ensure that these systems are sensitive to the societal norms, the legal frameworks, and cultural backgrounds.

A combination of the ethical aspect in the technical architectures and governance systems can help to improve the quality of decisions and protect the individual rights, social trust, and the overall well-being of the society. This would guarantee that the progress of cognitively aligned AI helps to create responsible, transparent, and human-oriented systems of decision-making.

9. FUTURE RESEARCH DIRECTIONS

Although significant advances have been made over the last several years, brain-inspired decision intelligence is still an interdisciplinary and dynamic area that has many open theoretical, methodological, and practical issues. These issues will be dealt with not only by improvements in the computational modelling but also in the way that human cognition works and how it can be translated into scalable and reliable decision systems. Neuromorphic computing is one of the potential directions of future studies because it attempts to realize the brain-inspired principles on the hardware level. Neuromorphic architectures with a neural-style of parallelism, event-based computation, and energy-efficient learning, can significantly increase the efficiency, scalability and real time responsiveness of decision systems. This kind of hardware, in the case of brain-inspired decision making, may allow ongoing learning and dynamism in unpredictable environments, especially resource-limited or time-dependent deployments. There is still a lot of effort required to match the capabilities of neuromorphic systems with high-level decision making, symbolic reasoning, and explainability. The other potential area of research is the combination of multimodal cognition and neurological data. Behavioural signals, physiological measurements, and contextual information will be useful in future decision systems to structure more profound and more detailed models of human decision states. The Multimodal integration can help models to consider emotional dynamics, attentional switches, and cognitive load which are a vital component in the real world decision-making process. The main problem is to create effective mechanisms for combining heterogeneous data sources and uncertainty, noise, and ethical limitations associated with data gathering and utilization.

The creation of scalable and generalisable cognitive architectures is another important area of research. Most of the available brain-like models are specific to one task or domain and thus have restricted generalizability. The new architectures should strike the right balance between the domain generality and task-specific adaptability so that the transfer learning, lifelong learning, and flexible reasoning might be applied in different decision-making situations. It is a balance that require the development of hybrid methods that will integrate neural, symbolic, and probabilistic representations in coherent decision frameworks.

Evaluation and validation are other problems that should be tackled in future studies. It is also common to assess the brain-inspired decision models using the behavioural plausibility, interpretability, human alignment, and performance measures in contrast to the traditional systems that are based on optimisation. It will be necessary to create standard benchmarks, experimental protocol, and longitudinal evaluation techniques to compare the models and the actual impact on the world. The additional way of enhancing the evaluation criteria to mirror the realistic human decision environment is the closer integration with decision science. Further advances in brain-inspired decision intelligence will be based on long-term cross-disciplinary cooperation between neuroscience, cognitive science, artificial intelligence, and decision science. This type of collaboration is essential for translating theoretical knowledge about the functioning of the brain, as well as human behaviour, to practical, ethically grounded decision support systems. Through the creation of integrative research work and mutual methodological bases, the field may progress to intelligent systems not merely powerful in computation but also cognitively consistent, trustworthy, and socially accountable.

10. CONCLUSION

The chapter involves an interdisciplinary and integrative analysis of brain-inspired models of artificial intelligence in intelligent decision-making. It has based its insights on neuroscience, cognitive science, and artificial intelligence to present major conceptual frameworks and design principles that can be used to develop decision intelligence systems that can perform well in highly complex, uncertain, and dynamic environments. The chapter has brought forth the role of cognitively informed models to show how human decision-making can be more accurately characterised and reflected by focusing on approaches that do not solely depend on optimisation. One of the main themes of this chapter has been the correspondence of the artificial decision process with the human cognitive processes, such as learning, attention, memory, emotion, and bias. This kind of alignment allows us to create not only adaptive and context-sensitive decision systems, but also transparent and interpretable ones. The brain-inspired solutions allow reasoning on uncertainty, continuous learning via feedback, and a combination of short-term and long-term goals, thereby enhancing the quality of the decisions and the strength of the system. Significantly, these features help to increase the level of trust and usability, which are critical to the successful use of intelligent systems in real environments. The practicality of brain-inspired decision intelligence in various areas of application including finances, health, and human-AI co-operative settings, has also been highlighted in the chapter. Cognitively compatible systems in such settings have the potential to supplement human knowledge, reduce established decision biases, and make more informed and responsible decisions. Brain-inspired models enhance human agency and cognitive abilities by positioning AI as a decision-support tool that compliments, rather than replaces, human judgment. The chapter has highlighted the significance of responsible design, transparency, and interdisciplinary collaboration in the consideration of ethical issues and future research directions. Within the system lifecycle, ethical and human-centred principles have to be incorporated in order to make sure that any improvements in the brain-inspired AI do not violate the autonomy of individuals and are supported by the values of society. Simultaneously, the current work on neuromorphic computing, multi-modal data integration, and scalable cognitive architecture provides promising opportunities to develop the sphere. Comprehensively, the chapter serves as a contribution to the edited book in that it provides a balanced approach between theory and practice, giving an all-encompassing and consistent view on brain-inspired artificial intelligence to intelligent decision-making. It offers a platform on which future research and practice can derive to create intelligent decision systems that are resilient, credible, and are intelligibly consistent with human cognition by combining cognitive and computational perspectives.

REFERENCES

Campos, A. C., Pinto, P., & Scott, N. (2019). Bottom-up factors of attention during the tourist experience: An empirical study. *Current Issues in Tourism, 23*(24), 3111–3133. DOI: 10.1080/13683500.2019.1681383

Cohen, M. X. (2006). Individual differences and the neural representations of reward expectation and reward prediction error. *Social Cognitive and Affective Neuroscience, 2*(1), 20–30. DOI: 10.1093/scan/nsl021 PMID: 17710118

Cortés, U., Sànchez-Marrè, M., Ceccaroni, L., R-Roda, I., & Poch, M. (2000). Artificial intelligence and environmental decision support systems. *Applied Intelligence, 13*(1), 77–91. DOI: 10.1023/A:1008331413864

Cowan, N. (2022). Working memory development in childhood. In *The Development of Memory in Infancy and Childhood* (pp. 111–145). Psychology Press., DOI: 10.4324/9781003016533-5

Diederen, K. M. J., & Fletcher, P. C. (2020). Dopamine, prediction error and beyond. *The Neuroscientist, 27*(1), 30–46. DOI: 10.1177/1073858420907591 PMID: 32338128

Domenech, P., & Koechlin, E. (2015). Executive control and decision-making in the prefrontal cortex. *Current Opinion in Behavioral Sciences, 1*, 101–106. DOI: 10.1016/j.cobeha.2014.10.007

George, J. M., & Dane, E. (2016). Affect, emotion, and decision making. *Organizational Behavior and Human Decision Processes, 136*, 47–55. DOI: 10.1016/j.obhdp.2016.06.004

Grossberg, S., & Gutowski, W. E. (1987). Neural dynamics of decision making under risk: Affective balance and cognitive-emotional interactions. *Psychological Review, 94*(3), 300–318. DOI: 10.1037/0033-295X.94.3.300 PMID: 3039561

Janjua, N. K., Hussain, F. K., & Hussain, O. K. (2012). Semantic information and knowledge integration through argumentative reasoning to support intelligent decision making. *Information Systems Frontiers, 15*(2), 167–192. DOI: 10.1007/s10796-012-9365-x

Kalaidjieva, M. A., & Swanson, G. A. (2004). Intelligence and living systems: A decision-making perspective. *Systems Research and Behavioral Science, 21*(2), 147–172. DOI: 10.1002/sres.550

Keiflin, R., & Janak, P. H. (2015). Dopamine prediction errors in reward learning and addiction: From theory to neural circuitry. *Neuron*, *88*(2), 247–263. DOI: 10.1016/j.neuron.2015.08.037 PMID: 26494275

Mahadeen, T., Galanakis, K., Samara, E., & Kilintzis, P. (2020). Heuristics and evidences decision (heed) making: A case study in a systemic model for transforming decision making from heuristics-based to evidenced-based. *Journal of the Knowledge Economy*, *12*(4), 1668–1693. DOI: 10.1007/s13132-020-00688-4

Mitchell, D. G. V. (2011). The nexus between decision making and emotion regulation: A review of convergent neurocognitive substrates. *Behavioural Brain Research*, *217*(1), 215–231. DOI: 10.1016/j.bbr.2010.10.030 PMID: 21055420

O'Callaghan, M. (2023). Decision intelligence framework for organizational decision-making. In *Decision Intelligence* (pp. 195–218). Chapman and Hall/CRC., DOI: 10.1201/b23322-7

Ochsner, K. N., & Schacter, D. L. (2000). A social cognitive neuroscience approach to emotion and memory. In *The Neuropsychology Of Emotion* (pp. 163–193). Oxford University PressNew York., DOI: 10.1093/oso/9780195114645.003.0007

Sakagami, M., Pan, X., & Uttl, B. (2006). Behavioral inhibition and prefrontal cortex in decision-making. *Neural Networks*, *19*(8), 1255–1265. DOI: 10.1016/j.neunet.2006.05.040 PMID: 16952442

Schultz, W. (2024). The dopamine reward prediction error signal. SSRN *Electronic Journal*. DOI: 10.2139/ssrn.4816666

. Waring, J. D., Payne, J. D., Schacter, D. L., & Kensinger, E. A. (2010). Emotion-induced memory trade-off measure. *PsycTESTS Dataset*. DOI: 10.1037/t31780-000

Zhang, R., Deng, H., & Xiao, X. (2024). The insular cortex: An interface between sensation, emotion and cognition. *Neuroscience Bulletin*, *40*(11), 1763–1773. DOI: 10.1007/s12264-024-01211-4 PMID: 38722464

KEY TERMS AND DEFINITIONS

Bounded Rationality: The concept that human decision-making is constrained by limited cognitive capacity, incomplete information, and time pressure, leading individuals to rely on heuristics and satisficing rather than optimal solutions.

Brain-Inspired Artificial Intelligence (BI-AI): An interdisciplinary approach to AI that draws on principles from neuroscience and cognitive science to model functional aspects of human cognition, such as learning, attention, memory, and adaptive decision-making, rather than replicating biological structures.

Cognitive Architectures: Computational frameworks that model the structured interaction of cognitive processes such as perception, memory, attention, learning, and reasoning to produce goal-directed and interpretable intelligent behavior.

Decision Intelligence: A human-centred AI paradigm focused on supporting and augmenting human decision-making in complex, uncertain, and dynamic environments through adaptive, transparent, and context-aware computational systems.

Explainability: The capacity of an AI system to provide human-understandable accounts of how and why decisions or recommendations are generated, supporting trust, accountability, and responsible use.

Human Centred AI: An approach to AI design that prioritises human values, agency, transparency, and ethical responsibility, positioning AI as a collaborative decision-support tool rather than a substitute for human judgement.

Neuro-Symbolic Models: Hybrid AI systems that integrate neural learning mechanisms with symbolic reasoning and knowledge representation to combine adaptability with structured, explainable decision processes.

Chapter 7
Attention, Memory, Reasoning, Judgment, and Metacognitive Control in AI-Based Educational Decisions:
Brain-Inspired Decision Intelligence for Education

Ashraf Alam
http://orcid.org/0000-0001-6178-1187
Department of Education, National Institute of Technology Calicut, Kozhikode, India

ABSTRACT

This chapter advances a conceptual-theoretical framework of brain-inspired Decision Intelligence for education, reframing AI not as prediction machinery but as a governance-gated socio-technical control system operating under partial observability, path dependence, and high-rights constraints. It synthesizes attentional gating, governed episodic-semantic-procedural memory, constraint-aware and causal-leaning reasoning, multi-objective judgment under harm asymmetry, and metacognitive control via calibrated uncertainty, selective prediction, abstention, and escalation. Hybrid architectures are specified across retrieval-centric, neuro-symbolic, causal planning, and human-AI teaming paradigms, coupled to procedurally defensible reasons-giving, contestability, and lifecycle accountability. The

DOI: 10.4018/979-8-3373-8988-2.ch007

chapter delivers design constitutions through five tables, operationalizing admissible evidence, salience ethics, anti-stigma retrieval, bounded adaptation, and audit-ready due process for globally deployable, rights-compatible educational decision systems.

1. INTRODUCTION

1.1 Education as a High-Uncertainty, High-Rights Decision Ecology

Education functions as a dense decision ecology in which micro-decisions and macro-decisions propagate across time, compounding into life-chance effects. Classroom triage, referral timing, disability accommodations, pathway guidance, assessment moderation, safeguarding escalation, scholarship allocation, and disciplinary adjudication are not isolated events, they are sequential commitments under partial observability, institutional scarcity, and contested legitimacy (O'Callaghan, 2023; Heilig & Scheer, 2023). This ecology is now globally strained by measurable structural shocks that intensify uncertainty and raise the cost of error. In low-income and middle-income settings, *Learning Poverty* has been estimated at 70 percent for ten-year-olds, indicating inability to read and understand age-appropriate text, while the global out-of-school population has been reported at 250 million children, creating a high-volume, high-stakes allocation problem for systems already capacity constrained. Climate disruption further injects volatility, with 242 million students in 85 countries reported to have experienced schooling disruption in 2024 due to extreme events, amplifying institutional risk and evidentiary noise.

1.2 Decision Intelligence as a Distinct Paradigm Beyond Educational Analytics

Decision Intelligence reframes educational AI from score production to end-to-end decision governance, spanning decision framing, evidentiary admissibility, inference, recommendation, action authorization, monitoring, and post-hoc accountability. Predictive analytics can forecast enrolment, identify attendance anomalies, or estimate progression probabilities, yet education decisions require more than prediction because they are constrained by rights, procedurally reviewable, and causally entangled with the very interventions they trigger (Al Humdan, 2025; Meng et al., 2025). A decision-centric framing treats thresholds as policy, escalation as safeguarding infrastructure, and monitoring as institutional self-regulation rather than a technical afterthought. This paradigm aligns with the practical reality that AI in education is already used for administrative forecasting, timetable optimization, resource allocation, and compliance automation, thereby reshaping workload distribution and accountability structures across the system. The relevant unit is not an isolated model output, but a decision chain with auditable handoffs, explicit risk budgets, and governance gates that prevent metric fixation, proxy dominance, and unreviewed automation of high-rights determinations.

1.3 Brain-Inspired Cognition as a Design Grammar for Educational Decision Systems

Brain-inspired cognition is used here as a functional design grammar, not as biological mimicry, specifying computational primitives that match how decisions must be made under uncertainty, overload, and normative constraint. *Attention* becomes resource allocation across competing alerts and cases, controlling salience and preventing institutional tunnel vision. *Memory* becomes structured precedent, policy knowledge, and case retrieval, enabling consistent decisions and reasons-giving that can survive contestation (Alam, 2022; Murthy et al., 2023). *Reasoning* becomes constraint-aware and causally oriented intervention selection, avoiding the category error of treating educational improvement as mere risk prediction. *Judgment* becomes explicit tradeoff management across equity, efficiency, wellbeing, safety, and legitimacy, using risk-sensitive and regret-aware policies rather than monolithic optimization. *Metacognitive Control* becomes calibrated uncertainty, selective prediction, abstention, and escalation, enabling epistemic humility by design. This grammar is policy salient because global governance bodies now foreground human-centred guardrails for generative systems in education, emphasizing immediate actions, long-horizon capacity building, and risk containment aligned to rights and accountability.

1.4 Scope, Unit of Analysis, and Analytic Commitments

The chapter is conceptual-theoretical and evaluates education through the unit of educational decisions rather than outcomes alone, because outcomes are temporally lagged, multi-causal, and frequently endogenous to institutional response. Decisions are differentiated by stake level, reversibility, evidentiary density, and rights intensity, for example formative feedback versus disciplinary exclusion, advising versus financial aid determination, and supportive referral versus safeguarding escalation. The analysis spans classroom, school, district or state, and higher education governance strata, recognizing that decision rights, documentation burdens, and procedural safeguards vary by level and jurisdiction. The scope excludes empirical benchmarking, product comparisons, and implementation case studies to preserve a mechanism-to-architecture-to-governance logic that remains portable across systems with different data infrastructures and legal regimes. The analytic stance is socio-technical, treating cognition as distributed across humans, artifacts, dashboards, rubrics, and accountability regimes, and rights-sensitive, treating *Procedural Justice*, *Epistemic Justice*, and contestability as first-class system requirements rather than downstream ethics statements.

1.5 Chapter Roadmap and Contribution Architecture

The chapter proceeds from foundations to mechanisms to architectures, then to reasons-giving and governance, yielding a compact design program for education-grade *Brain-Inspired Decision Intelligence*. Section 2 formalizes education as a constrained, multi-objective decision ecology and specifies why normativity must be operationalized as system constraint. Section 3 maps attention, memory, learning, reasoning, judgment, and metacognitive control into decision functions that are audit-ready and institution-compatible. Section 4 builds hybrid architectures that integrate retrieval, symbolic constraints, and causal planning with governance gates that encode decision rights and escalation rules. Section 5 reconceptualizes explainability as cognitively aligned reasons-giving that supports calibrated trust, contestation, and due process without creating privacy or gaming vulnerabilities. Section 6 operationalizes ethics and bias as lifecycle governance, including bounded adaptation, incident response, and capacity building for decision owners. Section 7 consolidates design principles and a research agenda focused on decision usefulness, abstention quality, and governance robustness under global volatility.

2. CONCEPTUAL FOUNDATIONS, DECISION INTELLIGENCE, COGNITION, AND EDUCATION AS A COMPLEX DECISION ECOLOGY

2.1 Formalizing Educational Decisions as Sequential, Constrained, Multi-Objective Systems

Educational decision-making is best specified as a temporally extended control problem in which actions produce delayed, path-dependent consequences, often under partial observability and institutional scarcity. Within a *Decision Intelligence* framing, decisions are not single outputs but chained commitments that require policy-grounded thresholds, evidentiary admissibility, and auditable handoffs across roles (Chohra & van der Wal, 2025; Kaneko, 2023a). Formal lenses such as *Partially Observable Markov Decision Processes*, *Contextual Bandits*, and constrained optimization clarify why education-grade decision systems must encode hard constraints, for example safeguarding non-negotiables and accommodation entitlements, while optimizing across soft objectives such as attainment, wellbeing, inclusion, fiscal efficiency, and institutional legitimacy. Crucially, the decision criterion cannot be predictive accuracy alone, it must be decision utility under rights constraints, with explicit management of uncertainty, reversibility, and harm asymmetry. This multi-objective reality anticipates the foundational lenses consolidated in Table 1, which operationalizes design imperatives that prevent proxy dominance, metric myopia, and unreviewed threshold drift in high-stakes educational ecologies

2.2 Cognitive Architecture of Professional Judgment in Schools and Universities

Educational professionals decide within bounded cognitive bandwidth, where time pressure, attentional fragmentation, and documentation burden shape what can be perceived, remembered, and acted upon. *Dual-Process* cognition foregrounds the oscillation between rapid heuristic appraisal and slower deliberative scrutiny, implying that decision systems must be ergonomically compatible with both fast triage and reflective review (Khachaturian et al., 2023; Rajak et al., 2025). *Naturalistic Decision-Making* and *Recognition-Primed* judgment highlight that expertise often operates through cue-based pattern completion rather than exhaustive search, which becomes brittle when institutional dashboards amplify specific cues and silence others. *Cognitive Load Theory* and working-memory constraints imply that decision support must compress complexity into decision-relevant representations, yet without inducing overconfidence or automation bias. Organizational cognition adds additional distortions, including anchoring on prior labels, compliance-induced satisficing, and authority effects that convert advice into de facto mandates. Table 1 anticipates these cognitive fragilities by translating them into design constraints such as calibrated uncertainty, dissent affordances, and procedural gating that preserves professional agency while improving decision reliability

2.3 Neuroscience-Informed Priors as Translational Constraints for Decision System Design

Brain-inspired design is most defensible when it functions as a mechanistic prior that shapes robustness and safety properties, rather than as literal neural emulation. *Predictive Processing* provides a principled frame for continual error monitoring and anomaly sensitivity, enabling decision systems to treat discordant evidence as a signal for escalation, not as noise to be smoothed away (S. Wu et al., 2025; Alam, 2023a). *Active Inference* supplies an architecture for epistemic action, where the system can request additional evidence, defer recommendations, or route cases to specialized human judgment when uncertainty remains irreducible. *Complementary Learning Systems* suggests a division between fast, context-bound case memory and slow, policy-stable semantic memory, reducing catastrophic interference and supporting reasons-giving grounded in precedent and protocol (Pratt et al., 2023; Kaneko, 2024). *Executive Control* models motivate governance gates, access controls, and arbitration mechanisms that constrain action selection in high-rights contexts. The integrative logic prefigured in Table 1 treats these priors as enforceable design constraints, ensuring that learning, adaptation, and automation are bounded by procedural legitimacy, harm asymmetry, and contestability expectations across diverse jurisdictions

2.4 Distributed Cognition, Socio-Material Infrastructures, and Institutional Rationalities

Educational cognition is distributed across humans, artifacts, routines, metrics, and accountability regimes, making decision-making a socio-material phenomenon rather than an isolated mental act. *Distributed Cognition* and *Activity-Theoretic* framing clarifies how forms, rubrics, dashboards, and workflow systems externalize memory and attention, thereby shaping what becomes visible, urgent, and actionable. Institutional rationalities, including performance governance, compliance cultures, and resource scarcity, modulate decision thresholds, escalation norms, and the perceived legitimacy of discretionary judgment (Qian, 2025; Vasuki et al., 2025). This creates a recursive ecology where decision systems not only support decisions but also reconfigure professional noticing, evidentiary standards, and organizational incentives. In such environments, the central failure mode is not only technical error but epistemic deformation, where what can be measured becomes what matters, and what is salient becomes what is treated as real. Table 1 consolidates the foundational lenses required to counter this deformation by specifying design imperatives that stabilize evidentiary integrity, preserve contextual interpretability, and protect against institutionalized tunnel vision through governance-bounded salience and memory controls.

Table 1. Foundational lenses and design implications for educational DI

Theoretical Lens Archetype	Core Mechanistic Claim	Decision-Ecology Translation	Pathologies Under Neglect	DI Design Imperative
Sequential Decision-Theoretic Control	Educational decisions operate as temporally coupled action sequences under uncertainty, where utility is contingent on constraints, reversibility, and asymmetric harm functions	Specify decision chains across time horizons, encode delayed outcomes and path dependence, and treat intervention sequencing as a controllable policy rather than a one-off recommendation	Reduction to static classification, unmanaged threshold drift across cohorts, and compounding error cascades that generate downstream exclusion, misallocation, or safeguarding failure	Implement constrained multi-objective policies, explicit escalation and de-escalation rules, and continuous post-decision surveillance for drift, harm signals, and policy misalignment
Bounded and Ecological Rationality	Decision agents satisfice under limited working memory and attentional bandwidth, relying on context-fit heuristics that trade optimality for tractability and speed	Design decision support for time-pressured cue integration, workload-aware evidence compression, and contextually meaningful heuristics that remain legible under institutional constraints	Automation bias, cue monopolies that suppress weak signals, and overconfident reliance on compressed proxies that displace professional noticing and situated judgment	Provide uncertainty-forward summaries, dissent-supporting interaction affordances, and cognitive forcing functions that trigger reflective review when stakes, ambiguity, or rights intensity are elevated
Predictive Processing and Active Inference	Cognition is organized around prediction-error minimization and epistemic action selection, where unresolved uncertainty should provoke inquiry, abstention, or escalation	Treat discordant indicators as diagnostically valuable, route cases into information-seeking workflows, and operationalize abstention as a safety-preserving decision outcome rather than a system failure	Over-smoothing of anomalies, premature labeling under noisy evidence, and failure to request additional information that could avert stigmatizing or irreversible actions	Deploy calibrated uncertainty estimation, selective prediction with abstention, boundary-condition detection, and inquiry prompts that request targeted evidence before issuing high-stakes recommendations

continued on following page

Table 1. Continued

Theoretical Lens Archetype	Core Mechanistic Claim	Decision-Ecology Translation	Pathologies Under Neglect	DI Design Imperative
Complementary Learning Systems Memory	Robust adaptation requires separable fast episodic storage and slow semantic consolidation, preventing interference while sustaining contextual recall and normative stability	Partition case retrieval from policy knowledge, enforce provenance and retention limits, and manage access through role-based controls aligned to decision rights and safeguarding boundaries	Stigma persistence through sticky labels, privacy leakage via unconstrained retrieval, and catastrophic interference when continual updates overwrite policy-relevant invariants	Build governed retrieval with privacy-preserving similarity, semantic policy graphs with versioning, and bounded adaptation protocols that lock decision thresholds to reviewed governance states
Distributed Cognition and Socio-Materiality	Cognition is enacted across people, artifacts, routines, dashboards, and accountability regimes that jointly shape what counts as evidence and what becomes actionable	Treat forms, rubrics, and workflow systems as cognitive infrastructure, evaluate how salience settings steer institutional attention, and align representations with role-specific authority and obligations	Metric myopia, performativity traps that reward gaming over learning, and institutional epistemic narrowing where measurable traces displace contextual realities	Audit salience effects and representational bias, embed governance gates at critical handoffs, and design interfaces that preserve contextual interpretability while constraining proxy dominance
Normative Legitimacy and Procedural Justice	Decisions in high-rights domains must be reasons-giving, contestable, and procedurally defensible, with safeguards that prevent discriminatory impact and unreviewable authority	Encode due process workflows, evidence admissibility standards, and appeal pathways as system constraints, ensuring that affected parties can understand and challenge decisions using intelligible criteria	Legitimacy collapse, discriminatory harm through opaque proxy use, and de facto automation of authority that erodes professional agency and rights protections	Require structured reasons-giving, contestability mechanisms, audit-ready decision logs, and explicit human authority boundaries for decisions that affect dignity, safety, access, or life chances

The table functions as a compact design constitution for education-grade *Decision Intelligence*, translating heterogeneous theoretical lenses into implementable constraints that can be encoded in pipelines, interfaces, and governance routines. Its integrative value lies in converting abstract constructs, such as bounded rationality, distributed cognition, and procedural justice, into engineering-relevant imperatives, such as abstention logic, governed retrieval, constraint solvers, and auditable handoffs. When read in conjunction with the decision-chain framing in 2.1 and the cognitive constraints in 2.2, Table 1 clarifies why education systems require hybrid architectures that can simultaneously optimize under uncertainty and remain rights-compatible under contestation. The table also anticipates downstream chapters by specifying which failure modes are structurally likely in institutional settings, including metric myopia, stigma persistence, and legitimacy collapse, and by prescribing concrete control surfaces that can be operationalized without collapsing normative commitments into proxy metrics

2.5 Normativity, Legitimacy, and Rights as First-Class Constraints in Educational Decision Intelligence

Education is a high-rights domain where decision legitimacy depends on procedural defensibility, contestability, and non-discrimination, not merely on aggregate performance improvement. Normativity is therefore not an ethical appendix but an architectural constraint that must be expressed in admissible evidence rules, decision-rights allocation, escalation protocols, and remedy pathways. *Procedural Justice* requires that affected individuals can understand the grounds of decisions and challenge them using intelligible standards, while *Epistemic Justice* requires that lived experience and contextual testimony are not structurally discounted by metric-centric pipelines (Jeyanthi et al., 2022; Wang & Bai, 2023). *Ethics of Care* adds relational obligations, emphasizing non-stigmatizing interventions, minimal intrusive data practices, and safeguarding primacy. These constraints imply that educational DI must include reasons-giving formats that support review, appeal, and institutional learning, while restricting automation in decisions that alter dignity, safety, or life chances. Table 1 consolidates this normative stance into enforceable design imperatives, ensuring that optimization remains bounded by legitimacy, rights protections, and governance that preserves professional authority while enhancing decision reliability across diverse global systems.

3. BRAIN-INSPIRED COGNITIVE MECHANISMS AS A DESIGN GRAMMAR FOR EDUCATIONAL DECISION INTELLIGENCE

3.1 Perception and Representation as Epistemic Engineering of Educational Evidence

Educational data should be treated as perceptual traces that are noisy, proxy-laden, and value-saturated, rather than as neutral observations. Decision-grade representation therefore requires psychometric and inferential discipline, including *construct validity*, *measurement invariance*, *differential item functioning*, *construct-irrelevant variance*, and robust handling of missing-not-at-random mechanisms that systematically distort vulnerability signals (Annu & Tripathi, 2024; Alam, 2023b). Representation is also a governance act, because feature selection determines what becomes legible to institutions and what remains structurally invisible, which triggers *Goodhart-type* degeneration when proxies become targets and surveillance replaces support. A brain-inspired stance operationalizes representation as uncertainty-aware evidence coding, using *signal detection theory* to manage false-negative versus false-positive asymmetry, and *Bayesian decision theory* to express evidentiary strength as calibrated belief rather than categorical labels. The representational primitive summarized in Table 2 is therefore not feature engineering, it is admissibility engineering, where provenance, sensitivity, and decision relevance are encoded as constraints consistent with the legitimacy safeguards consolidated earlier in Table 1

3.2 Attention and Salience as Resource Allocation Under Cognitive and Institutional Overload

Attention is the steering wheel of educational decision ecologies, because what the system renders salient becomes what staff notice, document, escalate, and ultimately govern. A brain-inspired framing treats attentional design as resource allocation under overload, using *attentional gating*, sparse prioritization, and *information foraging* principles to minimize cognitive friction while preserving epistemic breadth (Theodorakopoulos et al., 2025; Bratu, 2023). In operational terms, triage must behave like a calibrated queueing system, where caseload ordering integrates uncertainty, stakes, reversibility, and rights intensity, rather than ranking by predicted risk alone. Salience must also be ethically constrained, because repeated surfacing of the same proxies creates cue monopolies that amplify stigma persistence and suppress weak signals of distress that are qualitatively expressed. Decision interfaces should therefore implement progressive disclosure, uncertainty-weighted prominence, and diversity-aware alert portfolios that counter institutional tunnel vision. The mechanism-to-function translation in Table 2 specifies how attentional controls can be instrumented as auditable salience policies, aligned with the procedural defensibility conditions already crystallized in Table 1

3.3 Memory Architectures as Governed Precedent, Policy, and Workflow Substrates

Memory-augmented decision intelligence must emulate the functional pluralism of human memory by separating episodic case traces, semantic policy knowledge, and procedural workflow scripts, because each supports a distinct class of decision microfunctions. Episodic analogs enable *case-based reasoning* and analogical retrieval for intervention planning, yet require privacy-preserving similarity, retention limits, and contamination controls to prevent permanent labeling (Clark, 2025; Kaneko, 2023b). Semantic analogs operationalize policy and pedagogy as structured knowledge, using ontological schemas and knowledge-graph commitments that make constraints explicit, reviewable, and versionable. Procedural analogs encode escalation pathways, documentation obligations, and review cycles, ensuring that due process is not dependent on individual discretion. The integrated mapping in Table 2 expresses memory as a governed substrate, where provenance, access control, and rights-sensitive retrieval become enforceable design properties, consistent with the normative guardrails consolidated in Table 1

Table 2. Cognitive mechanisms mapped to educational decision functions and risks

Cognitive Mechanism Primitive	Computational Instantiation Pattern	Educational Decision Microfunctions	Dominant Failure Modes	Governance and Safety Controls
Representational Inference	Validity-aware feature pipelines, uncertainty-coded embeddings, and construct-aligned measurement models with explicit admissibility constraints	Convert heterogeneous traces into decision-relevant evidence, preserve interpretability for professional reasoning, and quantify evidentiary strength for reversible actions	Proxy dominance, label noise amplification, and construct drift that silently redefines what the institution treats as learning or risk	Provenance registers, admissible-feature policies, subgroup robustness checks, and periodic construct audits with versioned thresholds
Attentional Gating	Salience policies, uncertainty-weighted triage queues, sparse prioritization modules, and interface-level progressive disclosure tuned to workload	Prioritize safeguarding follow-ups, allocate support capacity, and surface weak signals without collapsing nuance into single-score urgency	Cue monopolies, alert fatigue, and institutional tunnel vision that repeatedly spotlights the same groups while missing quiet distress	Salience audits, diversity-aware alert portfolios, suppression of sensitive proxies, and mandatory escalation triggers under high uncertainty
Episodic Case Retrieval	Retrieval-augmented decision support using privacy-preserving similarity, de-identified exemplars, and bounded recall windows	Ground recommendations in analogous trajectories, support example-based reasons-giving, and enable counterfactual reflection using comparable cases	Stigma persistence, privacy leakage through retrieval, and overfitting to idiosyncratic precedents that distort fairness	Retention limits, role-based access, retrieval logging, similarity constraints, and retrieval diversification to avoid stereotyping

continued on following page

Table 2. Continued

Cognitive Mechanism Primitive	Computational Instantiation Pattern	Educational Decision Microfunctions	Dominant Failure Modes	Governance and Safety Controls
Semantic Policy Graph Memory	Knowledge graphs, ontologies, rule-constrained reasoning layers, and versioned policy schemas linked to audit trails	Enforce accommodations, safeguarding constraints, and curriculum obligations, while making constraint application contestable and reviewable	Ontological bias, rule brittleness, and policy-practice mismatch that converts compliance into rigid proceduralism	Participatory schema governance, exception-handling protocols, version control, and traceable rule-application logs for appeal workflows
Consolidation-Based Continual Learning	Drift detection, replay-based consolidation, modular adaptation, and change-point monitoring bounded by governance-approved ranges	Adapt to cohort shifts and regime changes without catastrophic forgetting, maintain calibration under distribution shift, and stabilize decision thresholds	Silent policy drift, catastrophic interference, and feedback-induced drift where interventions reshape the data generating process	Bounded adaptation policies, update review boards, calibration monitoring, subgroup drift alarms, and rollback-ready model versioning
Metacognitive Uncertainty Control	Calibration layers, selective prediction, abstention policies, escalation routers, and structured reasons-giving logs	Decide when to defer, request more evidence, or escalate to human authority, while documenting uncertainty and decision rationale	Overconfidence, forced outputs under ambiguity, and unreviewable automation that erodes due process and legitimacy	Calibrated abstention thresholds, mandatory human review for high-rights decisions, contestability workflows, and audit-ready decision logs

The table operationalizes cognitive primitives as engineering-ready control surfaces that can be embedded into decision pipelines, interfaces, and governance routines without reducing educational legitimacy to a proxy score. Its key contribution is the conversion of cognitive functions into auditable system behaviors, for example attentional gating becomes a salience policy, memory becomes governed retrieval plus versioned policy graphs, and metacognition becomes calibrated abstention with escalation routers. Read alongside Table 1, Table 2 also clarifies that robustness and rights-compatibility are co-dependent, because uncontrolled representation and retrieval generate stigma persistence, while uncontrolled adaptation generates silent threshold drift. The table therefore functions as a mechanism-level specification for later architectural synthesis, indicating which modules must be coupled, which must be bounded by decision rights, and which require lifecycle controls such as versioning, rollback, and contestability pathways to remain defensible under review

3.4 Learning and Adaptation as Drift-Resilient, Governance-Bounded Plasticity

Educational environments are intrinsically non-stationary, with cohort effects, policy reforms, assessment redesigns, migration shocks, and climate-linked disruptions producing distribution shift that makes static models epistemically hazardous. A brain-inspired design stance treats adaptation as controlled plasticity, governed by the *stability-plasticity dilemma* and protected against *catastrophic forgetting* through consolidation analogs such as replay, modular parameter isolation, and change-point detection (Zaimoglu et al., 2023; Sousa et al., 2022). Drift must be typed and monitored, including covariate drift in behavioral traces, label drift in administrative annotations, and feedback-induced drift where interventions change recorded outcomes, thereby corrupting naive performance metrics. Decision intelligence should therefore implement bounded adaptation, separating technical calibration updates from normative threshold changes that require institutional authorization and audit trails. The continual learning row in Table 2 specifies concrete controls, including rollback-ready versioning, subgroup drift alarms, and calibration monitoring that protect procedural legitimacy while enabling safe responsiveness to regime change, consistent with the constraint logic stabilized in Table 1

3.5 Reasoning, Judgment, and Metacognitive Control as Safety-Critical Executive Functions

Reasoning in education DI must be articulated as constraint-aware, causally oriented planning rather than predictive scoring, because interventions require counterfactual sensitivity, resource feasibility, and rights-compatible action selection. Executive reasoning can be operationalized through hybrid control stacks that integrate *constraint satisfaction*, policy graphs, and causal world models, enabling recommendations that respect accommodations, safeguarding non-negotiables, and evidentiary admissibility, while optimizing across multi-criteria objectives (Chinnaraju, 2025; Akcetin, 2024). Judgment is the tradeoff engine, requiring explicit risk preferences, harm asymmetry encoding, and regret-aware thresholding so that reversible supports are favored under ambiguity and irreversible sanctions are buffered by stronger evidence standards. Metacognitive control then governs epistemic humility through calibration, selective prediction, abstention, and escalation, making uncertainty a first-class output that routes cases to human authority when stakes or ambiguity exceed safe bounds. The metacognitive and executive primitives consolidated in Table 2 anchor these functions in auditable controls, ensuring alignment with the legitimacy and procedural defensibility imperatives consolidated earlier in Table 1

4. HYBRID COGNITIVE-COMPUTATIONAL ARCHITECTURES FOR AI-ENHANCED EDUCATIONAL DECISION INTELLIGENCE

4.1 Structural Necessity of Hybridization in Rights-Intensive Education Systems

Hybrid architectures are structurally necessary in education because the domain couples statistical uncertainty with non-negotiable normative constraints, heterogeneous context, and multi-stakeholder accountability, producing a design space where purely data-driven inference is under-specified and purely rule-driven automation is brittle. Educational decisions instantiate hard constraints, such as safeguarding imperatives, accommodation entitlements, and procedural defensibility, alongside soft objectives, such as attainment, wellbeing, inclusion, and fiscal sustainability, which demands *multi-objective optimization* under admissibility constraints rather than single-metric maximization (S. Z. Wu, 2025; Alam, 2023c). Hybridization operationalizes this requirement by fusing learned representations with *constraint satisfaction*, policy-encoded deontic commitments, and governance gates that regulate authority, override rights, and escalation. It also reduces epistemic fragility under distribution shift by combining pattern sensitivity with rule-grounded stability, thereby avoiding silent threshold drift and proxy substitution. The architecture families later taxonomized in Table 3 clarify how distinct hybrids trade expressivity, auditability, and adaptability under high-rights conditions

4.2 Governance-Gated Cognitive Pipelines as Executable Institutional Control Systems

A governance-gated pipeline treats decision intelligence as an executable institutional control system, where sensing, representation, salience, retrieval, reasoning, recommendation, explanation, authorization, and monitoring are coupled through explicit checkpoints that function as *executive control* for socio-technical decision chains. The critical move is to encode decision rights and evidence standards as controllable interfaces, including admissible-input filters, role-based access to sensitive signals, calibration thresholds that trigger abstention, and escalation routers that enforce separation of duties for high-stakes determinations (Nozari et al., 2025; Josephson et al., 2024). This pipeline aligns with *control theory* intuitions, where feedback loops must be designed for stability, not merely for responsiveness, and where drift detection must be connected to rollback, review, and policy re-approval rather than autonomous adaptation. In education, governance gates also serve procedural justice, because they guarantee that reasons-giving, documentation, and appealability are not discretionary add-ons. Table 3 positions this pipeline family alongside retrieval-centric and neuro-symbolic alternatives, enabling principled selection by risk tier, reversibility, and institutional capacity

4.3 Memory-Centric and Retrieval-Augmented Decision Architectures for Grounded Recommendations

Memory-centric architectures foreground retrieval as the primary substrate for actionable decision support, treating case traces, precedents, and policy artifacts as a governed evidence base that enables analogical reasoning, example-based justification, and consistency across time. Retrieval-augmented decision support is most defensible when it is bounded by privacy-preserving similarity, de-identification, retention limits, and diversification constraints that prevent stereotype reinforcement and stigma persistence (Wang et al., 2024; Zhao et al., 2023). In educational settings, retrieval can support intervention planning by surfacing trajectories that match contextual features, while simultaneously exposing boundary conditions and uncertainty through comparative exemplars rather than opaque scores. This architecture also improves contestability because stakeholders can interrogate the relevance of retrieved analogs, challenge mismatches, and introduce contextual evidence that re-anchors similarity judgments. The comparative taxonomy in Table 3 situates memory-centric designs within a broader architecture ecology, clarifying when retrieval should be dominant, when policy graphs should arbitrate, and when causal planning must supersede analogical evidence under long-horizon intervention selection

Table 3. Taxonomy of architecture families for education decision intelligence

Architecture Family Signature	Constituent Computational Subsystems	Institutional Fit and Decision Terrain	Predictable Failure Modes Under Stress	Governance Precondition for Safe Deployment
Governance-Gated DI Pipeline Architecture	Evidence admissibility filters, calibrated inference, explanation generators, authorization gates, drift monitors, audit logs tightly coupled as a control loop	Best fit for high-rights decisions with procedural review, multi-role accountability, and frequent escalation events across schools and universities	Gate circumvention through informal workarounds, compliance ritualism, and latent automation bias when gate outputs acquire de facto mandate status	Role-based decision rights, override documentation norms, escalation protocols, and rollback-ready version control institutionalized as policy
Retrieval-Centric Memory-Augmented Architecture	Privacy-preserving similarity search, de-identified episodic repositories, diversification constraints, exemplar-based justification layers for grounded recommendations	Best fit for advising, intervention selection, and support allocation where precedent and contextual comparability are decision-relevant	Stigma persistence via sticky labels, privacy leakage through reconstruction, and analogical overreach when similarity collapses causal heterogeneity	Retention limits, retrieval logging, access controls, similarity constraints, and contestable exemplar curation with periodic governance review
Neuro-Symbolic Constraint-Enforced Architecture	Learned representations coupled with rule engines, constraint solvers, and policy graphs that encode admissibility and compliance requirements	Best fit for accommodations, safeguarding workflows, curriculum constraints, and policy-heavy adjudication requiring auditability	Rule brittleness under novel context, ontological bias, and policy-practice mismatch that converts discretion into rigid proceduralism	Versioned knowledge governance, exception-handling protocols, traceable constraint application logs, and participatory schema maintenance
Causal World-Model and Planning Architecture	Structural causal models, counterfactual engines, treatment heterogeneity modules, sequential planners with resource constraints and uncertainty propagation	Best fit for long-horizon intervention planning, program allocation, and prevention strategies where prediction without causality misleads action	Model mis-specification, unobserved confounding, and brittle planning under sparse data that yields false certainty about intervention effects	Evidentiary standards for causal assumptions, human review of causal claims, continuous calibration, and conservative abstention thresholds

continued on following page

Table 3. Continued

Architecture Family Signature	Constituent Computational Subsystems	Institutional Fit and Decision Terrain	Predictable Failure Modes Under Stress	Governance Precondition for Safe Deployment
Human-AI Teaming Joint Cognitive Architecture	Decision orchestration interfaces, cognitive forcing functions, disagreement capture, authority boundaries, escalation routers, and reflective prompts	Best fit for decisions requiring situated expertise, qualitative evidence, and negotiated legitimacy across stakeholders	Authority drift toward the system, diffusion of responsibility, and superficial human oversight that rubber-stamps recommendations	Training for calibration literacy, enforced decision-rights separation, structured dissent capture, and accountability assignment at each handoff
Safety-Critical Selective-Prediction Architecture	Uncertainty quantification layers, selective prediction, abstention policies, risk-tiered escalation, and harm-weighted thresholding	Best fit for high-uncertainty contexts where errors are asymmetric and reversibility is limited, including safeguarding triage	Over-abstention that burdens staff, under-abstention that forces outputs, and calibration drift that silently alters risk posture	Risk-tiered operating procedures, calibration audits, periodic threshold re-authorization, and incident response playbooks with audit trails

The taxonomy in Table 3 functions as an architecture selection map, aligning computational form with decision terrain, institutional capacity, and legitimacy burden, thereby preventing a default drift toward monolithic prediction stacks. Its primary utility is in making architecture choice itself a governance decision, where reversibility, harm asymmetry, and due-process requirements determine whether retrieval, constraints, planning, or selective prediction should dominate the control stack. When read alongside the cognitive mechanism specification in Table 2 and the foundational guardrails in Table 1, the table clarifies that education-grade decision intelligence is not a single model, but an orchestrated system with distinct modules that must be bounded by authority, documentation, and contestability. The table also anticipates lifecycle requirements by naming predictable stress failures, such as gate circumvention and authority drift, which can be addressed only through institutionalized controls rather than interface aesthetics alone, thereby tightening the coupling between technical architecture and procedural legitimacy

4.4 Neuro-Symbolic Constraint Reasoning and Compliance-Executable Knowledge Structures

Neuro-symbolic architectures operationalize educational normativity by coupling learned representations with explicit constraint reasoning, enabling compliance-executable decision support that is legible under audit and contestation. The symbolic substrate can be structured as policy graphs, curriculum ontologies, accommodation schemas, safeguarding rule sets, and procedural workflows, which collectively function as a deontic layer that restricts action spaces, enforces admissible evidence, and records constraint application as traceable reasons (Zeng et al., 2023; Li et al., 2024a). This design is particularly valuable when decisions must be defensible in institutional review, because constraint satisfaction produces inspectable paths rather than probabilistic opacity. However, neuro-symbolic systems require epistemic humility about formalization, since ontological commitments can encode bias, and rule sets can become brittle under contextual novelty. The architecture taxonomy in Table 3 highlights these risks, including policy-practice mismatch and rule brittleness, which can be mitigated only through versioned governance, exception protocols, and participatory maintenance that treats knowledge structures as living institutional artifacts rather than static technical assets

4.5 Causal World Models, Planning Under Uncertainty, and Human-AI Teaming for Legitimate Action

Causal world models are indispensable when decision intelligence supports intervention selection rather than mere risk labeling, because action requires counterfactual sensitivity, treatment heterogeneity awareness, and resource-feasible sequencing across time. A planning-capable architecture integrates causal graphs with uncertainty propagation, enabling conservative, harm-weighted action selection that distinguishes correlation-driven alerts from intervention-relevant mechanisms (Schmidgall et al., 2024; Mehonic & Kenyon, 2022). Long-horizon educational planning also requires explicit modeling of constraints, including staffing capacity, budget envelopes, service availability, and legally mandated accommodations, so that recommendations do not become performative fantasies detached from implementability. Human-AI teaming then becomes the legitimacy substrate, operationalized through authority boundaries, disagreement capture, structured deliberation prompts, and escalation routers that preserve professional judgment where qualitative evidence and contextual interpretation are decisive. The architecture families in Table 3 clarify that causal planning and teaming must be coupled with selective prediction and abstention, because forcing confident outputs under confounding uncertainty produces procedural fragility, reputational harm, and irreversible misallocation in high-rights educational ecologies

5. EXPLAINABILITY, INTERPRETABILITY, AND COGNITIVELY ALIGNED REASONS-GIVING IN EDUCATIONAL DECISION SYSTEMS

5.1 Explainability as Procedural Reasons-Giving Under Institutional Accountability

Explainability in education should be treated as procedural reasons-giving rather than post-hoc model narration, because legitimacy hinges on whether a decision can be audited, contested, and remediated across stakeholders with asymmetric power and information. A defensible reasons-giving artifact must encode a minimal decision dossier, including decision trigger, admissible evidence summary, uncertainty profile, constraint ledger, and action authorization boundary, so that the institution can demonstrate *procedural justice* and *epistemic justice* without delegating authority to opaque scoring (Amrouch et al., 2022; Alam & Mohanty, 2022). Interpretability for developers targets parameter sensitivity and failure analysis, while reasons-giving for schools targets *simulatability*, *actionability*, and rights-compatible recourse. Explanations must therefore be decision-class specific, with different epistemic granularity for low-stakes nudges versus high-rights determinations such as exclusion, safeguarding escalation, or accommodation denial. Table 4 later in this section specifies stakeholder-tuned explanation modalities and safeguards, enabling globally portable reason templates that remain robust under diverse governance regimes and data infrastructures

5.2 Cognitive Fit and Explanation Ergonomics for Professional Decision Work

Cognitive fit requires that explanation formats align with human judgment under bounded attention and working-memory constraints, otherwise decision support becomes cognitive debt that increases error rates and compliance ritualism. A cognitively aligned interface should be engineered for dual-mode cognition, supporting rapid triage through compressed decision cards while enabling deliberative scrutiny via progressive disclosure of evidence, constraints, and uncertainty (Xu & Poo, 2023; Yang et al., 2022). Explanation ergonomics should respect *cognitive load theory*, *sensemaking* dynamics, and distributed work practices, where teachers, counselors, and leaders coordinate across asynchronous documentation. Practical design primitives include uncertainty-forward summarization, contrastive prompts that highlight decisive evidence, and structured reflection cues that prevent *automation bias* without inducing paralysis. Explanations should explicitly separate descriptive evidence from normative policy constraints, reducing category errors where statistical likelihood is misread as entitlement or culpability. The stakeholder mapping in Table 4 operationalizes cognitive fit by matching explanation modality to decision tempo, expertise, and accountability load, thereby stabilizing calibrated trust rather than rhetorical persuasion

5.3 Plural Explanation Families for Educational Decisions and their Governance Boundaries

Education requires a plural explanation repertoire because different decision types demand different epistemic commitments, causal granularity, and contestability affordances. Exemplar-based rationales support analogical reasoning when context similarity is salient, while counterfactual recourse narratives clarify what changes could alter a recommendation, which is critical for dignity-preserving feedback and actionable remediation (Jiao et al., 2023; Romeira et al., 2023). Causal explanations are necessary when interventions are recommended, because correlation-only rationales mis-specify mechanism and can misallocate scarce support. Procedural explanations expose applied constraints, including accommodation rules, safeguarding non-negotiables, and evidentiary admissibility, enabling contestation focused on policy interpretation rather than on hidden model behavior. Uncertainty explanations must distinguish epistemic uncertainty from irreducible variance, then route abstention and escalation as legitimate outcomes rather than system failure. Table 4 provides a decision-facing matrix that binds explanation families to stakeholder roles, mis-calibration risks, and validation safeguards, preventing a one-size explanation that either overwhelms or manipulates decision-makers

Table 4. Reasons-giving designs for trust-calibrated educational decisions across stakeholder roles

Stakeholder Role Anchor	Decision Terrain and Stakes Profile	Preferred Reasons-Giving Modality	Primary Miscalibration Risk Vector	Safeguard and Validation Criterion
Frontline Teacher	High-frequency micro-decisions on differentiation, attendance response, and classroom support triage under tight time budgets and incomplete context	Uncertainty-forward decision card with exemplar snippets, constraint reminders, and next-step suggestions framed as reversible actions	Overreliance under workload, cue monopolies that narrow noticing, and escalation drift where alerts become mandates	Mandatory uncertainty display, reversible-first action defaults, and structured disagreement capture logged for periodic review
Student Support Practitioner	Medium-frequency, medium-stakes referrals, intervention sequencing, and safeguarding screening requiring sensitive evidence integration and coordination	Contrastive explanation plus procedural workflow trace, including evidence admissibility, escalation thresholds, and boundary-condition warnings	Confirmation bias reinforcement, premature labeling, and privacy overreach through uncontrolled retrieval	Role-based access controls, retention limits, selective prediction with abstention, and audit-ready escalation logs
School Leader and Compliance Officer	Low-frequency, high-rights determinations on exclusions, placements, accommodations, and resource allocation with formal accountability exposure	Constraint ledger with policy graph trace, multi-criteria tradeoff summary, and reasoned authorization boundary with documented alternatives	Legitimacy inflation through institutional pressure, threshold drift, and diffusion of responsibility across committees	Two-key authorization for high-rights decisions, versioned thresholds, and appeal-triggered independent review protocol
Student and Youth Learner	Directly experienced outcomes on placement, support access, assessment consequences, and disciplinary processes affecting dignity and opportunity	Recourse-oriented counterfactual narrative, plain-language evidence synopsis, and rights-focused procedural explanation of review pathways	Perceived inevitability, opacity-induced disengagement, and stigma internalization from deficit framing	Dignity-preserving language constraints, actionable recourse options, and guaranteed contestability window with human review

continued on following page

Table 4. Continued

Stakeholder Role Anchor	Decision Terrain and Stakes Profile	Preferred Reasons-Giving Modality	Primary Miscalibration Risk Vector	Safeguard and Validation Criterion
Parent and Caregiver	High emotional salience decisions involving safeguarding, accommodations, pathway transitions, and resource eligibility with asymmetric information	Layered explanation with evidence provenance, uncertainty summary, and procedural roadmap for escalation, remedies, and documentation	Authority bias toward institutional outputs, mistrust from ambiguity, and adversarial escalation due to unclear standards	Standardized reason templates, evidence admissibility disclosure, and time-bound appeal pathway with documented response obligations
Auditor and Regulator	System-level oversight of fairness, safety, privacy, and due process across populations, vendors, and lifecycle updates	Audit dossier with decision logs, drift indicators, calibration diagnostics, and constraint application traceability at cohort scale	Formal compliance theater, metric gaming, and blind spots from incomplete logging or unscoped data use	Immutable audit trails, independent red-team testing, drift-triggered rollback readiness, and purpose-limited data governance checks

Table 4 functions as an operational design constitution for reasons-giving, converting abstract commitments to *contestability*, calibration, and procedural defensibility into stakeholder-tuned explanation modalities with explicit safeguards. The matrix is deliberately action-oriented, because it specifies who needs what kind of explanation, which cognitive and institutional risks are most probable, and which validation criteria are non-negotiable for safe use. When integrated into a governance-gated pipeline, the table enables explanation templates to be bound to decision classes, ensuring that high-rights determinations trigger constraint-led traces and two-key authorization, while low-stakes supports prioritize reversibility and uncertainty-forward nudges. The table also reduces cross-role friction by aligning teacher-facing brevity with auditor-facing traceability, thereby preventing explanation overload at the front line and explanation poverty at the oversight layer. This design stance complements Table 3 by aligning architecture selection with reasons-giving requirements, and complements Table 2 by binding cognitive mechanisms to accountability artifacts

5.4 Trust Calibration, Overreliance Dynamics, and the Sociology of Authority in AI-Supported Decisions

Trust in educational decision systems should be engineered as calibration, not as acceptance, because overtrust produces automation bias while undertrust yields workarounds that destroy governance visibility. Calibration requires explicit modeling of authority dynamics, where institutional hierarchies, audit anxiety, and performance regimes can transform advisory outputs into quasi-mandates. Effective countermeasures are structural rather than rhetorical, including disagreement capture as a first-class data object, mandatory second-look review for high-rights actions, and escalation routers that prevent single-actor ratification under uncertainty (Park et al., 2023; Al-Shamma & Baldwin, 2023). Decision support should embed cognitive forcing functions, for example brief reflective prompts that activate deliberative scrutiny when uncertainty is high, stakes are irreversible, or sensitive constraints are implicated. Trust should also be bounded by operational limits, with selective prediction and abstention treated as competence signals rather than as failure. Table 4 specifies role-sensitive safeguards, such as two-key authorization and appeal-triggered independent review, which create institutional friction at the right moments, thereby converting calibrated trust into durable legitimacy rather than fragile compliance

5.5 Transparency Tradeoffs, Privacy-Security Constraints, and Gaming-Resilient Explainability

Transparency in education is a constrained good because maximal openness can violate privacy, compromise safeguarding, and enable strategic gaming of assessment and allocation systems, while minimal transparency erodes due process and amplifies mistrust. A defensible transparency regime should therefore partition explainability into what is necessary for contestability and what must be protected by purpose limitation, least-privilege access, and controlled disclosure. Privacy-preserving explanation requires careful redaction of sensitive proxies, bounded retention of case memories, and governance controls over retrieval and logging so that explanations do not become surveillance instruments (Hayashi, 2022; Gong et al., 2023). Security requires threat modeling for prompt injection, data exfiltration, and model inversion risks in AI-assisted workflows, particularly when generative interfaces are used to summarize cases or draft recommendations. Gaming resistance requires that explanation templates disclose decision logic without revealing exploit-ready thresholds, and that monitoring detects anomalous behavioral adaptation following transparency changes. Table 4 anticipates these tradeoffs through stakeholder-specific safeguards, ensuring that auditability and contestability are achieved without converting educational decision systems into vulnerability amplifiers or incentive-distorting score factories

6. ETHICS, BIAS, AND RESPONSIBLE GOVERNANCE FOR BRAIN-INSPIRED EDUCATIONAL DECISION INTELLIGENCE

6.1 Education as a High-Rights Domain with Non-Negotiable Design Constraints

Education-grade *Decision Intelligence* must be engineered as a high-rights socio-technical system where legitimacy is produced through constraint compliance, procedural defensibility, and dignity preservation, not through performance rhetoric. A rights-sensitive architecture operationalizes *proportionality*, *purpose limitation*, and *data minimization* as hard constraints, ensuring that informational intrusiveness scales with stakes and reversibility rather than with model convenience (Alam & Mohanty, 2023; Gandolfi et al., 2022). *Capability-oriented* design shifts evaluation from narrow attainment proxies toward opportunity sets, safeguarding that supports expand agency rather than intensify surveillance. *Ethics of Care* adds relational non-derogation, requiring that triage and intervention pathways avoid stigmatizing classifications, deficit framings, and irreversible labeling. Harm asymmetry must be explicitly encoded through conservative thresholds for high-rights determinations, paired with reversible-first action policies under ambiguity. These constraints become operational only when mapped into enforceable controls, therefore the governance logic later consolidated in Table 5 should be treated as a system constitution rather than as compliance ornamentation

6.2 Bias as a Lifecycle Socio-Technical and Cognitive Phenomenon

Bias in educational decision systems is a lifecycle phenomenon generated by measurement choices, sampling regimes, labeling conventions, institutional incentives, and deployment conditions, which jointly create structural distortion even when model internals appear calibrated. Measurement bias emerges when constructs are operationalized through proxies that are differentially valid across groups, while selection bias emerges when visibility and documentation differ by class, language, disability, geography, or institutional track. Labeling bias arises when administrative categories encode disciplinary cultures rather than learner states, and deployment bias appears when interventions change the data generating process, creating feedback loops that amplify differential surveillance (Li et al., 2024b; Liu et al., 2023). Cognitive distortions compound these pipeline biases, because *anchoring*, *availability effects*, and *confirmation dynamics* make salient flags self-reinforcing within distributed decision work. Aggregation bias also appears when system-level optimization masks local harm, producing equity externalities that are invisible to global metrics. Table

5 provides a structured taxonomy of these bias vectors by linking them to cognitive mechanisms, predictable failure signatures, and governance instruments that can be audited and re-authorized under changing institutional regimes

6.3 Mitigation as Mechanism Design using Attention, Memory, Uncertainty, and Constraints

Mitigation in brain-inspired educational DI is best framed as mechanism design, where the system's choice architecture is engineered to reduce harm likelihood, contain harm magnitude, and preserve contestability under uncertainty. Attention mechanisms must be throttled and diversified so that salience does not collapse into proxy-driven cue monopolies, while memory mechanisms must be governed through retention limits, privacy-preserving retrieval, and anti-stigma controls that prevent sticky labels from becoming institutional destiny (H. Wu et al., 2025; Wang et al., 2025). Uncertainty control should operationalize *selective prediction*, calibrated abstention, and escalation routing, converting epistemic humility into a formal safety function rather than a discretionary gesture. Constraint reasoning must encode admissibility, due-process requirements, and rights-protective action boundaries, ensuring that high-stakes recommendations are procedurally fenced and auditable. Table 5 operationalizes this mitigation logic by mapping risk domains to cognitive couplings, failure signatures, and evidence artifacts that support accountability under review-ready governance

Table 5. Governance controls linking cognitive mechanisms to education risk mitigation

Risk Domain Vector	Cognitive Mechanism Coupling	Failure Mode Signature	Control and Oversight Instrument	Accountability Evidence Artifact
Construct and Measurement Integrity	*Representational inference* coupled with validity-aware encoding, subgroup sensitivity, and uncertainty coding for noisy proxies	Proxy dominance, construct drift, and differential validity that silently redefines need, merit, or risk across cohorts	Admissible-feature governance, construct audits with versioned definitions, and subgroup robustness stress-testing under shift	Feature provenance registers, construct change logs, and decision-threshold re-authorization records aligned to policy
Salience Steering and Attention Harm	*Attentional gating* coupled with triage queues, alert portfolios, and workload-aware interface ergonomics	Cue monopolies, alert fatigue, and institutional tunnel vision that amplifies surveillance while missing quiet distress	Salience audits, diversity-aware prioritization, suppression of sensitive proxies, and risk-tiered escalation triggers	Salience configuration histories, alert outcome traces, and disagreement capture logs linked to case review cycles
Memory, Privacy, and Stigma Persistence	*Episodic retrieval* coupled with de-identified exemplars, similarity constraints, and bounded recall windows	Sticky labeling, privacy leakage through reconstruction, and analogical overreach that hardens stereotypes into policy	Retention limits, role-based access, retrieval logging, and privacy-preserving similarity with diversification constraints	Retrieval audit trails, access logs, retention schedules, and incident reports documenting remediation and rollback
Continual Learning and Threshold Drift	*Consolidation-based adaptation* coupled with drift detection, replay, and calibration maintenance under regime change	Silent policy drift, catastrophic interference, and feedback-induced drift that makes outcomes endogenous to intervention	Bounded adaptation policies, update review boards, calibration monitoring, and rollback-ready version control	Model version dossiers, calibration drift dashboards, update approvals, and rollback decisions with rationale trails
Authority, Explainability, and Overreliance	*Metacognitive control* coupled with reasons-giving, abstention, and reflective prompting under high uncertainty	Automation bias, de facto mandates, and diffusion of responsibility across committees and workflows	Mandatory uncertainty display, two-key authorization for high-rights actions, and structured dissent capture	Explanation templates, authorization logs, override rationales, and audit-ready decision dossiers supporting contestation
Due Process, Recourse, and Contestability	Constraint-led *executive control* coupled with appeal workflows, evidence admissibility, and procedural traceability	Opaque outcomes, inaccessible appeals, and legitimacy collapse when affected parties cannot challenge decisions	Contestability pathways, time-bound review obligations, independent escalation routes, and remedy protocols	Appeal records, response timelines, evidence admissibility disclosures, and remedial action logs tied to governance review

Table 5 functions as a governance blueprint that converts ethical commitments into executable institutional controls, ensuring that mitigation is not reduced to statistical parity rhetoric. By tying each risk vector to cognitive mechanisms, the table clarifies where harms originate, how they propagate through attention, memory, and adaptation, and where governance gates must intervene to preserve rights compatibility. The evidence artifacts column is deliberately operational, because accountability collapses when controls cannot be audited, versioned, and re-authorized under changing policy. When read alongside Table 2, Table 5 makes clear that cognitive mechanisms are not neutral, they are institutional steering devices whose failure modes must be anticipated as predictable stress responses. This framework also supports global portability, because the control instruments are expressed as governance primitives that can be localized to different legal regimes without weakening procedural defensibility

6.4 Contestability, Due Process, and Epistemic Justice as Functional Requirements

Contestability must be treated as a functional requirement that shapes system architecture, because educational legitimacy depends on the ability of affected parties to understand, challenge, and remediate decisions without needing technical expertise. Due process in AI-supported education should be operationalized through admissible-evidence standards, structured reasons-giving, time-bound review obligations, and remedy pathways that are not discretionary, ensuring that procedural guarantees remain stable under institutional pressure (Y. Wu et al., 2022; Yordanova, 2026). *Epistemic justice* requires that contextual testimony, lived constraints, and qualitative evidence have standing within decision workflows, preventing metric-centric pipelines from structurally discounting non-quantified realities. Recourse should be engineered as actionability, providing feasible pathways to change outcomes through support, documentation, or review, while preventing explanations from becoming persuasive narratives that mask uncertainty (Seok et al., 2024; Alam & Mohanty, 2024). High-rights decisions should trigger separation of duties, independent review, and explicit authority boundaries so advisory outputs cannot morph into unreviewable mandates. Table 5 specifies these contestability primitives as governance instruments and evidence artifacts, enabling consistent implementation across jurisdictions with different administrative norms and accountability capacities

6.5 Lifecycle Governance Architecture and Capacity Building for Durable Compliance

Responsible educational DI requires lifecycle governance that spans procurement, configuration, deployment, monitoring, incident response, and retirement, because harms often arise after launch through drift, policy change, and informal workarounds. A durable governance architecture includes risk-tiering by decision class, role-based decision rights, version-controlled thresholds, calibration audits, and red-team stress testing for privacy leakage, gaming susceptibility, and prompt-driven manipulation in human-facing interfaces (Pedro, 2023; Feng et al., 2024). Drift monitoring must be coupled to rollback readiness and threshold re-authorization, otherwise adaptive systems become ungoverned policy engines. Governance must also include socio-technical capacity building, because institutions without calibration literacy, explanation literacy, and bias literacy convert even well-designed systems into compliance theater or coercive instrumentation. Training should focus on uncertainty interpretation, dissent capture, and evidence admissibility, enabling professional agency and accountable overrides (Atashgahi et al., 2022; Xiong et al., 2023; Alohali et al., 2022). Decision logs must be treated as institutional memory for learning and remediation, not as surveillance exhaust. Table 5 provides the minimal governance primitives that can be instantiated in resource-rich and resource-constrained contexts, preserving rights-compatible decision-making under global volatility and institutional scarcity

7. CONCLUSION

7.1 Integrative Synthesis of a Brain-Inspired Design Logic for Educational Decision Intelligence

This chapter has specified a coherent design logic in which *attention*, *memory*, *reasoning*, *judgment*, and *metacognitive control* operate as mutually constraining primitives for education-grade *Decision Intelligence* in high-uncertainty, high-rights environments. The synthesis is not neuro-mimetic, it is functionally prescriptive, translating cognitive mechanisms into auditable control surfaces that stabilize decision chains under partial observability, delayed effects, and institutional scarcity. *Attentional gating* becomes salience governance that prevents cue monopolies and metric myopia, while preserving weak-signal detectability in safeguarding and support allocation. *Memory* becomes governed precedent and policy substrate, separating episodic retrieval from semantic constraint knowledge and procedural workflows, thereby enabling consistency without stigma persistence. *Reasoning* is repositioned as constraint-aware and causally oriented intervention planning, while *judgment* is formalized as multi-objective tradeoff management under harm asymmetry, reversibility gradients, and legitimacy burdens. *Metacognitive control* becomes calibrated uncertainty, selective prediction, abstention, and escalation routing, ensuring epistemic humility by design. Across the chapter, the central claim is that education DI must be architected as a socio-technical control system whose technical modules and institutional gates jointly produce robustness, contestability, and rights compatibility

7.2 Core Conceptual Contributions and System-Level Design Principles

The core contribution is a mechanism-to-architecture-to-governance chain that treats technical design, institutional accountability, and normative legitimacy as inseparable dimensions of the same system. At the mechanism level, the chapter has reframed representational inference, attentional salience, memory retrieval, consolidation-based adaptation, and metacognitive calibration as safety-critical functions rather than as performance optimizations. At the architectural level, it has justified hybridization, demonstrating why governance-gated pipelines, retrieval-centric memory architectures, neuro-symbolic constraint enforcement, causal world-model planning, and human-AI teaming are different but complementary archetypes that must be selected by decision terrain, rights intensity, and institutional capacity. At the governance level, it has operationalized accountability through versioning, audit trails, admissible-evidence rules, contestability workflows, two-key authorization for high-rights decisions, and bounded adaptation that prevents silent threshold drift. The design principles implied by this chain are concrete, including reversible-first action defaults under ambiguity, uncertainty-forward interaction ergonomics, traceable constraint ledgers, privacy-preserving retrieval with retention limits, drift-aware rollback readiness, and structured dissent capture that protects professional agency while increasing decision reliability across global education systems

7.3 Research and Design Agenda for Globally Deployable, Rights-Compatible Educational DI

A forward agenda must shift evaluation from predictive accuracy to decision usefulness, which requires formal metrics for reversibility-sensitive harm, calibrated abstention quality, escalation timeliness, and contestability efficacy under realistic institutional workloads. The next design frontier is the rigorous coupling of causal planning with resource feasibility, ensuring that intervention recommendations propagate through staffing constraints, service capacity, and legally mandated accommodations, rather than remaining score-driven abstractions. Equally critical is standardizing explanation as procedural reasons-giving, with globally portable templates that preserve due process while respecting privacy, safeguarding, and gaming resistance, particularly in generative interfaces where exfiltration and manipulation risks are non-trivial. Governance research must prioritize lifecycle controls that scale to resource-constrained settings, including lightweight auditability, threshold re-authorization routines, and incident-response playbooks that do not presuppose mature data infrastructures. Finally, capacity building must be treated as an institutional technology, developing calibration literacy, bias literacy, and explanation literacy so that human authority boundaries remain credible rather than ceremonial. The chapter's closing position is that brain-inspired primitives offer a principled vocabulary for engineering decision systems that support, rather than supplant, responsible human judgment, while remaining defensible under rights scrutiny and resilient under global volatility and regime change

REFERENCES

Akcetin, E. (2024). Integrating Scientific Instruments and Intelligence Frameworks for Digital Enterprises. In *Game Theory-Computational Aspects and Applications*. IntechOpen. DOI: 10.5772/intechopen.1007794

Al Humdan, E. (2025). Enhancing transportation agility through neuroadaptive AI and behavioural decision intelligence. *Transportation Research Interdisciplinary Perspectives*, *31*, 101468. DOI: 10.1016/j.trip.2025.101468

Al-Shamma, S., & Baldwin, D. (2023). Neuro-Informatics Approaches for Human-Machine Interaction and Brain-Inspired Artificial Intelligence. *Frontora*, *1*(1), 13–20.

Alam, A. (2022). Impact of university's human resources practices on professors' occupational performance: empirical evidence from India's higher education sector. In *Inclusive businesses in developing economies: Converging people, profit, and corporate citizenship* (pp. 107–131). Springer International Publishing. DOI: 10.1007/978-3-031-12217-0_6

Alam, A. (2023a). Media multitasking with M-learning technology in real-time classroom learning: Analysing the dynamics in formal educational settings for the future of E-learning in India. In *2023 2nd International Conference on Smart Technologies and Systems for Next Generation Computing (ICSTSN)* (pp. 1-6). IEEE.

Alam, A. (2023b). The secret sauce of student success: Cracking the code by navigating the path to personalized learning with educational data mining. In *2023 2nd International Conference on Smart Technologies and Systems for Next Generation Computing (ICSTSN)* (pp. 1-8). IEEE.

Alam, A. (2023c). Intelligence unleashed: An argument for AI-enabled learning ecologies with real world examples of today and a peek into the future. In *International Conference on Innovations in Computer Science, Electronics & Electrical Engineering*-2022 (Vol. 2717, No. 1, p. 030001). AIP Publishing LLC. DOI: 10.1063/5.0129803

Alam, A., & Mohanty, A. (2022, December). Facial analytics or virtual avatars: competencies and design considerations for student-teacher interaction in AI-powered online education for effective classroom engagement. In *International Conference on Communication, Networks and Computing* (pp. 252-265). Cham: Springer Nature Switzerland.

Alam, A., & Mohanty, A. (2023, January). Learning on the move: a pedagogical framework for state-of-the-art mobile learning. In *International Conference on Data Management, Analytics & Innovation* (pp. 735-748). Singapore: Springer Nature Singapore. DOI: 10.1007/978-981-99-1414-2_52

Alam, A., & Mohanty, A. (2024). Integrated constructive robotics in education (ICRE) model: A paradigmatic framework for transformative learning in educational ecosystem. *Cogent Education, 11*(1), 2324487. DOI: 10.1080/2331186X.2024.2324487

Alohali, M. A., Al-Wesabi, F. N., Hilal, A. M., Goel, S., Gupta, D., & Khanna, A. (2022). Artificial intelligence enabled intrusion detection systems for cognitive cyber-physical systems in industry 4.0 environment. *Cognitive Neurodynamics, 16*(5), 1045–1057. DOI: 10.1007/s11571-022-09780-8 PMID: 36237400

Amrouch, H., Imani, M., Jiao, X., Aloimonos, Y., Fermuller, C., Yuan, D., & Sutor, P. (2022, October). Brain-inspired hyperdimensional computing for ultra-efficient edge ai. In *2022 International Conference on Hardware/Software Codesign and System Synthesis (CODES+ ISSS)* (pp. 25-34). IEEE. DOI: 10.1109/CODES-ISSS55005.2022.00017

Annu, & Tripathi, R. (2024). A bibliometric and systematic review analysis of adopting decision intelligence analytics for rational decision-making. *Qualitative Research in Financial Markets, 16*(5), 915-936.

Atashgahi, Z., Pieterse, J., Liu, S., Mocanu, D. C., Veldhuis, R., & Pechenizkiy, M. (2022). A brain-inspired algorithm for training highly sparse neural networks. *Machine Learning, 111*(12), 4411–4452. DOI: 10.1007/s10994-022-06266-w

Bratu, S. (2023). Brain-inspired artificial intelligence and machine learning-based navigation algorithms, spatial data visualization and digital twin modeling tools, and emotion detection and recognition technologies in the decentralized and interconnected metaverse. *Review of Contemporary Philosophy*, (22), 51–67.

Chinnaraju, A. (2025). Quantum computing in consumer behavior: A theoretical framework for market prediction and decision analytics. *Quantum : the Open Journal for Quantum Science, 5*(2).

Chohra, A., & van der Wal, C. N. (2025, October). Integrating Emotional, Personal, and Social Intelligences in Complex Collective Decision-Making. In *Proceedings of the Future Technologies Conference* (pp. 176-193). Cham: Springer Nature Switzerland.

Clark, L. F. B. (2025). Multimodal AI–Driven Real-Time Decision Intelligence for Secure Healthcare Cloud and SAP Data Networks. [IJEETR]. *International Journal of Engineering & Extended Technologies Research, 7*(4), 10277–10283.

Feng, J., Jirsa, V., & Lu, W. (2024). Human brain computing and brain-inspired intelligence. *National Science Review, 11*(5), nwae144. DOI: 10.1093/nsr/nwae144 PMID: 38742233

Gandolfi, D., Puglisi, F. M., Serb, A., Giugliano, M., & Mapelli, J. (2022). Brain-inspired computing: Neuroscience drives the development of new electronics and artificial intelligence. *Frontiers in Cellular Neuroscience, 16*, 1115395. DOI: 10.3389/fncel.2022.1115395 PMID: 36605614

Gong, C., Jing, C., Chen, X., Pun, C. M., Huang, G., Saha, A., Nieuwoudt, M., Li, H.-X., Hu, Y., & Wang, S. (2023). Generative AI for brain image computing and brain network computing: A review. *Frontiers in Neuroscience, 17*, 1203104. DOI: 10.3389/fnins.2023.1203104 PMID: 37383107

HAYASHI, R. (2022). Perspective of Brain-inspired AI Research and Multiomics Data Analysis Approach. *Kiso Shinrigaku Kenkyu, 41*(1).

Heilig, T., & Scheer, I. (2023). *Decision intelligence: Transform your team and organization with AI-Driven decision-making*. John Wiley & Sons.

Jeyanthi, P. M., Choudhury, T., Hack-Polay, D., Singh, T. P., & Abujar, S. (2022). *Decision Intelligence Analytics and the Implementation of Strategic Business Management*. Springer International Publishing. DOI: 10.1007/978-3-030-82763-2

Jiao, L., Huang, Z., Liu, X., Yang, Y., Ma, M., Zhao, J., You, C., Hou, B., Yang, S., Liu, F., Ma, W., Li, L., Chen, P., Feng, Z., Tang, X., Guo, Y., Zhang, X., Quan, D., Wang, S., & Feng, J. (2023). Brain-inspired remote sensing interpretation: A comprehensive survey. *IEEE Journal of Selected Topics in Applied Earth Observations and Remote Sensing, 16*, 2992–3033. DOI: 10.1109/JSTARS.2023.3247455

Josephson, C. B., Aronica, E., Beniczky, S., Boyce, D., Cavalleri, G., Denaxas, S., French, J., Jehi, L., Koh, H., Kwan, P., McDonald, C., Mitchell, J. W., Rampp, S., Sadleir, L., Sisodiya, S. M., Wang, I., Wiebe, S., Yasuda, C., & Youngerman, B. (2024). Big data research is everyone's research—Making epilepsy data science accessible to the global community: Report of the ILAE big data commission. *Epileptic Disorders, 26*(6), 733–752. DOI: 10.1002/epd2.20288 PMID: 39446076

Kaneko, T. (2023). Naturally decision intelligence: Perfect algorithm generated by the hypothetical and synchronizing model for life system. *Intelligent Decision Technologies, 17*(1), 195–210. DOI: 10.3233/IDT-220231

Kaneko, T. (2023, March). Decision Intelligence into Synchronization. In *2023 15th International Conference on Computer and Automation Engineering (ICCAE)* (pp. 326-332). IEEE. DOI: 10.1109/ICCAE56788.2023.10111454

Kaneko, T. (2024, August). Comparison of decision-making technologies and basic concepts of Naturally Decision Intelligence. In *International Conference on Artificial Intelligence-empowered Software Engineering* (pp. 181-188). Cham: Springer Nature Switzerland.

Khachaturian, A. S., Cassin, B., & Finney, G. R. (2023). Using clinical decision intelligence applications to improve pathways for earlier detection of underrecognized cognitive disorders. *JAR Life*, *12*, 14–17. DOI: 10.14283/jarlife.2023.3 PMID: 37234274

Li, G., Deng, L., Tang, H., Pan, G., Tian, Y., Roy, K., & Maass, W. (2024). Brain-inspired computing: A systematic survey and future trends. *Proceedings of the IEEE.* DOI: 10.1109/JPROC.2024.3429360

Li, G., Huang, Q., Liu, C., Wang, G., Guo, L., Liu, R., & Liu, L. (2024). Fully automated diagnosis of thyroid nodule ultrasound using brain-inspired inference. *Neurocomputing*, *582*, 127497. DOI: 10.1016/j.neucom.2024.127497

Liu, Z., Gan, E., & Tegmark, M. (2023). Seeing is believing: Brain-inspired modular training for mechanistic interpretability. *Entropy (Basel, Switzerland)*, *26*(1), 41. DOI: 10.3390/e26010041 PMID: 38248167

Mehonic, A., & Kenyon, A. J. (2022). Brain-inspired computing needs a master plan. *Nature*, *604*(7905), 255–260. DOI: 10.1038/s41586-021-04362-w PMID: 35418630

Meng, L., Zhang, W., Xu, F., Han, L., Guo, S., Chen, T., . . . Wang, F. Y. (2025, November). Brain-Inspired Parallel Data Intelligence: a Generality-Specificity Fused Approach. In *2025 21st IEEE International Conference on Mechatronic and Embedded Systems and Applications (MESA)* (pp. 71-77). IEEE. DOI: 10.1109/MESA68091.2025.11278866

Murthy, B. K., Reddy, B. V. R., Hasteer, N., & Van Belle, J. P. (2023). Decision Intelligence. *InCITe, 1.*

Nozari, H., Abdi, H., & Szmelter-Jarosz, A. (2025). A Neuromorphic and Bio-Inspired Framework for Optimization in Multi-Echelon Supply Networks. *Transformative Science*, *1*(1), 9–17.

O'Callaghan, M. (2023). *Decision intelligence: human–machine integration for decision-making*. CRC Press. DOI: 10.1201/b23322

Park, T. J., Deng, S., Manna, S., Islam, A. N., Yu, H., Yuan, Y., Fong, D. D., Chubykin, A. A., Sengupta, A., Sankaranarayanan, S. K. R. S., & Ramanathan, S. (2023). Complex oxides for brain-inspired computing: A review. *Advanced Materials*, *35*(37), 2203352. DOI: 10.1002/adma.202203352 PMID: 35723973

Pedro, F. (2023). Theoretical Analysis of the Brain and Artificial Intelligence. *Journal of Robotics Spectrum, 1*, 024-035.

Pratt, L., Bisson, C., & Warin, T. (2023). Bringing advanced technology to strategic decision-making: The Decision Intelligence/Data Science (DI/DS) Integration framework. *Futures*, *152*, 103217. DOI: 10.1016/j.futures.2023.103217

Qian, S. U. N. (2025, June). Research on the Application and Development Trend of Artificial Intelligence Based Decision Making Intelligence. In *2025 5th International Conference on Artificial Intelligence, Big Data and Algorithms (CAIBDA)* (pp. 1370-1373). IEEE.

Rajak, M. K., Dey, S., Kumar, V., Pratihar, D. K., Singh, V. P., & Islam, S. (2025). Foundations of decision intelligence in healthcare: Integrating machine learning, robotics, and biomedical analytics. In *Advancing Healthcare through Decision Intelligence* (pp. 1–7). Academic Press. DOI: 10.1016/B978-0-443-26480-1.00010-2

Romeira, B., Adão, R., Nieder, J. B., Al-Taai, Q., Zhang, W., Hadfield, R. H., Wasige, E., Hejda, M., Hurtado, A., Malysheva, E., Dolores Calzadilla, V., Lourenço, J., Castro Alves, D., Figueiredo, J. M. L., Ortega-Piwonka, I., Javaloyes, J., Edwards, S., Davies, J. I., Horst, F., & Offrein, B. J. (2023). Brain-inspired nanophotonic spike computing: Challenges and prospects. *Neuromorphic Computing and Engineering*, *3*(3), 033001. DOI: 10.1088/2634-4386/acdf17

Schmidgall, S., Ziaei, R., Achterberg, J., Kirsch, L., Hajiseyedrazi, S., & Eshraghian, J. (2024). Brain-inspired learning in artificial neural networks: A review. *APL Machine Learning*, *2*(2), 021501. DOI: 10.1063/5.0186054

Seok, H., Lee, D., Son, S., Choi, H., Kim, G., & Kim, T. (2024). Beyond von Neumann architecture: Brain-inspired artificial neuromorphic devices and integrated computing. *Advanced Electronic Materials*, *10*(8), 2300839. DOI: 10.1002/aelm.202300839

Sousa, M. J., Dal Mas, F., Osório de Barros, G., & Tavares, N. (2022). Artificial intelligence: technologies, applications, and policy perspectives. Insights from Portugal. In *Decision Intelligence Analytics and the Implementation of Strategic Business Management* (pp. 69–84). Springer International Publishing. DOI: 10.1007/978-3-030-82763-2_6

Theodorakopoulos, L., Theodoropoulou, A., & Halkiopoulos, C. (2025). Cognitive bias mitigation in executive decision-making: A data-driven approach integrating big data analytics, AI, and explainable systems. *Electronics (Basel)*, *14*(19), 3930. DOI: 10.3390/electronics14193930

Vasuki, M., Kumar, A. D., Celestin, M., & Alghazali, T. A. H. (2025). Mathematical Logic as the Engine of Autonomous Decision Intelligence in Digital Systems. *International Journal of Applied and Advanced Scientific Research*, *10*(2), 97–108.

Wang, G., Bao, H., Liu, Q., Zhou, T., Wu, S., Huang, T., Yu, Z. F., Lu, C. W., Gong, Y. H., Zhang, Z. X., & He, S. (2024). Brain-inspired artificial intelligence research: A review. *Science China. Technological Sciences*, *67*(8), 2282–2296. DOI: 10.1007/s11431-024-2732-9

Wang, K., & Bai, D. (2023, August). Overview of Game Decision Intelligence. In *2023 IEEE 11th International Conference on Information, Communication and Networks (ICICN)* (pp. 835-843). IEEE. DOI: 10.1109/ICICN59530.2023.10392489

Wang, Y., Yan, T., Li, S., Cai, Y., Zhang, P., Cao, M., Wang, F., Zhan, X., & Wang, Z. (2025). Brain-inspired neural networks: Neuromorphic devices and their practical applications. *Nanotechnology*, *36*(47), 472001. DOI: 10.1088/1361-6528/ae1840 PMID: 41151104

Wu, H., Shao, H., Gao, X. Z., Yan, M., Heng, J., & Li, Z. (2025). Brain-inspired Network with Distinct Brain Mechanisms for Short-term Load Forecasting. *IEEE Internet of Things Journal.*

Wu, S., Oltramari, A., Francis, J., Giles, C. L., & Ritter, F. E. (2025). Cognitive LLMs: Toward human-like artificial intelligence by integrating cognitive architectures and large language models for manufacturing decision-making. *Neurosymbolic Artificial Intelligence*, *1*, 29498732251377341. DOI: 10.1177/29498732251377341

Wu, S. Z. (2025). Emerging Next-Generation AI Techniques. In *The AI City* (pp. 237–245). Springer Nature Singapore. DOI: 10.1007/978-981-96-2560-4_15

Wu, Y., Zhao, R., Zhu, J., Chen, F., Xu, M., Li, G., Song, S., Deng, L., Wang, G., Zheng, H., Ma, S., Pei, J., Zhang, Y., Zhao, M., & Shi, L. (2022). Brain-inspired global-local learning incorporated with neuromorphic computing. *Nature Communications*, *13*(1), 65. DOI: 10.1038/s41467-021-27653-2 PMID: 35013198

Xiong, H., Chu, C., Fan, L., Song, M., Zhang, J., Ma, Y., ... & Jiang, T. (2023). The digital twin brain: A bridge between biological and artificial intelligence. Intelligent Computing, 2, 0055.

Xu, B., & Poo, M. M. (2023). Large language models and brain-inspired general intelligence. *National Science Review*, *10*(10), nwad267. DOI: 10.1093/nsr/nwad267 PMID: 37942481

Yang, X., Yan, J., Wang, W., Li, S., Hu, B., & Lin, J. (2022). Brain-inspired models for visual object recognition: An overview. *Artificial Intelligence Review*, *55*(7), 5263–5311. DOI: 10.1007/s10462-021-10130-z

Yordanova, Z. (2026). Merging AIoE technology With Brain-Inspired Computing in Education. In *AI-Augmented Creativity in Learning Analytics* (pp. 273–292). IGI Global Scientific Publishing.

Zaimoglu, A. K., Pratt, L., & Fisher, B. (2023). Epistemological role of human reasoning in data-informed decision-making. *Frontiers in Communication*, *8*, 1250301. DOI: 10.3389/fcomm.2023.1250301

Zeng, Y., Zhao, D., Zhao, F., Shen, G., Dong, Y., Lu, E., Zhang, Q., Sun, Y., Liang, Q., Zhao, Y., Zhao, Z., Fang, H., Wang, Y., Li, Y., Liu, X., Du, C., Kong, Q., Ruan, Z., & Bi, W. (2023). Braincog: A spiking neural network based, brain-inspired cognitive intelligence engine for brain-inspired ai and brain simulation. *Patterns (New York, N.Y.)*, *4*(8), 100789. DOI: 10.1016/j.patter.2023.100789 PMID: 37602224

Zhao, L., Zhang, L., Wu, Z., Chen, Y., Dai, H., Yu, X., Liu, Z., Zhang, T., Hu, X., Jiang, X., Li, X., Zhu, D., Shen, D., & Liu, T. (2023). When brain-inspired ai meets agi. *Meta-Radiology*, *1*(1), 100005. DOI: 10.1016/j.metrad.2023.100005

Chapter 8
Explainable and Interpretable Decision Intelligence for Human-Centered AI Systems:
Bridging Cognitive Computing and Transparent AI Through Brain-Inspired Mechanisms

Firdaus Alamsyah
http://orcid.org/0009-0009-8539-3416
Esa Unggul University, Indonesia

ABSTRACT

As AI systems increasingly drive critical decisions across healthcare, finance, and policy domains, transparency and interpretability have become essential requirements. This chapter explores explainable AI and decision intelligence through brain-inspired cognitive mechanisms that prioritize human understanding and trust. Drawing from neuroscience and cognitive science, we examine how human-centered AI systems can mirror the brain's natural decision-making processes while maintaining interpretability. We present a comprehensive framework balancing model performance with human comprehensibility, investigating XAI techniques including attention mechanisms, feature importance, and counterfactual explanations grounded in cognitive principles. The chapter explores neuro-symbolic AI in creating hybrid systems that combine neural networks' pattern recognition with symbolic reasoning's

DOI: 10.4018/979-8-3373-8988-2.ch008

interpretability. Through practical case studies, we demonstrate how brain-inspired explainable AI enhances human-AI collaboration and builds stakeholder trust while addressing interpretability challenges.

1. INTRODUCTION

The proliferation of artificial intelligence systems in critical decision-making domains has created an urgent need for transparency and interpretability. From medical diagnosis to financial credit scoring, from criminal justice sentencing to autonomous vehicle navigation, AI systems increasingly influence outcomes that profoundly affect human lives. However, many of these systems operate as "black boxes," making decisions through complex mathematical operations that remain opaque even to their creators(Doshi-Velez & Kim, 2017)

This opacity presents significant challenges across multiple domains. Healthcare professionals struggle to trust AI diagnostic recommendations when they cannot understand the reasoning behind them, leading to hesitation in adopting potentially life-saving technologies. Financial institutions face stringent regulatory requirements to explain loan decisions to applicants, yet deploy sophisticated models whose internal workings defy simple explanation. Autonomous systems make split-second decisions with life-or-death consequences, yet their decision processes remain largely inscrutable to safety investigators when accidents occur.

The field of explainable AI (XAI) has emerged to address these challenges, developing techniques and frameworks to make AI decision-making more transparent and interpretable. However, early XAI approaches often focused narrowly on technical solutions without considering the cognitive and contextual needs of human users. A post-hoc explanation generated by an algorithm may be technically accurate yet fail to provide meaningful insight to domain experts who need to validate, debug, or improve the system. (Balasubramanian, 2023)

This chapter advocates for a paradigm shift toward human-centered explainable AI that draws inspiration from neuroscience and cognitive science. By understanding how the human brain processes information, makes decisions, and constructs explanations, we can design AI systems that communicate their reasoning in ways that align with natural human cognition. This brain-inspired approach to decision intelligence represents a convergence of multiple disciplines: machine learning provides the computational substrate, neuroscience offers insights into biological intelligence, cognitive psychology illuminates how humans understand and trust explanations, and human-computer interaction design shapes the interface through which AI systems communicate with users.

1.1 The Explainability Challenge in Modern AI

The challenge of explainability has grown more acute as AI systems have become more powerful. Traditional machine learning models like decision trees and linear regression offered inherent interpretability through their simple structure. A decision tree's branching logic could be traced from root to leaf, and a linear model's coefficients directly indicated feature importance. However, these simpler models often lacked the capacity to capture complex patterns in high-dimensional data. (Minh et al., 2022)

The deep learning revolution dramatically improved predictive performance across domains from computer vision to natural language processing. Neural networks with millions or billions of parameters achieved superhuman performance on tasks previously thought impossible for machines. (Mumuni & Mumuni, 2025) Yet this performance came at the cost of interpretability. A deep neural network's decision emerges from the collective interaction of countless neurons across multiple layers, creating what researchers call a "distributed representation" that resists straightforward interpretation. (Minh et al., 2022)

This trade-off between performance and interpretability is sometimes called the "interpretability-accuracy frontier," suggesting that gains in one dimension necessitate losses in the other. (Minh et al., 2022) However, this framing may be overly pessimistic. Recent research suggests that with careful design, we can develop models that achieve both high performance and meaningful interpretability, particularly when we ground our approach in principles of human cognition. (Mumuni & Mumuni, 2025)

1.2 Decision Intelligence as a Framework

Decision intelligence represents an emerging discipline that integrates AI, decision theory, and social sciences to improve organizational and individual decision-making. Unlike narrow AI focused solely on prediction accuracy, decision intelligence emphasizes the entire decision-making pipeline: understanding the decision context, gathering relevant information, generating options, evaluating trade-offs, and supporting implementation. (Balasubramanian, 2023)

This holistic perspective makes decision intelligence a natural framework for human-centered AI. A decision intelligence system must not only produce accurate predictions but also communicate uncertainty, surface relevant contextual factors, and support human judgment rather than replacing it. Explainability is not an optional feature but a core requirement for effective decision support. (Tripathi et al., 2025)

Figure 1. Conceptual framework showing the relationship between decision intelligence, explainable AI, and human-centered AI, highlighting their intersections and key components within each domain

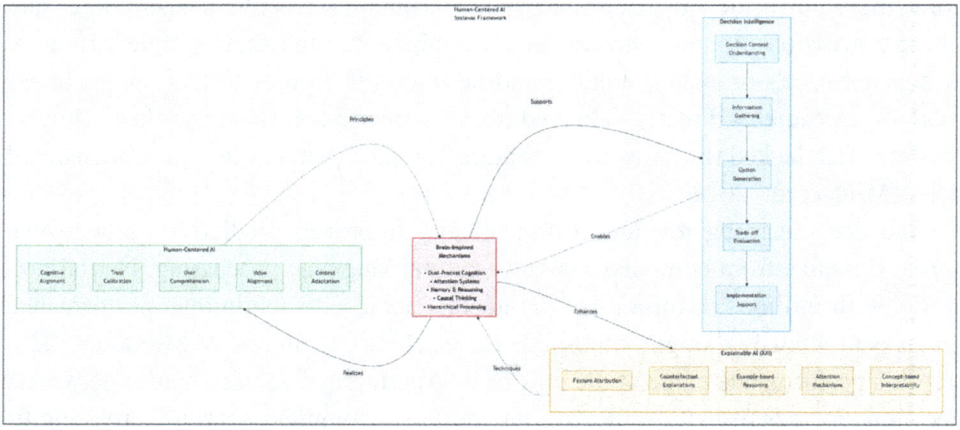

1.3 Chapter Objectives and Organization

This chapter pursues three primary objectives. First, we examine how insights from neuroscience and cognitive science can inform the design of explainable AI systems that align with natural human reasoning processes. Second, we present a comprehensive framework for developing decision intelligence systems that balance predictive performance with human comprehensibility. Third, we demonstrate through case studies how brain-inspired explainable AI can enhance human-AI collaboration across diverse application domains.

The remainder of this chapter is organized as follows. Section 2 reviews foundational concepts in explainable AI and decision intelligence, establishing key terminology and surveying the landscape of existing approaches. Section 3 explores brain-inspired mechanisms for AI explainability, drawing connections between neuroscientific findings and machine learning techniques. Section 4 presents our framework for human-centered decision intelligence systems. Section 5 examines specific XAI techniques through the lens of cognitive alignment. Section 6 discusses evaluation methodologies for explainable AI that account for human understanding. Section 7 presents case studies demonstrating practical applications. Section 8 addresses challenges and limitations, and Section 9 concludes with future research directions.

2. FOUNDATIONS OF EXPLAINABLE AI AND DECISION INTELLIGENCE

2.1 Defining Explainability and Interpretability

The terms "explainability" and "interpretability" are often used interchangeably in AI literature, yet some researchers draw meaningful distinctions between them. Interpretability typically refers to the degree to which a human can understand the cause of a decision, often associated with the inherent transparency of a model's structure. (Doshi-Velez & Kim, 2017) A linear regression model is interpretable because its coefficients directly indicate how input features influence predictions. Explainability, in contrast, often refers to post-hoc techniques that provide insight into already-trained models, regardless of their internal. (Minh et al., 2022)

For the purposes of this chapter, we adopt a unified perspective that encompasses both concepts under the umbrella of "transparent AI." What matters is not whether understanding comes from inherent model structure or post-hoc analysis, but whether humans can develop appropriate trust and effectively collaborate with AI systems. This pragmatic stance aligns with the goals of human-centered. (Zhang et al., 2024)

2.2 The Spectrum of Interpretability

AI models exist along a spectrum of interpretability. At one end lie inherently interpretable models whose structure directly reveals their reasoning. Linear models, decision trees, and rule-based systems fall into this category. These models offer "global interpretability" because their entire decision-making logic can be comprehended. (Minh et al., 2022)

At the opposite end of the spectrum lie complex black-box models like deep neural networks and large ensemble methods. These models typically require post-hoc explanation techniques to provide any insight into their decision-making. However, they can still offer "local interpretability" by explaining individual predictions, even if their overall behavior remains opaque. (Minh et al., 2022)

Between these extremes lie models with intermediate interpretability. Attention-based neural networks, for instance, learn to focus on specific input features when making predictions, providing some insight into their reasoning while maintaining the flexibility of deep learning. (Mumuni & Mumuni, 2025) Similarly, neural-symbolic hybrid systems combine the pattern recognition capabilities of neural networks with the logical transparency of symbolic reasoning. (Mumuni & Mumuni, 2025)

Figure 2. Spectrum of model interpretability ranges from highly interpretable approaches, such as decision trees and linear models, to less interpretable models, including deep neural networks and ensemble methods, with hybrid approaches positioned in between

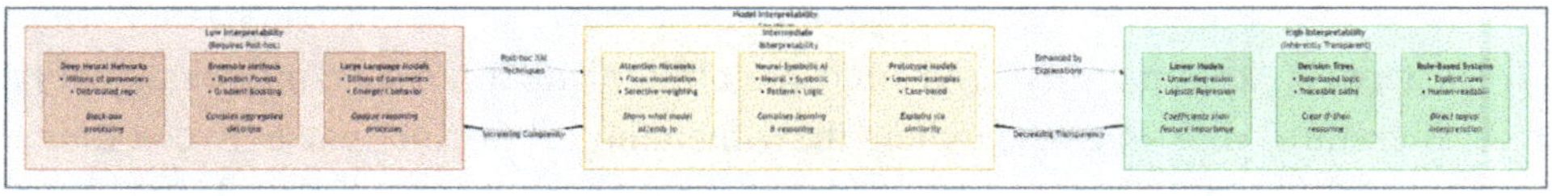

2.3 Types of Explanations

XAI research has identified several categories of explanations that serve different purposes and address different stakeholder needs. Feature importance explanations identify which input variables most strongly influenced a prediction, answering "what mattered most?". Example-based explanations provide similar past cases, answering "what does this remind you of?". (Mumuni & Mumuni, 2025) Counterfactual explanations describe how inputs would need to change to produce a different outcome, answering "what would need to be different?". (Minh et al., 2022)

Rule-based explanations extract human-readable if-then logic from trained models, providing procedural understanding. Attention visualizations highlight which parts of complex inputs like images or text the model focused on during prediction. (Mumuni & Mumuni, 2025) Concept-based explanations identify high-level concepts the model has learned, bridging the gap between low-level features and human-understandable abstractions. (Minh et al., 2022)

The appropriate type of explanation depends on the use case and audience. A radiologist reviewing an AI diagnosis may benefit from attention maps highlighting relevant image regions, while a loan applicant denied credit needs a counterfactual explanation indicating how their application could be improved. (Zhang et al., 2024)

2.4 Decision Intelligence Core Principles

Decision intelligence extends beyond prediction to encompass the full decision-making process. Several core principles distinguish decision intelligence from conventional machine learning. First, decision intelligence maintains a clear focus on actionable outcomes rather than abstract accuracy metrics. A model's value is measured not by test set performance but by its impact on real-world. (Balasubramanian, 2023)

Second, decision intelligence embraces uncertainty quantification as essential rather than optional. Decision-makers need to understand not just what the model predicts but how confident that prediction is and where uncertainty originates. (Kurniawan et al., 2023) Third, decision intelligence prioritizes causal understanding over mere correlation. Understanding causal relationships enables prediction under intervention and supports reasoning about "what if" scenarios.

Fourth, decision intelligence systems must account for human values and preferences. Technical optimization alone cannot resolve inherently value-laden choices about trade-offs between competing objectives. (Tripathi et al., 2025) Finally, decision intelligence recognizes that AI systems operate within organizational and social contexts. System design must consider implementation challenges, stakeholder buy-in, and alignment with existing workflows. (Balasubramanian, 2023)

2.5 The Role of Trust in Human-AI Systems

Trust plays a central role in effective human-AI collaboration. Appropriate trust calibration means users rely on AI systems when they perform well and override them when they err. (Alufaisan et al., 2020) Both overtrust and undertrust create problems. Overtrust leads users to accept incorrect recommendations without scrutiny, while undertrust prevents beneficial AI adoption. (Kurniawan et al., 2023)

Explainability directly influences trust by enabling users to evaluate system reliability. When users understand how an AI system reaches conclusions, they can better judge when to trust its recommendations and when to seek additional information or override its suggestions. (Alufaisan et al., 2020) However, the relationship between explainability and trust is nuanced. Explanations can increase trust even in incorrect predictions if users lack expertise to evaluate them critically, highlighting the importance of explanation quality over quantity. (Alufaisan et al., 2020)

3. BRAIN-INSPIRED MECHANISMS FOR AI EXPLAINABILITY

3.1 Cognitive Architecture and Decision-Making

Understanding human decision-making provides valuable guidance for designing explainable AI systems. Cognitive psychology distinguishes between System 1 and System 2 thinking, concepts popularized by Daniel Kahneman. System 1 operates quickly and intuitively, handling pattern recognition and routine decisions with minimal conscious effort. System 2 engages in slower, more deliberate reasoning for novel or complex problems requiring logical. (Sreedharan et al., 2024)

This dual-process framework suggests that effective AI explanations should support both cognitive modes. For routine decisions where users develop intuition through experience, AI systems might emphasize fast, salient visualizations that enable at-a-glance assessment. For high-stakes or novel decisions requiring deliberation, systems should provide detailed analytical explanations supporting systematic evaluation. (Sreedharan et al., 2024)

Neuroscience reveals that human decision-making involves distributed neural networks rather than localized processing centers. The prefrontal cortex plays a key role in executive function and deliberate reasoning, while subcortical structures like the amygdala contribute emotional valuation, and sensory cortices process perceptual. (Minh et al., 2022) This distributed architecture suggests that effective AI systems might similarly integrate multiple specialized components, each addressing different aspects of decision-making.

3.2 Attention Mechanisms as Cognitive Models

Attention mechanisms in neural networks draw direct inspiration from human visual and cognitive attention. Human attention selectively focuses cognitive resources on relevant information while filtering distractions, enabling efficient processing of complex environments. (Minh et al., 2022) Similarly, attention mechanisms in neural networks learn to weight different input features or regions based on their relevance to the task at hand.

This parallel between artificial and biological attention provides a natural pathway to interpretability. By visualizing attention weights, we can see which input regions the model considers most relevant, mimicking how humans might explain their focus of attention when making judgments. (Mumuni & Mumuni, 2025) Attention maps in image classification can highlight diagnostic features a radiologist should examine, while attention weights in language models can indicate which words most influenced sentiment analysis. (Minh et al., 2022)

However, attention-based explanations have limitations. Attention weights indicate correlation with model decisions but do not necessarily reveal causation. (Minh et al., 2022) Multiple attention patterns may yield similar predictions, and post-hoc rationalization differs from true mechanistic understanding. Despite these caveats, attention mechanisms represent a promising bridge between black-box neural networks and human-understandable. (Mumuni & Mumuni, 2025)

Figure 3. Comparison of human visual attention mechanisms with neural network attention models, illustrating parallels between biological attention patterns and attention representations in AI systems

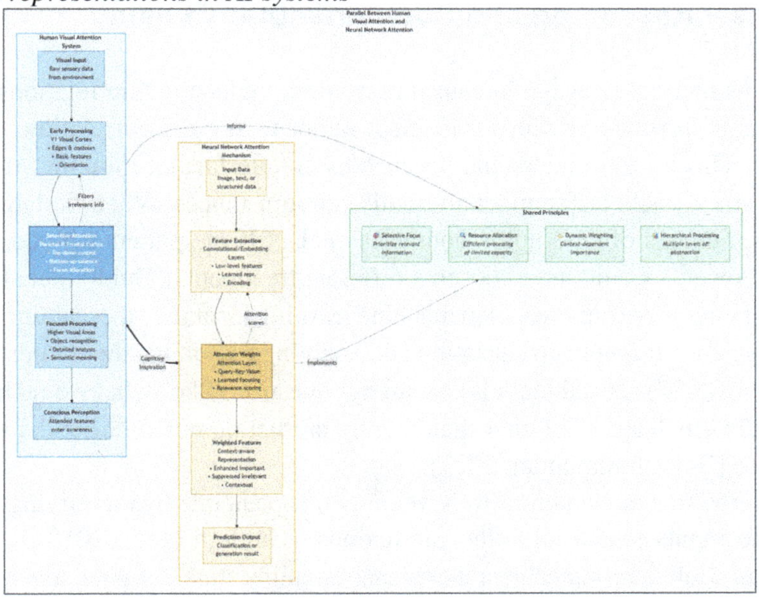

3.3 Memory Systems and Case-Based Reasoning

Human memory operates through multiple interacting systems. Episodic memory stores specific past experiences, semantic memory contains general knowledge and concepts, and working memory maintains information for immediate cognitive tasks. (Minh et al., 2022) These memory systems enable humans to reason by analogy, drawing on past experiences to interpret novel. (Mumuni & Mumuni, 2025)

Case-based reasoning (CBR) in AI mirrors this cognitive process by solving new problems through adaptation of past solutions. (Guidotti et al., 2018) CBR systems maintain a library of past cases with their solutions and retrieve relevant examples when facing new problems. This approach offers natural interpretability because explanations can reference concrete past instances rather than abstract patterns. (Guidotti et al., 2018)

Example-based explanations leverage this principle by presenting similar training instances that support a prediction. When an AI system classifies a medical image as showing a tumor, showing similar past cases provides intuitive justification that physicians can evaluate based on their experience. (Balasubramanian, 2023)

Prototype-based models extend this idea by learning representative examples that capture essential features of each class. (Mumuni & Mumuni, 2025)

3.4 Causal Reasoning and Counterfactual Thinking

Humans naturally engage in causal reasoning, understanding relationships between events in terms of cause and effect rather than mere correlation. (Minh et al., 2022) This causal understanding enables counterfactual thinking, imagining how outcomes would differ under alternative circumstances. When explaining their decisions, humans often employ counterfactuals: "If the patient's blood pressure had been lower, I would have chosen a different treatment".(Guidotti et al., 2018)

Causal inference frameworks in machine learning formalize these intuitions, distinguishing correlation from causation and enabling reasoning about interventions. (Minh et al., 2022) Causal models can answer questions like "what would happen if we changed this feature?" rather than merely predicting outcomes under observed conditions. (Balasubramanian, 2023)

Counterfactual explanations in XAI apply this principle by identifying minimal changes to inputs that would alter predictions. (Guidotti et al., 2018) For a loan application denial, a counterfactual explanation might state: "If your annual income were $5,000 higher or your debt-to-income ratio were 5% lower, you would have been approved." This format aligns with natural human explanation preferences and provides actionable guidance. (Lai et al., 2023)

3.5 Hierarchical Processing and Abstraction

The human brain processes information hierarchically, with lower levels handling basic features and higher levels integrating these into abstract concepts. In vision, early processing detects edges and textures, intermediate stages recognize object parts, and higher regions identify complete objects and scenes. (Minh et al., 2022) This hierarchical organization enables efficient processing and supports reasoning at multiple levels of abstraction. (Mumuni & Mumuni, 2025)

Deep neural networks exhibit similar hierarchical organization, with early layers learning low-level features and deeper layers combining these into increasingly abstract representations. (Mumuni & Mumuni, 2025) This parallel suggests opportunities for interpretability at multiple abstraction levels. Rather than explaining predictions solely through input features, we can identify intermediate concepts the network has learned. (Minh et al., 2022)

Concept-based interpretability methods leverage this hierarchy by identifying human-understandable concepts in model representations. Network dissection techniques probe individual neurons to discover what concepts they respond to.

(Minh et al., 2022) Concept activation vectors quantify how much a model's prediction depends on specific high-level concepts rather than low-level features. These approaches bridge the semantic gap between machine representations and human understanding. (Minh et al., 2022)

4. FRAMEWORK FOR HUMAN-CENTERED DECISION INTELLIGENCE

4.1 Design Principles

Building on cognitive and neuroscientific insights, we propose a framework for human-centered decision intelligence systems grounded in five core design principles. First, cognitive alignment ensures that system behavior and explanations match natural human reasoning patterns. This includes structuring information to support both intuitive and analytical thinking, providing explanations that reference familiar concepts, and respecting cognitive limitations like working memory capacity. (Tripathi et al., 2025)

Second, contextual adaptation recognizes that explanation needs vary across users, tasks, and situations. A domain expert requires different information than a novice, urgent decisions demand faster explanations than reflective analysis, and different cultural contexts shape explanation expectations. (Zhang et al., 2024) Systems should adapt their communication style to match user expertise, task demands, and organizational culture.

Third, transparency with parsimony balances completeness against cognitive overload. While comprehensive information supports thorough analysis, excessive detail overwhelms users and obscures key. (Alufaisan et al., 2020) Effective systems present essential information prominently while making additional details available on demand, implementing progressive disclosure that reveals complexity gradually as needed. (Zhang et al., 2024)

Fourth, uncertainty communication explicitly conveys confidence levels and identifies sources of uncertainty. Decision-makers need to know not just what the model predicts but how reliable that prediction is and where caution is warranted. (Kurniawan et al., 2023) This includes distinguishing epistemic uncertainty from model limitations versus aleatory uncertainty from inherent randomness. (Minh et al., 2022)

Fifth, value alignment ensures systems respect human values and support rather than supplant human judgment. AI should empower users to make better-informed decisions rather than coerce specific choices. (Tripathi et al., 2025) This includes making assumptions and trade-offs explicit, acknowledging value-laden aspects of decisions, and maintaining meaningful human control over outcomes. (Sreedharan et al., 2024)

4.2 Stakeholder Analysis

Effective decision intelligence systems must serve diverse stakeholders with varying needs and expertise. End users who interact directly with the system require explanations that support their specific tasks and integrate with their workflows. (Lai et al., 2023) A radiologist using AI-assisted diagnosis needs different information than a patient receiving that diagnosis, and both differ from what a hospital administrator monitoring system performance requires.

Domain experts bring deep knowledge that AI systems should leverage rather than ignore. Explanations for experts can reference domain-specific concepts and assume technical background, while explanations for novices must build understanding from foundational concepts. (Teso et al., 2023) Regulatory auditors and compliance officers need explanations that demonstrate adherence to legal requirements and ethical standards. (Tripathi et al., 2025)

Developers and maintainers require different forms of interpretability focused on debugging, validation, and improvement. They need to understand model failure modes, identify sources of bias, and diagnose performance issues. (Guidotti et al., 2018) Business stakeholders making adoption decisions need explanations of system capabilities, limitations, and expected impact that inform strategic choices. (Balasubramanian, 2023)

This diversity of stakeholders highlights the need for multi-level explanations tailored to different audiences and purposes. A single explanation type cannot serve all stakeholders equally well. (Zhang et al., 2024)

4.3 System Architecture

Our proposed architecture for human-centered decision intelligence systems comprises five interconnected components. The perception module processes raw inputs, extracting relevant features and patterns. This component might employ deep neural networks for unstructured data like images and text, or structured data processing for tabular inputs. (Mumuni & Mumuni, 2025)

The reasoning module generates predictions, recommendations, or decisions based on processed inputs. This is where core machine learning models operate, whether neural networks, ensemble methods, or hybrid approaches. Importantly, the reasoning module should maintain information necessary for explanation generation, such as attention weights, feature importance, or activated concepts. (Mumuni & Mumuni, 2025)

The explanation module generates human-understandable accounts of system reasoning. Rather than a single explanation type, this module should produce multiple explanation modalities adapted to user needs: feature importance for analytical evaluation, example-based comparisons for intuitive understanding, counterfactuals for actionable guidance. (Zhang et al., 2024) The explanation module interfaces with the reasoning module to extract relevant information and with the dialogue module to present explanations appropriately.

The dialogue module manages interaction with users, presenting information and explanations through an interface suited to the task context. This includes visualizations for complex data, natural language for conversational interaction, and interactive elements enabling users to query the system and explore alternative scenarios. (Sreedharan et al., 2024) The dialogue module should support progressive disclosure, initially presenting essential information and revealing additional detail as users request it.

The monitoring module tracks system performance, user interactions, and explanation effectiveness over time. This enables continuous improvement through feedback loops. When users frequently override specific types of recommendations, this signals areas where explanations may be insufficient or model performance questionable. (Teso et al., 2023) The monitoring module feeds insights back to other components to support model refinement and explanation adaptation.

4.4 Explanation Generation Process

Generating effective explanations requires a structured process that transforms model internals into human-understandable communication. The first stage involves explanation planning, where the system determines what type of explanation to generate based on user characteristics, task context, and available. (Zhang et al., 2024) This planning considers user expertise, time pressure, decision stakes, and explanation purpose (e.g., justification, learning, debugging).

The second stage extracts relevant information from the reasoning module. For feature importance explanations, this might involve computing SHAP values or other attribution measures. For counterfactual explanations, this requires searching for minimal input modifications that change. (Guidotti et al., 2018) For example-based explanations, this involves retrieving similar cases from training data or learned prototypes. (Guidotti et al., 2018)

The third stage involves abstraction, translating low-level model information into concepts meaningful to users. Rather than reporting that pixel values in specific image regions influenced classification, abstraction identifies semantic concepts like "irregular border" or "unusual texture" that domain experts recognize. (Minh et al., 2022) This stage bridges the semantic gap between machine representations and human understanding.

The fourth stage addresses presentation, formatting explanations for effective communication. This includes choosing appropriate visualization types, structuring information hierarchically, and using language suited to the user's background. (Zhang et al., 2024) Visual explanations might employ saliency maps, bar charts, or interactive diagrams depending on the data type and user preferences. (Guidotti et al., 2018)

The final stage involves validation and calibration. Generated explanations should be faithful to model behavior rather than post-hoc rationalizations that diverge from actual decision-making. (Minh et al., 2022) This requires testing whether explanations accurately represent model reasoning and revising them when discrepancies emerge. Additionally, explanations should be calibrated to promote appropriate trust, neither understating nor overstating model reliability. (Kurniawan et al., 2023)

5. EXPLAINABILITY TECHNIQUES THROUGH A COGNITIVE LENS

5.1 Feature Attribution Methods

Feature attribution methods quantify how much each input feature contributed to a specific prediction, addressing the question "what mattered most?" Local Interpretable Model-agnostic Explanations (LIME) approximates complex models locally with interpretable linear models. For a given prediction, LIME perturbs the input, observes how predictions change, and fits a simple model to approximate behavior in that local region. This approach aligns with human reasoning about local causation and provides intuitive feature importance rankings.

SHapley Additive exPlanations (SHAP) grounds feature attribution in cooperative game theory, distributing prediction credit among features based on their contributions across all possible feature coalitions. SHAP values satisfy desirable mathematical properties including local accuracy, missingness, and consistency, providing theoretically sound attribution. (Minh et al., 2022)

From a cognitive perspective, feature attribution aligns with how humans explain decisions by identifying salient factors. When a doctor explains a diagnosis, they highlight key symptoms that drove their conclusion. However, human explanations often focus on a few critical factors rather than exhaustively listing all contributions. (Sreedharan et al., 2024) This suggests that feature attribution methods should prioritize presenting the most important features rather than overwhelming users with complete rankings.

Integrated Gradients represents another principled attribution method that satisfies implementation invariance and sensitivity properties. The technique computes gradients along a path from a baseline input to the actual input, accumulating attribution along this trajectory. This approach provides fine-grained attribution for deep neural networks and has been successfully applied in domains from natural language processing to genomics. (Minh et al., 2022)

Challenges remain in feature attribution methods. Different techniques can produce contradictory attributions for the same prediction, raising questions about which explanation to trust. (Minh et al., 2022) Attribution methods may also be vulnerable to manipulation, where models can be trained to produce misleading explanations while maintaining prediction accuracy. (Minh et al., 2022) These limitations underscore the importance of using multiple explanation approaches and validating explanation faithfulness.

5.2 Counterfactual and Contrastive Explanations

Counterfactual explanations answer "what would need to be different for a different outcome?" by identifying minimal input changes that alter predictions. This format mirrors human contrastive reasoning, where explanations often take the form "this rather than that because of X".(Sreedharan et al., 2024) When asked why a particular decision was made, humans typically explain by contrasting it with plausible alternatives rather than providing comprehensive causal accounts.

Diverse Counterfactual Explanations (DiCE) generates multiple counterfactual examples showing different ways to achieve a desired outcome. For a loan denial, DiCE might show that approval could be achieved by either increasing income by a specific amount, reducing debt, or improving credit score. This diversity respects that different individuals have different capacities to change various features and provides actionable guidance. (Minh et al., 2022)

From a decision intelligence perspective, counterfactual explanations directly support improvement and intervention planning. They transform explanations from passive justifications into active guidance for achieving desired outcomes. (Tripathi et al., 2025) However, generating realistic counterfactuals requires causal understanding to ensure proposed changes would actually produce the predicted outcome in practice. (Minh et al., 2022)

Contrastive explanations extend counterfactual reasoning by explaining why a prediction occurred instead of a specific alternative the user had in mind. (Guidotti et al., 2018) Rather than comparing against all possible alternatives, contrastive explanations focus on the specific foil the user cares about. A patient might ask not just "why this diagnosis?" but specifically "why this diagnosis rather than the one

I expected?" Answering this targeted question provides more relevant information than a general explanation.

5.3 Example-Based and Prototype Learning

Example-based explanations leverage human preference for reasoning by analogy and concrete cases. Rather than abstract feature importance scores, these methods present similar examples from training data that support a prediction. (Guidotti et al., 2018) When a diagnostic AI classifies a medical image, showing similar past cases provides intuitive justification that physicians can evaluate based on their clinical experience.

Influence functions identify training examples that most strongly influenced a specific prediction. By tracing predictions back to their training data roots, influence functions support debugging and bias detection. If a model's predictions consistently trace back to mislabeled or unrepresentative training examples, this signals data quality issues requiring attention. (Guidotti et al., 2018)

Prototype learning methods constrain models to make predictions by comparing inputs to learned prototypical examples of each class. (Mumuni & Mumuni, 2025) This architectural constraint builds interpretability directly into the model rather than adding it post-hoc. Prototypes serve as anchors for classification decisions, and prediction explanations can reference similarity to relevant prototypes. (Mumuni & Mumuni, 2025)

Case-based reasoning systems extend example-based approaches by adapting past solutions to new problems. (Guidotti et al., 2018) This approach particularly suits domains where expertise develops through experience with specific cases rather than abstract principles. Medical diagnosis, legal reasoning, and design often involve case-based reasoning where experts draw on their experience with similar past situations. (Balasubramanian, 2023)

Challenges in example-based explanation include ensuring retrieved examples are truly representative and relevant. Similarity metrics may not align with human perception of similarity, leading to confusing or misleading example sets. (Minh et al., 2022) Additionally, privacy concerns arise when showing actual training examples that might contain sensitive information. (Tripathi et al., 2025)

5.4 Attention Mechanisms and Saliency Maps

Attention mechanisms provide a window into what neural networks focus on during prediction. In computer vision, attention maps highlight image regions the model weighted most heavily. (Mumuni & Mumuni, 2025) In natural language processing, attention weights show which words or phrases most influenced outputs.

(Mumuni & Mumuni, 2025) These visualizations create intuitive explanations by showing where the model "looked" when making decisions.

Gradient-based saliency maps compute input gradients with respect to outputs, identifying which input changes would most impact predictions. Class Activation Mapping (CAM) and its variants like Grad-CAM produce coarse localization maps showing important regions for specific predictions. These techniques have been widely adopted in medical imaging, where highlighting diagnostically relevant image regions helps clinicians validate AI recommendations. (Balasubramanian, 2023)

From a neuroscientific perspective, attention mechanisms mirror biological selective attention processes that filter sensory information. (Minh et al., 2022) However, the correspondence is imperfect. Biological attention involves complex dynamics across brain regions and integrates top-down goals with bottom-up salience, while artificial attention typically implements simpler weighting functions. (Minh et al., 2022)

Recent research has questioned whether attention weights provide faithful explanations of model behavior. Attention weights may not uniquely determine predictions, and models can maintain similar predictions with substantially different attention patterns. (Minh et al., 2022) Additionally, attention mechanisms can be adversarially manipulated to produce misleading explanations. (Minh et al., 2022) These findings suggest attention visualizations should be interpreted cautiously and validated against other explanation methods.

5.5 Concept-Based Interpretability

Concept-based methods bridge the semantic gap between low-level features and high-level human understanding. Rather than explaining predictions through raw input features like pixel values, concept-based approaches identify human-meaningful concepts the model has learned. (Minh et al., 2022) In medical imaging, relevant concepts might include "irregular borders," "asymmetric shape," or "unusual texture" rather than specific pixel patterns.

Testing with Concept Activation Vectors (TCAV) quantifies how much a model's predictions depend on user-defined concepts. TCAV works by training linear classifiers to distinguish concept examples from random examples in the model's activation space, then measuring how much predictions change when activations are perturbed in the concept direction. This enables testing whether models use concepts humans consider relevant, supporting validation and debugging.

Automated concept discovery methods identify interpretable concepts without requiring manual specification. (Minh et al., 2022) These techniques cluster activation patterns and attempt to characterize clusters with human-understandable descriptions. Network dissection systematically probes individual neurons to discover what visual concepts they respond to. (Mumuni & Mumuni, 2025) Such automated discovery can reveal unexpected concepts models have learned, including spurious correlations that undermine reliability. (Minh et al., 2022)

Concept-based explanations align with hierarchical processing in both biological and artificial neural networks. Explanations at different abstraction levels serve different purposes: low-level features for detailed technical analysis, mid-level concepts for domain expert validation, high-level concepts for communication with non-.(Mumuni & Mumuni, 2025) Effective explanation systems should provide access to multiple abstraction levels based on user needs.

6. EVALUATION OF EXPLAINABLE AI SYSTEMS

6.1 Technical Evaluation Metrics

Evaluating explainability requires metrics beyond traditional ML performance measures. Faithfulness measures whether explanations accurately represent model reasoning rather than providing misleading post-hoc rationalizations. (Minh et al., 2022) One approach involves removing or perturbing features identified as important and verifying that predictions change as the explanation suggests. If removing a supposedly critical feature has minimal impact, the explanation lacks faithfulness.

Stability or robustness assesses whether explanations remain consistent for similar inputs. (Minh et al., 2022) Unstable explanations that change dramatically for minor input perturbations undermine user trust and suggest the explanation method is unreliable. However, some instability may be appropriate when inputs cross decision boundaries, so stability must be evaluated in context.

Consistency measures agreement between different explanation methods applied to the same model. (Minh et al., 2022) While different explanation types naturally emphasize different aspects of model behavior, dramatic inconsistencies signal potential problems. For instance, if feature attribution methods and counterfactual explanations suggest completely different factors drive predictions, this warrants investigation.

Computational efficiency matters for real-time decision support. Some explanation methods require expensive optimization or repeated model evaluations, making them impractical for time-sensitive applications. (Minh et al., 2022) Efficient approximations that sacrifice some accuracy for speed may be preferable in such contexts, highlighting trade-offs between explanation quality and practical constraints.

6.2 Human-Centered Evaluation

Technical metrics alone cannot assess whether explanations serve human needs. Human-centered evaluation requires user studies examining how explanations affect understanding, trust, and decision quality. (Zhang et al., 2024) Comprehensibility measures whether users can understand explanations, often assessed through questions testing explanation content or by asking users to predict model behavior after viewing explanations. (Lai et al., 2023)

Trust calibration evaluates whether explanations help users develop appropriate trust, relying on AI when it performs well and questioning it when it errs. (Alufaisan et al., 2020) Both overtrust and undertrust are problematic: overtrust leads users to accept incorrect recommendations uncritically, while undertrust prevents beneficial AI adoption. Effective explanations should promote trust proportional to system reliability.

Decision quality ultimately matters most for decision intelligence applications. Do explanations help users make better decisions? This can be evaluated by comparing decision outcomes with and without AI assistance, and with different explanation types. (Alufaisan et al., 2020) However, isolating explanation effects from overall system design presents methodological challenges, as explanations interact with interface design, task context, and user characteristics.

User satisfaction and perceived utility provide subjective measures of explanation value. (Zhang et al., 2024) While users may appreciate explanations without necessarily making better decisions, or vice versa, satisfaction matters for adoption and continued use. Qualitative feedback through interviews and think-aloud protocols can reveal explanation strengths and weaknesses not captured by quantitative metrics. (Zhang et al., 2024)

6.3 Task-Specific Evaluation

Explanation needs vary across tasks and domains, necessitating task-specific evaluation criteria. For medical diagnosis support, appropriate metrics might include whether explanations help physicians identify diagnostic errors, whether they support learning by residents, and whether they satisfy medicolegal documentation requirements. (Balasubramanian, 2023) For financial lending, relevant criteria include whether explanations satisfy regulatory disclosure requirements and whether they enable applicants to improve their creditworthiness. (Tripathi et al., 2025)

Debugging and model improvement tasks require explanations that reveal failure modes and bias sources. (Guidotti et al., 2018) Evaluation should measure whether explanations successfully identify mislabeled training data, spurious correlations, and distribution shift. Teaching and knowledge transfer applications need explanations that support learning transferable insights rather than memorizing specific cases. (Balasubramanian, 2023)

Contested decisions like loan denials or hiring recommendations require explanations that enable challenge and appeal. (Tripathi et al., 2025) Evaluation criteria include whether affected individuals can understand decisions well enough to contest them and whether explanations provide actionable guidance for improvement. Scientific discovery applications demand explanations that generate testable hypotheses and novel insights. (Minh et al., 2022)

6.4 Longitudinal and Ecological Evaluation

Most XAI evaluation studies involve short-term laboratory experiments, but real-world deployment requires understanding long-term effects. Longitudinal studies examine how explanation effectiveness evolves as users gain experience with systems. (Alufaisan et al., 2020) Novice users may require detailed explanations that become unnecessary as they develop intuition. Conversely, initial explanation helpfulness may diminish if users develop overtrust and stop critically evaluating recommendations.

Ecological validity demands evaluating systems in realistic usage contexts rather than artificial laboratory tasks. (Zhang et al., 2024) Real-world constraints like time pressure, multitasking demands, and organizational context influence how explanations are used. Field deployments reveal usage patterns and challenges that laboratory studies miss, though they sacrifice experimental control.

Organizational-level evaluation considers system impact beyond individual users. (Balasubramanian, 2023) Does explainability facilitate stakeholder buy-in and smooth implementation? Does it support knowledge sharing across teams? Does it enable compliance with governance requirements? These broader questions matter for successful deployment but require evaluation methods beyond standard user studies.

7. CASE STUDIES AND APPLICATIONS

7.1 Healthcare: AI-Assisted Medical Diagnosis

Medical diagnosis represents a critical application domain for explainable AI, where incorrect decisions have serious consequences and physicians require understanding to responsibly adopt AI tools. Deep learning models have achieved impressive performance in radiology, dermatology, pathology, and other specialties.

(Balasubramanian, 2023) However, physician adoption has been slower than technical capability would suggest, largely due to opacity concerns. (Kurniawan et al., 2023)

A representative case study involves AI-assisted chest X-ray interpretation for detecting pneumonia, lung nodules, and other abnormalities. (Balasubramanian, 2023) The system employs a deep convolutional neural network trained on hundreds of thousands of radiographs. To provide interpretability, the system generates multiple explanation types: attention maps highlighting diagnostically relevant image regions, similar historical cases with confirmed diagnoses, textual descriptions of observed abnormalities, and confidence scores indicating prediction uncertainty.

Attention maps align with how radiologists naturally examine images, directing attention to concerning areas. However, clinician feedback revealed that attention maps alone were insufficient. Radiologists wanted semantic descriptions of what abnormalities the AI detected rather than just spatial localization. (Balasubramanian, 2023) This motivated integrating concept-based explanations identifying specific abnormality features like “irregular opacity” or “ground-glass pattern” that radiologists recognize.

Example-based explanations proved particularly valuable for rare conditions where radiologists had limited personal experience. Seeing similar confirmed cases helped build confidence in AI recommendations and supported learning. (Balasubramanian, 2023) However, privacy concerns required careful anonymization of historical cases. The system also communicated uncertainty explicitly, distinguishing cases where the model was highly confident from ambiguous cases requiring additional testing.

Evaluation through a clinical trial showed that AI assistance with these explanations improved diagnostic accuracy, particularly for less experienced radiologists. (Balasubramanian, 2023) Importantly, explanations helped calibrate trust: radiologists were more likely to accept correct AI recommendations and question incorrect ones compared to using AI without explanations. Qualitative feedback indicated that explanations made radiologists feel more comfortable adopting the technology and better able to document their reasoning in medical records.

7.2 Finance: Credit Risk Assessment and Loan Decisions

Financial services face both strong incentives to adopt AI for improved risk assessment and strict regulatory requirements for transparency. Credit scoring determines access to financial resources that profoundly affect people's lives, making explainability both ethically necessary and legally mandated. (Tripathi et al., 2025) Traditional credit scoring relied on interpretable models like logistic regression, but modern approaches employ more complex methods that offer better predictive performance at the cost of interpretability. (Guidotti et al., 2018)

A case study of a large bank's credit decisioning system illustrates how explainable AI can balance performance with regulatory compliance. (Tripathi et al., 2025) The bank deployed a gradient boosting model that significantly outperformed their legacy system but faced challenges explaining decisions to loan officers, applicants, and regulators. They implemented a multi-faceted explainability approach combining several techniques.

Feature attribution using SHAP values identified the most important factors influencing each credit decision. (Guidotti et al., 2018) For approved applications, this provided documentation of favorable factors. For denials, it satisfied regulatory adverse action notice requirements by identifying primary reasons. However, raw SHAP values required translation into language understandable to non-technical applicants: “debt-to-income ratio” rather than “feature 17.”

Counterfactual explanations addressed applicant questions about how to improve their creditworthiness. (Tripathi et al., 2025) The system generated actionable guidance like “Reducing your outstanding credit card debt by $3,000 or increasing your annual income by $5,000 would likely result in approval.” These counterfactuals were constrained to realistic, actionable changes rather than suggesting impossible modifications like changing age or past payment history.

The system also provided loan officers with case-based explanations showing similar historical applications and their outcomes. (Guidotti et al., 2018) This helped officers spot potential data errors or unusual circumstances not captured by model features. For regulatory compliance, the bank maintained detailed logs of model decisions and explanations, enabling auditors to verify decisions followed established policies without illegal discrimination.

Evaluation revealed nuanced outcomes. While the explainable AI system satisfied regulatory requirements and improved applicant satisfaction with the transparency of decisions, some loan officers initially resisted the new system. (Balasubramanian, 2023) They perceived it as constraining their professional judgment, highlighting the importance of framing AI as decision support rather than replacement. Training programs emphasizing that officers retained final authority while gaining powerful analytical tools improved acceptance.

7.3 Autonomous Systems: Explainable Robot Decision-Making

Autonomous robots operating in human environments must make real-time decisions involving safety trade-offs, yet their deep reinforcement learning controllers typically function as black boxes. (Sreedharan et al., 2024) This opacity creates barriers to public acceptance, complicates safety certification, and hinders debugging when robots behave unexpectedly. (Sreedharan et al., 2024)

A case study of an autonomous delivery robot navigating sidewalks illustrates brain-inspired explainability approaches. (Sreedharan et al., 2024) The robot uses deep reinforcement learning for navigation, making real-time decisions about path planning, obstacle avoidance, and pedestrian interaction. To provide interpretability, researchers implemented several explanation mechanisms operating at different timescales.

For real-time transparency, the robot displays visual indicators of its current intentions: directional arrows showing planned path, color-coded hazard awareness indicating detected obstacles, and speed indicators. (Sreedharan et al., 2024) These simple signals enable pedestrians to predict robot behavior and coordinate shared space usage without requiring technical understanding. This mimics how humans use body language and eye contact for implicit communication during navigation.

For post-hoc analysis after unexpected behaviors or near-misses, the system generates detailed explanations of decision sequences leading to events. (Sreedharan et al., 2024) Attention mechanisms highlight which sensor inputs most influenced decisions at critical moments. Counterfactual analysis identifies how different sensor readings or environmental conditions would have altered behavior. This supports debugging and helps engineers refine controller policies.

For interaction with non-technical users, the robot can provide natural language explanations through a mobile app: “I stopped because I detected a person entering my path from the left” or “I chose this route because it avoids the crowded area ahead”.(Sreedharan et al., 2024) These explanations balance technical accuracy with accessibility, translating sensor data and policy decisions into everyday language.

The system also learns from human feedback, implementing a form of interactive machine learning where users can query decisions and provide corrections. (Sreedharan et al., 2024) When users indicate that a robot decision was inappropriate, the system logs this feedback along with context to identify systematic issues. This creates a feedback loop supporting continuous improvement grounded in human values and expectations.

Evaluation involved both technical testing and public deployment. Technical metrics showed that explanation generation added minimal computational overhead, maintaining real-time performance. (Sreedharan et al., 2024) Public feedback indicated that visible intent signaling and explanation capabilities increased comfort with robot presence and reduced navigation conflicts. Safety analysis revealed that explanation mechanisms facilitated faster identification and resolution of edge cases where the robot behaved suboptimally.

8. CHALLENGES AND LIMITATIONS

8.1 The Interpretability-Accuracy Trade-Off

A persistent challenge in explainable AI is the apparent tension between model complexity and interpretability. Simpler models offer transparency but may lack the capacity to capture intricate patterns in complex data. (Minh et al., 2022) Deep neural networks achieve superior performance on many tasks but resist straightforward interpretation. This trade-off creates difficult choices, particularly in high-stakes domains where both accuracy and interpretability matter.

However, characterizing this as a fundamental trade-off may be overly pessimistic. Recent research demonstrates that carefully designed architectures can achieve competitive performance while maintaining interpretability. (Mumuni & Mumuni, 2025) Attention mechanisms, sparse models, and modular architectures incorporate interpretability constraints without necessarily sacrificing accuracy. The key is building interpretability into system design from the outset rather than adding it post-hoc. (Mumuni & Mumuni, 2025)

Moreover, the appropriate balance between interpretability and accuracy depends on application context. For low-stakes decisions like product recommendations, prioritizing accuracy may be reasonable. For high-stakes decisions affecting human rights and wellbeing, interpretability requirements appropriately constrain model choices even at some accuracy cost. (Tripathi et al., 2025) Framing this as a principled context-dependent choice rather than an unfortunate technical limitation helps navigate trade-offs.

8.2 Explanation Faithfulness and Manipulation

A critical concern is whether explanations faithfully represent model reasoning or provide misleading post-hoc rationalizations. Some explanation methods identify features correlated with predictions without establishing causal relationships. (Minh et al., 2022) Models can exploit spurious correlations while explanations highlight seemingly sensible features, creating false confidence in model robustness.

More concerning is that models can be explicitly trained to produce misleading explanations while maintaining prediction accuracy. (Minh et al., 2022) An image classifier might be trained to highlight contextually sensible image regions through attention mechanisms while actually relying on imperceptible adversarial perturbations for classification. Such "explanation laundering" undermines the trust-building purpose of XAI. (Minh et al., 2022)

Addressing these concerns requires multiple strategies. Using diverse explanation types helps detect inconsistencies that signal faithfulness problems. (Minh et al., 2022) Formal verification methods can provide guarantees about explanation properties, though at computational cost. (Minh et al., 2022) Adversarial testing that deliberately attempts to fool explanation methods reveals vulnerabilities before deployment. (Minh et al., 2022)

8.3 Cognitive Limitations and Information Overload

While explanations aim to help humans understand AI systems, they must respect cognitive limitations to be effective. Human working memory capacity limits how much information people can process simultaneously. (Alufaisan et al., 2020) Overly complex explanations overwhelm users, defeating their purpose. Yet oversimplified explanations may omit critical information needed for sound judgment.

This tension between completeness and parsimony has no universal solution. Appropriate explanation complexity depends on user expertise, available processing time, and decision stakes. (Zhang et al., 2024) Progressive disclosure interfaces that present essential information initially while making additional detail available on demand offer one approach. (Zhang et al., 2024)

Cognitive biases also affect how humans interpret explanations. Confirmation bias may lead users to accept explanations that align with preexisting beliefs while questioning contradictory explanations. (Sreedharan et al., 2024) Availability heuristics might cause vivid explanation examples to unduly influence trust. Anchoring effects mean initial explanations shape interpretation of subsequent information. Effective explanation design must account for these systematic deviations from ideal rational reasoning. (Zhang et al., 2024)

8.4 Standardization and Evaluation Gaps

The XAI field lacks standardized evaluation methodologies, making it difficult to compare approaches or assess progress. (Balasubramanian, 2023) Different studies use different metrics, datasets, and user populations, limiting generalizability. Technical metrics like faithfulness have multiple proposed definitions that may not correlate with each other. (Minh et al., 2022)

Human-subject studies face reproducibility challenges common to behavioral research. Small sample sizes, heterogeneous participant populations, and task-specific findings limit generalization. Lab studies may not reflect real-world usage where time pressure, expertise differences, and organizational context influence explanation effectiveness. Longitudinal field studies are expensive and rare despite their ecological validity. (Zhang et al., 2024)

Developing standardized benchmarks, common evaluation protocols, and shared datasets would advance the field. (Balasubramanian, 2023) However, explanation needs vary across domains and tasks, making universal standards challenging. Perhaps the solution lies in developing domain-specific evaluation frameworks that share methodological principles while adapting metrics to context. (Tripathi et al., 2025)

8.5 Sociotechnical and Ethical Considerations

Explainability intersects with broader questions of AI ethics, fairness, and accountability. Explanations might reveal that models rely on protected attributes like race or gender, either directly or through proxies. (Tripathi et al., 2025) While such transparency is necessary to detect bias, it creates uncomfortable questions about whether biased but explainable models are preferable to opaque but seemingly unbiased ones. (Tripathi et al., 2025)

Explainability might also create false sense of accountability. If organizations can provide explanations for algorithmic decisions, this might satisfy legal requirements without genuinely ensuring decisions are fair and appropriate. (Tripathi et al., 2025) Meaningful accountability requires not just ex-post explanations but genuine understanding and oversight throughout development and deployment. (Tripathi et al., 2025)

Power asymmetries affect who demands and who provides explanations. Institutions deploying AI typically have more resources and expertise than affected individuals, creating imbalances in explanation quality and comprehensibility. (Tripathi et al., 2025) Explanations targeted at technical auditors may be inaccessible to affected laypersons, while simplified explanations for the public may omit details necessary for genuine oversight. (Tripathi et al., 2025)

8.6 Practical Implications for Design and Deployment

Beyond theoretical contributions, explainable and interpretable decision intelligence carries significant practical implications for the design, deployment, and governance of AI systems in real-world environments. Organizations adopting AI-driven decision support tools must recognize that explainability is not solely a technical feature, but a socio-technical requirement that influences usability, trust, accountability, and long-term sustainability.

From a system design perspective, explainability should be integrated from the earliest stages of development. Retrofitting explanations onto fully developed black-box systems often leads to superficial or misleading interpretations that fail to reflect true model behavior. Explainability-by-design approaches encourage developers to select architectures, features, and training objectives that naturally support. (Mumuni & Mumuni, 2025) This includes modular architectures, constrained optimization objectives, and hybrid symbolic–neural designs.

In deployment contexts, explainable decision intelligence supports more effective human-AI collaboration. Users are more likely to adopt AI systems when explanations align with their domain expertise and decision-making workflows. For example, in clinical environments, explanations that mirror diagnostic reasoning processes enhance acceptance and reduce resistance to automation. (Balasubramanian, 2023) Similarly, in organizational decision-making, transparent AI systems facilitate cross-functional communication between technical teams, management, and regulators.

Explainability also plays a critical role in post-deployment monitoring and governance. Transparent systems enable organizations to audit decisions, detect bias, and respond to errors more effectively. When unexpected outcomes occur, interpretable models and faithful explanations accelerate root-cause analysis and corrective action. This capability is essential for maintaining accountability and complying with evolving regulatory. (Tripathi et al., 2025)

8.7 Educational and Knowledge Transfer Implications

Explainable decision intelligence has important implications for education and professional training. Rather than treating AI systems as opaque tools, interpretable models can function as learning aids that expose underlying patterns, causal relationships, and decision logic. This is particularly valuable in domains where expertise development traditionally relies on apprenticeship and experiential learning.

In medical education, explainable AI systems can support residents and trainees by explicitly highlighting diagnostic cues, reasoning pathways, and uncertainty factors. (Balasubramanian, 2023) In finance and policy analysis, transparent decision intelligence systems can help students understand complex trade-offs and systemic interactions that are difficult to grasp through static textbooks alone.

Moreover, explainable AI can facilitate interdisciplinary knowledge transfer. Decision intelligence often sits at the intersection of data science, domain expertise, and organizational strategy. Explanations serve as a shared language that bridges these communities, enabling more effective collaboration and mutual understanding. (Zhang et al., 2024)

8.8 Limitations of Brain-Inspired Analogies

While brain-inspired mechanisms provide valuable guidance for explainable AI, caution is warranted when drawing analogies between biological cognition and artificial systems. Neural networks are, at best, highly simplified abstractions of biological neural processes. Overstating the similarity risks misleading interpretations and unjustified claims of cognitive plausibility. (Mumuni & Mumuni, 2025)

Biological cognition is shaped by embodiment, emotion, social interaction, and developmental learning, factors that most AI systems do not fully capture. Consequently, explanations that appear cognitively aligned may still diverge from genuine human reasoning in important ways. Recognizing these limitations helps prevent anthropomorphism and supports more rigorous evaluation of explanation effectiveness. (Tripathi et al., 2025)

9. FUTURE DIRECTIONS AND RESEARCH OPPORTUNITIES

9.1 Neurosymbolic Integration

Neurosymbolic AI combines neural networks' pattern recognition with symbolic reasoning's interpretability, representing a promising direction for human-centered decision intelligence. (Mumuni & Mumuni, 2025) Rather than treating neural and symbolic components as separate modules, future research should explore deeper integration where symbolic knowledge guides neural learning and neural representations inform symbolic reasoning. (Mumuni & Mumuni, 2025)

Differentiable logic and reasoning systems enable end-to-end learning while maintaining logical structure. (Mumuni & Mumuni, 2025) Incorporating common-sense knowledge graphs into neural models can ground learning in structured world knowledge. (Mumuni & Mumuni, 2025) Learning symbolic rules directly from data while preserving human interpretability remains an open challenge. Recent approaches explore extracting logical rules from trained neural networks or jointly learning neural representations and symbolic constraints. (Mumuni & Mumuni, 2025) These methods aim to combine the flexibility of data-driven learning with the transparency of rule-based reasoning.

In decision intelligence contexts, neurosymbolic systems offer particular promise. Symbolic components can encode domain knowledge, regulations, and ethical constraints, while neural components handle perception and pattern recognition from noisy real-world data. This division of labor aligns with human cognition, where intuitive perception is complemented by explicit reasoning rules. (Sreedharan et al., 2024) Future research should explore scalable architectures that maintain interpretability while handling complex, high-dimensional inputs.

9.2 Interactive and Conversational Explanations

Static explanations may be insufficient for complex decisions. Humans naturally ask follow-up questions, challenge assumptions, and explore alternatives when reasoning about decisions. Interactive and conversational explanation systems enable users to engage in dialogue with AI, refining explanations based on evolving information needs. (Sreedharan et al., 2024)

Conversational explanations allow users to ask "why," "why not," and "what if" questions, supporting deeper understanding and sense-making. For example, a decision-maker might ask why a particular option was recommended over another, or how changes in assumptions would affect outcomes. These interactions reflect natural human explanatory practices and align well with decision intelligence objectives. (Sreedharan et al., 2024)

Research challenges include maintaining explanation faithfulness across multiple conversational turns, managing user misconceptions, and balancing responsiveness with computational constraints. Integrating large language models with structured decision intelligence systems presents both opportunities and risks, particularly regarding hallucination and overconfidence. (Mumuni & Mumuni, 2025)

9.3 Learning from Human Feedback

Human feedback offers a powerful signal for improving both model performance and explanation quality. Interactive machine learning approaches incorporate user corrections, preferences, and critiques to refine system behavior over time. (Teso et al., 2023) In explainable AI, feedback can reveal whether explanations are misleading, incomplete, or misaligned with user mental models.

Preference learning techniques can be used to optimize explanation selection, prioritizing explanation types that users find most helpful in specific. (Lai et al., 2023) Reinforcement learning from human feedback (RLHF) offers another pathway for aligning AI behavior with human values and expectations, though applying it reliably to decision intelligence remains an active research area. (Sreedharan et al., 2024)

9.4 Standardization, Benchmarks, and Governance

Progress in explainable decision intelligence will benefit from shared benchmarks, standardized evaluation protocols, and clearer governance frameworks. Current XAI benchmarks often focus on narrow tasks or synthetic datasets that fail to capture real-world decision complexity. (Balasubramanian, 2023) Developing benchmarks grounded in realistic decision-making scenarios would enable more meaningful comparison across approaches.

Governance frameworks should define when and how explanations are required, who the intended audience is, and what level of detail is appropriate. Regulatory initiatives such as the EU AI Act highlight explainability as a core requirement but leave many practical questions unanswered. (Tripathi et al., 2025) Bridging the gap between technical research and policy implementation remains a critical interdisciplinary challenge.

9.5 Toward Truly Human-Centered Decision Intelligence

Ultimately, the goal of explainable decision intelligence is not merely to open black boxes, but to support better human decisions. This requires shifting research emphasis from model-centric interpretability toward outcome-oriented human-centered design. Explanations should be evaluated based on whether they improve understanding, support learning, enable accountability, and foster appropriate. (Tripathi et al., 2025)

Future systems should treat explainability as a dynamic, context-sensitive process rather than a static feature. By integrating insights from neuroscience, cognitive science, human-computer interaction, and ethics, decision intelligence systems can evolve toward genuinely collaborative human-AI partnerships.

10. CONCLUSION

As artificial intelligence systems increasingly influence high-stakes decisions across healthcare, finance, policy, and autonomous systems, the need for transparency and interpretability has become both a technical and societal imperative. This chapter has argued that explainability should not be treated as an afterthought or mere compliance requirement, but as a core component of human-centered decision intelligence.

By drawing inspiration from brain-inspired mechanisms and cognitive principles, we have shown how explainable AI can be designed to align more closely with natural human reasoning processes. Attention mechanisms, case-based reasoning, counterfactual explanations, and concept-based interpretability each reflect aspects of how humans perceive, remember, and explain decisions. When integrated thoughtfully, these techniques can bridge the gap between powerful machine learning models and human understanding.

We presented a comprehensive framework for human-centered decision intelligence that emphasizes cognitive alignment, contextual adaptation, uncertainty communication, and value-sensitive design. Through case studies in healthcare, finance, and autonomous systems, we demonstrated how explainable AI can enhance trust, improve decision quality, and support effective human-AI collaboration in real-world settings.

Despite significant progress, substantial challenges remain. Ensuring explanation faithfulness, avoiding information overload, standardizing evaluation, and addressing sociotechnical and ethical concerns require continued interdisciplinary research. Future directions such as neurosymbolic integration, interactive explanations, and learning from human feedback offer promising pathways toward more transparent and trustworthy AI systems.

In conclusion, explainable and interpretable decision intelligence represents a crucial step toward AI systems that do not merely optimize predictions, but genuinely support human judgment. By grounding AI explanations in cognitive and neuroscientific insights, we can move closer to systems that are not only intelligent, but also understandable, accountable, and aligned with human values.

REFERENCES

Alufaisan, Y., Marusich, L. R., Bakdash, J. Z., Zhou, Y., & Kantarcioglu, M. (2020). Does explainable artificial intelligence improve human decision-making? *Proceedings of the Human Factors and Ergonomics Society Annual Meeting*, *64*(1), 137–141. DOI: 10.1177/1071181320641025

Balasubramanian, S. (2023). Literature review on developing explainable artificial intelligence for trustworthy decision-making. *International Journal of Design and Manufacturing Technology*, *14*(1), 1–10.

Doshi-Velez, F., & Kim, B. (2017). Towards a rigorous science of interpretable machine learning. *arXiv preprint arXiv:1702.08608*. /arXiv.1702.08608DOI: 10.48550

Guidotti, R., Monreale, A., Ruggieri, S., Turini, F., Giannotti, F., & Pedreschi, D. (2018). A survey of methods for explaining black box models. *ACM Computing Surveys*, *51*(5), 93. Advance online publication. DOI: 10.1145/3236009

Kurniawan, D., Triyanto, D., Wahyudi, M., & Pujiastuti, L. (2023). Explainable artificial intelligence (XAI) for trustworthy decision-making. *Jurnal Teknik Informatika C.I.T Medicom, 15*(2), 45–54.

Lai, V., Zhang, Y., Chen, C., Liao, Q. V., & Tan, C. (2023). Selective explanations Leveraging human input to align explainable AI. *arXiv preprint arXiv:2301.09656* /arXiv.2301.09656DOI: 10.48550

Minh, D., Wang, H. X., Li, Y. F., & Nguyen, T. N. (2022). Explainable artificial intelligence: A comprehensive review. *Artificial Intelligence Review*, *55*(5), 3503–3568. DOI: 10.1007/s10462-021-10088-y

Mumuni, F., & Mumuni, A. (2025). Explainable artificial intelligence (XAI): From inherent explainability to large language models. *arXiv preprint arXiv:2501.09967*. /arXiv.2501.09967DOI: 10.48550

Sreedharan, S., Kulkarni, A., & Kambhampati, S. (2024). Explainable human-AI interaction: A planning perspective. *arXiv preprint arXiv:2405.15804*. /arXiv.2405.15804DOI: 10.48550

Teso, S., Alkan, Ö., Stammer, W., & Daly, E. (2023). Leveraging explanations in interactive machine learning: An overview. *Frontiers in Artificial Intelligence*, *6*, 113–129. DOI: 10.3389/frai.2023.1066049 PMID: 36909207

Tripathi, D. R., Urkude, A. S., & Parkhi, V. P. (2025). Explainable artificial intelligence (XAI): Interpretable models for ethical decision-making. *International Journal for Research in Applied Science and Engineering Technology*, *13*(1), 102–110. DOI: 10.22214/ijraset.2025.74356

Zhang, Y., Liao, Q. V., & Bellamy, R. K. E. (2024). From explainable to interactive AI: A literature review on current trends in human-AI interaction. *International Journal of Human-Computer Studies*, *182*, 103140. DOI: 10.1016/j.ijhcs.2023.103140

Chapter 9
From Algorithms to Cognition:
Brain-Inspired Mechanisms for AI-Enhanced Decision Intelligence

Salam Al E'mari
http://orcid.org/0000-0002-2134-4158
University of Petra, Jordan

Yousef Sanjalawe
http://orcid.org/0000-0002-4442-1865
University of Jordan, Jordan

Fadi Fataftah
http://orcid.org/0000-0002-3299-0972
University Canada West, Canada

ABSTRACT

Artificial intelligence systems are increasingly used for complex decision-making in domains characterized by uncertainty, scale, and dynamic change. While traditional algorithm-based approaches have achieved strong performance in data-driven tasks, they remain limited in flexibility, contextual awareness, and interpretability when compared with human cognition. At the same time, advances in neuroscience and cognitive science have clarified the mechanisms of perception, learning, memory, attention, and reasoning that underlie human decision-making. This chapter examines how such cognitive principles can be translated into computational models that bridge statistical learning and cognitive reasoning. Drawing on cognitive computing and neuro-symbolic AI, it proposes an integrative framework for AI-supported decision

DOI: 10.4018/979-8-3373-8988-2.ch009

intelligence that combines data-driven learning with brain-inspired architectures, emphasizing transparency, trust, and cognitive responsibility in the design of interpretable and context-sensitive intelligent systems.

1. INTRODUCTION: THE COGNITIVE TURN IN DECISION INTELLIGENCE

Today, the landscape of decision-making is undergoing a revolution driven by this new paradigm. For the past few decades, the "intelligent" benchmark for systems has been based on brute processing power: the capability to analyze an enormous number of data points at a speed that would have been impossible for human cognitive ability. But with the migration of Artificial Intelligence from the well-ordered environment of the lab into the unpredictable world of medicine, finance, and automated infrastructure, the shortcomings of such statistics-based methodologies have made themselves comically obvious. This is what I term the "Cognitive Turn", the recognition that true decision intelligence must utilize the more sophisticated methods performed by the human brain.

In contemporary society, AI is no longer merely an adjunct decision-maker and is now an integral part of high-stakes decision-making cycles. In the medical industry, AI interprets medical images to suggest treatment strategies; in the logistics industry, AI controls worldwide supply chains in real time; and in public policy, the trend is towards the use of AI in modeling economic outcomes. According to recent studies on the matter, the implementation of AI in these areas is due to the demand for "Decision Intelligence," which is the "use of machine learning for improving action outcome.

Nevertheless, the "intelligence" in such systems may be fragile. For example, although the intelligence in one such system may be better than the human intelligence associated with the identification of a specific pathology from an X-ray, the former may not possess the cognitive ability to grasp the socio-economic background of the patient, which would, in turn, be assessed by the doctor on an intuitive basis.

The "Connectionist" or deep learning approach primarily thrives on the idea that the world is a stationary place, i.e., the future will look statistically similar to the past. This gives way to some fundamental errors in complex decision-making spaces:

1. Contextual Blindness: Today's models have the problem of "out-of-distribution" data. If the environment suddenly changes (for instance, market crash or a new pandemic outbreak), statistical models will not work because they do not have a conceptual understanding of why things happen.

2. The Black Box Problem: Very often, the decisions reached by deep neural networks cannot be explained. In critical settings, "the algorithm said so" is clearly not a satisfactory answer to a problem of great significance (although it is a more recent topic in terms of cognitive alignment, as addressed in (Rai, 2020).
3. Data Inefficiency: Although it takes a human brain only one example to learn a rule for decision making (few-shot learning), it requires millions of data points for AI to perform equally well.

In their pioneering research on neuroscience-inspired artificial intelligence, Hassabis and other researchers pointed out that the absence of "common sense" as well as causal knowledge in computer models is a serious challenge that has continued to hinder fully realized autonomy (Whittington & Bogacz, 2022).

However, the brain remains the existence proof of a system that performs complex decision-making in the presence of high uncertainty, operating with very low power dissipation (around 20 watts). A promising avenue for developing more robust AI formulations might be to explore neuroscience. For instance, we might benefit from «brain-inspired» concepts, which imply adopting architecture principles that make the human brain so robust.

For example, one such principle under research is the Global Workspace Theory (GWT), in which specialized brain modules "broadcast" information to the workspace, enabling decisions to be made. Such principles could be integrated into AI systems, allowing the different data streams, such as inputs, memory, and logic, to work in harmony.

Moreover, a shift has been observed in Neuro-symbolic AI, aimed at connecting the capabilities of neural networks with pattern recognition and classical AI with symbolic processing, reflecting human capabilities that can merge intuition and logical processing of information. According to an argument made by the authors of a 2020 publication, followed by a new model introduced in 2021, this needs to be achieved because they explain that systems should be able to explain their decisions that align with human reasoning.

Ultimately, the need for this turn in our thinking is not purely academic. It is a necessity that corresponds to a future in which we increasingly delegate more consequential decisions to computers. These computers increasingly require the same adaptability, transparency, and ethical foundations that are hallmarks of human wisdom. The remainder of this chapter suggests that the path from statistical calculation to true decision-making power lies in translating cognitive principles into computational models. Figure 1 illustrates the conceptual transition from traditional statistical AI systems toward cognitively inspired decision intelligence frameworks.

Figure 1. From statistical AI to cognitive decision intelligence.

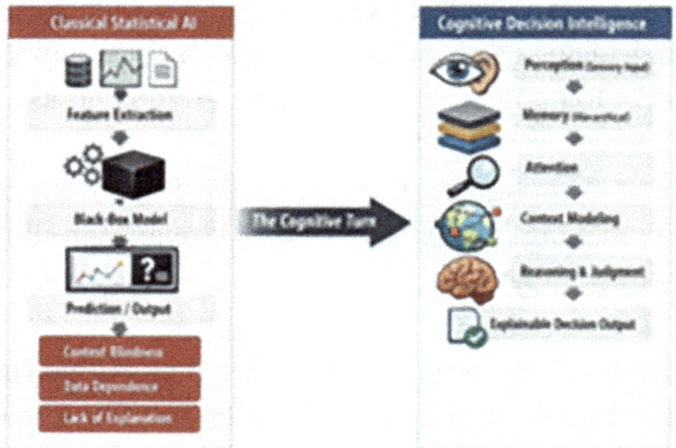

2. CLASSICAL EPISTEMOLOGIES AND THE PROBLEM OF CERTAINTY

Traditional epistemologies, founded on the ambition to develop knowledge through justified true beliefs, have always contributed to the methodological bases for the social sciences. Also, representation, evidence, and the ideal of objective correspondence between belief and reality have always been intertwined with interpretive authority in traditional approaches. Nevertheless, the situations under which the certainty of knowledge claims in the social sciences have always been full of complex elements. This section will investigate how traditional approaches to representation and evidence have implied interpretive authority, the place of uncertainty in scientific practice, and the limitations of deterministic and simple probabilistic approaches to representing the complexity of knowledge.

Classical approaches to knowledge in epistemology have frequently been adopted from a philosophical tradition that stressed justified true belief or similar approaches to knowledge. Within this tradition, evidence serves as a foundation for belief, and it mediates the researcher's ability to interpret observations in theoretical terms. Within this approach, evidence is presumed to represent a feature of a stable and real-world social fact, and accordingly, the function of interpretive authority within this tradition is closely tied to scientific expertise.

However, recent literature has demonstrated the limitations of the evidence-to-representations paradigm for the production of knowledge. The literature on social epistemology has broadened by exploring the roles of epistemic trust, authority, and social structures in characterizing knowledge as believable within the scientific community (Sinaga, Manugeren, & Barus, 2025). In a systematic review of the literature on social epistemology, the authors have established that scientific knowledge cannot be determined solely by observational means but rather by the collective practice and power associated with consensus and interpretation within the scientific community.

In addition, it draws attention to epistemic injustice in the form of exclusionary or marginal knowledge production, which is overlooked by traditional epistemologies. This occurs when the means of knowledge production marginalize alternative knowledge epistemologies, such as those grounded in underrepresented communities within knowledge discourses (Sinaga et al., 2025).

In contrast to the simplistic narrative that treats the treatment of uncertainty as a temporary methodological question, history reveals that uncertainty has played a significant role in epistemology. Right from the earliest forms of probability in philosophical epistemology to the present-day debates in the philosophy of science, uncertainty has determined not only the content of knowledge but the process of inquiry itself.

These early probabilistic philosophers, such as Leonelli, understood the complexity of the logic of probability in terms of claims of certainty; probability does not equate with truth but with the strength of belief based upon knowledge (Leonelli, 2019). More contemporary authors contend that uncertainty should be recognized as an enabling factor in scientific progress, rather than an impediment (Chen, 2022). This approach runs counter to the premise that the sole function of science is the reduction of uncertainty. Rather, uncertainty, far from being exclusionary or problematic within scientific discourse, can serve an enabling function by pinpointing the boundaries of knowledge, raising new questions, or keeping ideas open.

This historical shift is also connected to the evolution of Work —the philosophy that all claims to knowledge are corrigible, even when supported by strong evidence. This philosophy argues that there is no such thing as certain knowledge that cannot be reversed; rather, one can hold a belief even while acknowledging the possibility of being mistaken about it (Work, 2025). Uncertainty is thereby seen as a characteristic rather than a flaw in scientific inquiry.

Classical views on certainty usually imply that the world functions under stable laws that can be correctly mapped and forecasted. A similar determinist premise therefore underlies many positivists and early logical empiricist approaches to social science: if phenomena can be represented correctly, then knowledge closely approximates truth. Deterministic frameworks, however, have been attacked from both philosophical and practical standpoints for underestimating the complexity, contingency, and context sensitivity of social phenomena. More contemporary work within the philosophy of science illustrates that even probabilistic assumptions, while more flexible than strict determinism, maintain their own limits. A recently developed unified theory of truth approximation highlights that traditional accounts of verisimilitude focus on deterministic formulations of truth and, for this reason, exclude fundamentally probabilistic truths or representations from their scope (Cevolani & Festa, 2021). The attempt at unification undertaken by probabilistic and deterministic accounts shows that neither is entirely able to capture how knowledge claims represent their domains unless they explicitly account for the uncertainty inherent in observations and models.

Furthermore, probabilistic models' use within the social sciences comes with the challenge associated with their application regarding representation; the description-experience gap is one such challenge: "People differ in their interpretation of the same probabilistic information as a function of whether they learned about that information descriptionally as opposed to experiencing it" (Nie et al., 2024). This shows that even as uncertainty is incorporated into probabilistic models' predictions on outcomes, the processing of uncertainty by agents is a challenge that limits such models.

Taking together, these criticisms point to an underlying limitation of classical epistemologies: while these epistemologies privilege representability and evidence to a high standard, they downplay or misunderstand the role of uncertainty and negotiation in the production of knowledge. Rather, they point towards approaches that understand knowledge as contextual and more or less uncertain.

3. GENERATIVE SYSTEMS AND THE PRODUCTION OF SYNTHETIC MEANING

Modern trends in artificial intelligence, especially those associated with massive-scale generative models, raise questions about the conventional notion of knowledge production, given that they can produce outputs that function as knowledge yet fail to meet the conventional epistemological criteria of knowledge as representation and empirical correspondence. Knowledge, as a conventional epistemological notion, is grounded in justification, a true belief supported by empirical evidence; conversely, contemporary generative models emphasize probability distributions, pattern-based regularities, and the possibility/probability within algorithms as a foundation for meaning production. This section discusses how generative models produce knowledge-related outputs, the manner in which these models produce a form of meaning based on statistical patterns related to probability inference, and the phenomenon of hallucination as a form of epistemic uncertainty. Figure 2 presents the probabilistic process through which generative AI systems construct synthetic meaning from statistical associations.

Figure 2. How generative AI produces synthetic meaning

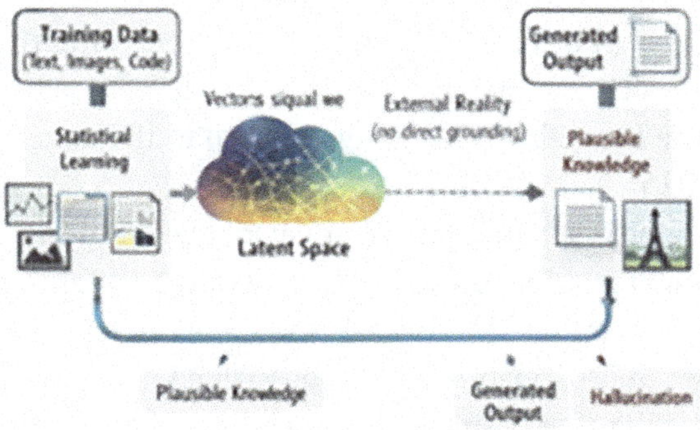

Large-scale generative models, such as transformer language models and diffusion models, have been trained on large amounts of text data, learning to predict probability distributions over the data. During training, the model adjusts its parameters to predict the next element in the sequence given the previous context. During the inference stage, the model samples a probability distribution to predict the tasks, aiming to forecast the most probable sequence following the input sequence. This predictive process does not depend on the prior definition of meaning through symbolic interpretation. Rather, it depends on the statistical properties of the data and can establish high-dimensional associations.

The epistemological status of these outputs has received attention in AI research and philosophy of science. One criticism of these generative outputs is that they should not be considered knowledge in the classical sense because they lack an intentional reference to external reality and engage in deductive reasoning (Bender et al., 2021). However, from a functional perspective, some of the outputs of a generative model may also exhibit characteristics of knowledge upon performance, such as explanatory cohesiveness, relevance, and task-specific adaptation, across areas including summarization, translation, and reasoning support. The outputs are employed in discovery methods in scientific and creative fields because they are timely and convey information that appears valid and useful, despite lacking empirical support. This upends epistemological standards that value representation based on procedural coherence.

The heart of all generative systems, then, is the Semantic Transition from traditional semantic representation through probabilistic pattern. Rather than representing context as vectors in the latent space, where closeness is measured by semantic truth, the internos model uses vectors in the latent space, where closeness is measured by statistical similarity (Huang, 2025). Meaning, then, is no longer based on the notion of referent, but pattern, probability, and believability.

In epistemology, this is important because plausibility is not truth or fact. A generative model may generate a sentence that is semantically coherent from a structural point of view, but false or lacking a concrete, empirical root. Such behavior is known as 'hallucination' in AI, as it refers to generated content which is fluent but not intended to refer to anything external to the generator (Anh-Hoang, Tran, & Nguyen, 2025).

This tension between plausibility and truth places generative models in a liminal, or threshold, position in relation to knowledge: models generate material that serves as knowledge in particular knowledge workflows, but do not necessarily satisfy the traditional conditions of justification in relation to that knowledge either empirically or justificatory speaking that assumes correspondence with things as they are in the external world in a non-mediated way in relation to statistical correspondence

3.1 Hallucination, Creativity, and Epistemic Ambiguity

The hallucination phenomenon has now come to the center of debates concerning the limits and potentials of generative AI. Hallucinations are those outputs that seem coherent but factually incorrect, inconsistent, or entirely fabricated. Importantly, scholars stress that hallucination is not merely a technical failure but also a feature of how generative systems synthesize patterns in the absence of direct referential grounding.

From an epistemic perspective, hallucination reveals something essential about the production of synthetic meaning: it shows that generative systems do not access truth conditions in the semantic sense but instead produce text that is plausible given the statistical properties of the training data. This brings ambiguity: output can be persuasive or epistemically useful without being reliably true or falsifiable. The same mechanism that allows for novel combinations of ideas and creative outputs also produces epistemic distortions that defy conventional judgments of credibility.

Others have suggested redefining hallucinations from a strictly erroneous phenomenon to one that reflects creative abilities within probabilistic models. Generative AIs are capable of unexpected combinations even for their users with expertise, which can energize a fresh approach or reframe a challenge in a manner that a human researcher may never have otherwise contemplated—one that is less about automating a process and more about supplementing creativity, as suggested by some researchers (Cai & Gao, 2025). Nonetheless, while its creativity is far-reaching, there are dangers and epistemological pitfalls as well.

These two forms of hallucination and synthetic creativity suggest there is an underlying ambivalence of knowledge production itself: generative models create meaning that is both creatively productive of new connections and vulnerable to knowledge errors. The authority of interpretation is thus distributed among users, training data contexts, algorithmic organization, and society-wide practices for validating those predictions. Sociology needs to engage with this paradigm change by working to develop forms of trust calibration, interpretive reflexivity, and algorithm-specific procedures of validation informed by synthetic aspects of generative predictions.

4. QUANTUM-INSPIRED INFERENCE AND NON-CLASSICAL REASONING

The quantum-inspired inference models are on the rise in cognitive science, decision theory, and modeling studies, as investigators face empirical observations that challenge classical probabilistic and causal models of uncertainty and cognition. Instead of deterministic epistemologies, which conceive observation as revealing a stable state and comparably conventional Bayesian models of uncertainty as subjective probabilities over well-defined alternatives, quantum models conceive uncertainty as irreducible, contextual, and non-commutative. The implication of quantum models of inference among social studies is that social systems are "quantum physical," and by implication, quantum models of inference are merely useful as a language of rigor in reasoning when observation affects a phenomenon and when meanings exist as non-separable, non-stable variables as indicated in current journal literature on quantum models of inference and how they can be applied in modeling cognition, social behavior, and human-machine hybrid settings of cognition and inference.

In quantum-inspired inference, uncertainty is not merely a matter of a lack of information about a pregiven state; it is a result of how the state is constituted by the context of inquiry. A key construct is superposition: in the formal system, it represents a situation in which, until an observation (measurement) is made, a system occupies multiple potential states simultaneously. Translated into social inquiry, superposition offers a means to model situations where agents' preferences, beliefs, or intentions are not well represented as fixed variables but instead exist as coexisting potentials that become determined through elicitation, framing, or interaction. Figure 3 demonstrates a quantum-inspired model of contextual inference in which cognitive states collapse through observation and interaction.

Figure 3. Quantum-inspired inference and contextual collapse

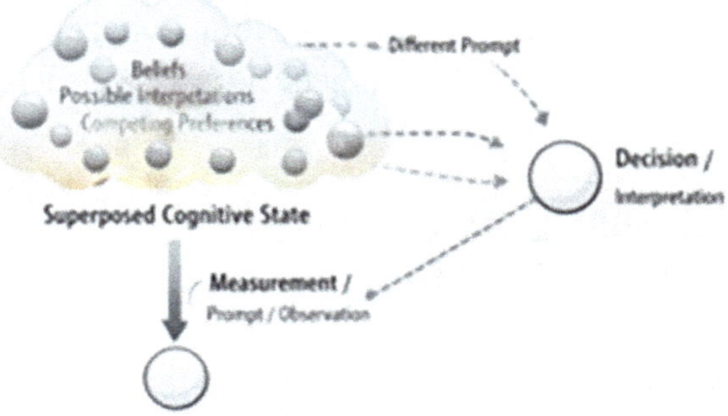

This epistemic reading is developed explicitly in work on quantum-like cognition and decision-making, where questions posed to respondents are treated as measurements that can perturb cognitive states. Fuyama and colleagues stress that quantum-like modeling is not a superficial borrowing of quantum vocabulary; it requires careful attention to measurement theory, and especially to forms of measurement that update states in ways not captured by classical conditioning (Fuyama et al., 2025).

Entanglement—the modeling of non-separable dependencies—wrests a new epistemic transformation from this legacy, contesting the idea that social variables should or can be reduced to independent components. In quantum models, entangled systems are those in which correlations cannot be decomposed into marginal distributions that are separately attributable across different variables. Social analogies exist in relational phenomena wherein meanings and identities are made in and as a consequence of their relational constitution and cannot be adequately captured as additive properties of individuals. A specific modeling example of this is quantum-like modeling of Bayesian networks that employ entanglement to model socially embedded bias and affective relations, such as friendships and hostilities, as features that recast dependencies on decisions (Meghdadi et al., 2022). Instead of modeling dependencies as noise, this model formulates them as generatively productive of outcome structures—a crucial shift in interpretive and computational studies of society broadly.

4.1 Beyond Linear Causality and Stable Observation

Traditional inference in the social sciences is typically grounded in the assumption of stable observations and linear models of causality, in which variables are assumed to exist prior to observation and in which the objective of explanation is the identification of their isolated effects. Quantum inference contests these assumptions on both counts. First, observation may be a transformation rather than a disclosure of state; and second, explanation may be context-dependent and non-linear in the sense that the conditions of elicitation shape what is observed.

There is a challenge to apply mathematics correctly, as made clear by Khrennikov's recent work in a refereed journal, summarized by the model of contextual measurement, which holds that the relationship between Hilbert-space variables and psychological/social phenomena should be tackled as a modeling problem rather than taken as obvious (Khrennikov, 2024). The message here is one of methodology: what represents "the same" entity may vary from context to context, and the study itself becomes an object. This corresponds to deep-seated interpretive findings in the social sciences, but what's new here begins with the mathematization of contextuality and the role of non-commuting observation sequences.

Quantum-inspired reasoning also equips us with ways to understand interference phenomena—where the probability of an outcome cannot be modeled as a sum of independent transitions. These are increasingly employed in the modeling of collective decision-making. For instance, Jiang & Liu develop a quantum model of group decision-making in which the beliefs of decision-making experts are represented in superposition, generating interference during consensus-building (Jiang & Liu, 2022). The epistemic implication here is that not only is the uncertainty summed over individuals, but there is also an active transformation of uncertainty that generates a group dynamic that cannot easily be modeled by a sum of independent transitions.

4.2 Implications for Explanation and Inference in Social Research

Quantum-inspired inference rearranges explanatory work by moving the focus from the application of causal inference in order to identify invariant causal relationships between variables to the application of inferring state dynamics in relation to the conditions under which "outcomes collapse into determinate patterns." This does not mean the end of explanation, but rather an "explanation about" something else. Explanation becomes an account of (a) the structure of potentiality, (b) the measurement context by which some possibilities, but not others, are realized, and (c) the relational structure by which the whole resists decomposition into causal factors.

This immediately impacts the area of AI-enhanced research. Formal models inspired by quantum mechanics are increasingly cited as conceptual tools for human-AI decision-making systems, in which the significance of the result is determined by the iterative prompting, interaction, and updating process. In the perspective article on quantum probability theory and human-AI decision-making systems, the author suggests that "quantum probability models may better describe the order effects or contextual changes that emerge in a human-in-the-loop environment, where decisions develop through an iterative process of prompting or updating rather than through prior commitment" (Humr, 2025).

Finally, quantum-inspired methods also interact with the computational progress in quantum machine learning simulations for inference. Although most of this work focuses on physical systems, it can also be seen as part of the epistemic drift toward treating inference as a process of modeling distributions over likely generative models under conditions of limited observability and complexity. Recently, an article in the peer-reviewed journal Frontiers discusses Bayesian inference via simulation methods based on parameterized quantum circuits as simulated generative models for likelihood-free inference, systematically defining Bayesian distributions in terms of epistemic uncertainty rather than error (Nikoloska et al., 2024). In relation to the social sciences, this findings takeaway might be framed in relation to its methodological implication—the process of inference becomes more about contrasting real data with simulated data based upon generative models whose underlying dynamics can be opaque.

In conclusion, quantum-inspired inference is more than a metaphor. Quantum-inspired inference presents a well-articulated, non-classical notion of epistemic grammar for situations in which (1) states are not stable objects independent of measurement, (2) meanings and decision-making are contextually defined, and (3) dependencies have a relational, irreducible character.

5. FROM ERROR TO CONDITION: RETHINKING UNCERTAINTY

Uncertainty has classically manifested itself as a problem of scientific or social scientific research as an error to correct, an obstacle to reduce, or a kind of noise that must be controlled. However, with the growing prevalence of generative AIs, data-driven computation, or complex adaptive systems as features of our current epistemic environment, the understanding of uncertainty as a problem of irritation or statistical residue seems less and less plausible as a methodological challenge or a challenge to statistical inference and modeling. From being a mischief within scientific methodology or a statistical trace without it, uncertainty increasingly comes into prominence as a defining aspect or condition of scientific knowledge that must be ontologically and methodologically identified and characterized as such. From a concern with a traditional understanding of uncertainty as a manifestation of error, bias, or probabilistic variation, a considerable body of philosophical and interdisciplinary philosophical work, that is, studies on philosophical theories and methods of research in science studies and studies of science and technology, has emerged with a conception of uncertainty as a territory and a field of inquiry.

In the traditional research paradigm, uncertainty is generally treated synonymously with, albeit related to, concepts of bias, noise, and error. By bias is meant systematic deviation produced either through research design or the prejudices of the researcher, while noise is regarded as the variability that hides the signal, and error refers to those outcomes that are different in problematic ways, often in defiance of expectations of the expected outcome, for example, in terms of non-reproducibility in experiments and wrong predictions.

Uncertainty, on the other hand, refers to the epistemological boundaries of knowledge that are possible within a given model, context, or assumption set. It is significant that uncertainty cannot be viewed solely as error/noise, since it is, more often than not, a consequence of the structural properties of the observed phenomenon rather than any deficiencies in methodology or data. When viewed within the context of complex systems, for example, uncertainty is generated by interrelationships, emergence, feedback, etc., that cannot be captured within error terms.

In socio-ecological modeling, the recognition of uncertainty arises from its multifaceted nature, which cannot be quantitatively diminished, regardless of the availability of more data or enhanced modeling complexity (Rounsevell et al., 2021). This also underscores that uncertainty is more about the nature of the interaction between representations and open systems than about unpolitic variations around the mean.

Similarly, the literature on methodology focuses on how uncertainty cannot be reduced to noise without ignoring its importance for epistemology. Indeed, recent studies on uncertainty quantification in AI claim that uncertainty should be assessed alongside AI predictions, rather than diminished (Zhang et al., 2025). From a viewpoint of ignoring uncertainty versus uncertainty converted into noise or error, ignoring uncertainty can lead people to treat uncertainties in interpretation as erroneous methodologies, concealing how these methodologies involve different possibilities for interpreting data appropriately.

Increasingly, there is a body of research suggesting that uncertainty is not simply a result of knowledge but is itself a property of phenomena—that is, it is ontological, not simply epistemological. This is most apparent in research about complex adaptive systems, which are inherently unstable with regard to initial conditions.

Uncertainty within socio-ecological systems, that is, systems of interplay between human and ecological systems—becomes ingrained in the system's own evolutionary dynamics. Uncertainty, as opposed to the lack of information or the lack of sufficient representation through samples or observations, becomes the possible states the system is capable of reaching on its own (Rounsevell et al., 2021). This directly affects research design by requiring that the research employ methods capable of handling uncertainty.

This is one area in which philosophical workplaces strong emphasis on how uncertainty reveals the limitations of classical representation and predictive idealization. Indeed, foundational work on frameworks of uncertainty demonstrates that the multiple conceptualizations (risk, ignorance, ambiguity) each adopted resist a singular synthesis precisely because uncertainty cannot be fully captured by a single universal metric or a single taxonomy. This would imply that uncertainty must be treated as irreducible methodologically; for this reason, researchers would have to detail what kind, or type, or combination of uncertainties they are dealing with and how they influence interpretation.

The idea that uncertainty has ontological significance also resonates with approaches that treat research phenomena as wicked problems, in which complexity, conflicting values, and interdependent processes render unique solutions or stable predictions impossible (Agustian, 2025). In such settings, uncertainty is epistemic, or about knowledge states, and ontological, or about the intrinsic openness of the world.

5.1 Productive Indeterminacy in Complex Social Systems

Recognizing uncertainty as a fundamental feature creates space for what has been termed productive indeterminacy—the notion that uncertainty can be a source of insight, innovation, and methodological transformation rather than a deficit to be overcome. In this view, uncertainty guides researchers to attend to what is not yet fully articulated within phenomena, inviting iterative model refinement, pluralistic methods, and reflexive interpretation.

In fields such as scenario planning, foresight studies, and futures research, uncertainty is treated not as a problem to be eliminated but as the starting point for exploring alternative pathways and mapping contingencies. These approaches emphasize that uncertainty can reveal latent structures and decision trade-offs that stable predictions would otherwise hide.

Productive indeterminacy also intersects with recent discussions on the epistemic use of generative systems: instead of conceiving model uncertainty as algorithmic error, researchers are beginning to consider how models generate multiple plausible representations that can be compared, contested, and enriched through human interpretation. This reframing aligns with broader methodological shifts toward reflexive analysis, where uncertainty becomes part of the interpretive work of theory building rather than a residual phase of statistical adjustment.

Practically, embracing uncertainty as a methodological condition encourages researchers to design inquiry that is adaptive, iterative, and open to transformation. Rather than privileging static validation criteria, methodologies can incorporate uncertainty diagnostics, scenario diversity, and robustness checks that do not assume a single "true" outcome. In doing so, uncertainty becomes not only something to measure but a resource for critical inquiry and epistemic humility.

6. MEANING-MAKING IN HUMAN–MACHINE ASSEMBLAGES

Generative artificial intelligence has not only brought new tools to social research and knowledge work; it has enabled a transformation in the architecture of sense-making itself. Today, in the digital environment, sense-making is no longer a human cognitive activity confined to the human brain; it now takes place in assemblages of human and non-human agents, such as models, platforms, interfaces, and data infrastructure. Such assemblages have enabled the distribution of sense-making agency, in that they have brought into play human and non-human agents in ways unknown in human sense-making.

Such a shift requires a re-conceptualization of interpretation from being a cognitive task to a socio-technical endeavor. There has been a recent trend in science and technology studies and digital epistemology to recognize that algorithmic networks play an increasingly important role as mediators of meaning, in that they decide what gets seen, what gets said, and what gets interpreted as meaningful. Generative AI networks not only learn and compute data; they construct, in effect, the semantic space wherein interpretation of reality takes place, hence co-producing meaning/knowledge (Mayer et al., 2025; Beer, 2018).

6.1 Distributed Interpretation Across Humans, Models, and Platforms

Accordingly, the construction of meaning in human–machine assemblages occur through the recursive interaction between humans and algorithmic systems. When a human sends a query to a generative model, it automatically receives a response in the form of statistically generated text, based on patterns encoded in the algorithmic systems' training data and architectures.

This is consistent with descriptions of socio-technical mediation in STS theories, in which technologies not only transmit but also shape meanings. According to Mayer et al. (2025), Today's AI systems may be seen as epistemic infrastructures in the sense that they shape what counts as legitimate claims, which questions count as relevant ones, and which answers count as plausible ones. In generative models of AI, these infrastructures work in a probabilistic rather than in a rule-based fashion and map a space of meanings as fluid and dependent on the design of the models.

Another factor, which also becomes a deciding point here, is the role of platforms. Search engines, productivity software, or social media platforms are beginning to integrate AI as a new interpretive layer, where information received by users will be filtered or interpreted through these means. This, according to Beer (2018), leads to the formation of algorithmic sense-making environments where Reality itself becomes interpretatively mediated through computationally mediated representations, as opposed to direct access to information, where meaning anyhow gets mediated through machine-made narratives, summaries, or framings.

6.2 Prompting, Steering, and Interpretive Intervention

As generative AI models lack innate intentions and goals, the primary steering force behind them is the human prompting mechanism, which essentially guides them by providing operational definitions of the semantic domain and normative direction of their responses. Here, the role of prompting becomes similar to that of operationalization in conventional social science analysis or framing in qualitative analysis.

Recent empirical work reveals that the same model can produce radically different outputs depending on the prompt, even for the same topic (Zamfirescu-Pereira et al., 2023). Such variations support the notion that generative models are not containers of predetermined meanings but generate meanings on the spot through interaction. The user issuing the query is not only accessing a knowledge base but also actively taking a step toward building an interpretation message. However, this epistemology also enables new knowledge and new knowledge holders, or expertise. Those who know how to question effectively—to specify constraints, styles, and goals—are in a uniquely powerful position within an epistemological system. Thus, literacy in the question becomes a form of epistemic capital in the sense that it determines who can effectively act in human-machine sense-making (Seaver, 2021).

Further, the use of prompting not only directs output but also weaves human interpretation, values, and beliefs into the process of text creation. The model reproduces these interpretations in magnified or distorted form, thereby establishing a feedback loop between the human interpretive frame of meaning and the generated machine text. Here, meaning is a result of co-evolution rather than a mapping relationship between data and reality.

6.3 The Shifting Locus of Sense-Making Authority

One of the most significant aspects of generative AI is the way it upends knowledge authority. In traditional knowledge systems, knowledge authority lay largely with human experts, with organizational or institutional knowledge, or with knowledge methodological standards. In human-machine arrangements, knowledge authority is increasingly vested in computer-generated systems that produce fluent, confident, and highly persuasive statements.

Empirical research reveals that users often ascribe epistemic credibility to generative models despite being aware of their limitations. According to a study by Jago & Carroll (2024), users conflate AI-generated text with more objective, authoritative, and, particularly, informative text than similar text written by humans. Thus, there's a danger in which synthetic meaning can be confused with validated knowledge, underwriting epistemic overconfidence instead of rigorous examination.

However, it is important to note that these models do not actually 'know' anything in a classical sense; rather, they create meaning via probabilistic pattern recognition. This creates a paradox in which accountability is granted to systems that lack epistemic accountability. Epistemic accountability, therefore, must be reattributed to human agents and to institutionally mediated decision-making structures that create and interpret these systems.

For example, as assemblages, these practices require new modes of epistemic accountability, where models being used will be less about whether or not it is 'correct' but more about how its results are produced, who views are included within the training data, or what reinterpretations of its use mean for practices of interpretation (Beer, 2018; Mayer et al., 2025).

7. VALIDATION, INTERPRETATION, AND EPISTEMIC RESPONSIBILITY

As generative and quantum-inspired models become ever more embedded in the process of social knowledge production, the problem of how claims are validated and accounted for takes center stage. Traditional models of validation in the form of replication, the significance level of p values, peer review, and the transparency of evidence were framed for a world in which the production of observation statements and models was sharply distinct. In AI-enhanced research practice, however, claims for knowledge typically co-emerge as the results of human and computational processes of inference and sense-making. Hybrid paradigms of knowledge production, such as this, blur the boundaries of familiar notions of rigor and demand a critical redefinition of frameworks of traceability and epistemic accountability grounded in concepts of traceability and interpretive plurality.

In today's research settings, generative systems have been capable of delivering an output that can be fluent, coherent, and contextually relevant, yet for which the task of stabilizing knowledge or making a determination about whether an output is evidential, explanatory, or insightful, as being verifiable and referentially grounded, becomes problematic. Here, instead of asking whether a particular output is "true," it has become common practice to evaluate robustness across a series of tasks, contexts, or data slices.

Recent studies in AI evaluation illustrate the crucial role of distributional and epistemic uncertainty in evaluating AI system outputs. Therefore, Amodei et al. (2021), among others, in reviewed journals such as AI & Society and Nature Machine Intelligence, contend that system success scores should be supplemented with uncertainty-inclusive validation approaches that specifically evaluate the sensitivity of system outputs to variations in framing and sampling of inputs. In social studies, therefore, a generative output qualifies epistemically not on the account of its convincing alone but on its coherence and informativeness in different independent inquiries.

This trend can be understood in relation to open science and computational reproducibility research environments that are also being transformed to establish assertions by code, data, or analytical workflow rather than by rhetorical persuasive ability alone. In cases of generating literature synthesis, in addition to hypothesis formulation or qualitative analysis, using these artificial intelligence approaches, it is essential to provide procedural transparency to explain how the particular claim was arrived at (Mayer et al., 2025).

Studies in the Journal Big Data & Society and Information, Communication & Society make it clear that traceability is also a political-ethical issue. If meaning is mediated by systemic meaning, those who can reconstruct it can contest, reinterpret, or challenge it (Beer, 2018). Without traceability, synthetic meaning can come across as authoritative yet unaccountable, thereby perpetuating power inequalities among those who design the system, those who own the platform, and those who use knowledge.

Reflexivity complements traceability by foregrounding the situatedness of interpretation. Reflexive practices ask researchers to articulate their assumptions, values, and positionalities, as well as how these shape their interactions with generative systems. Recent STS-informed studies show that users often project intentions, neutrality, or objectivity onto AI outputs, obscuring the social and technical contingencies embedded in those outputs (Jago & Carroll, 2024). Reflexivity counters this tendency by insisting that AI-mediated interpretations be treated as provisional constructions rather than privileged views from nowhere.

Interpretive pluralism follows from these commitments. If generative models can produce multiple plausible narratives from the same data, then no single output should be treated as definitive. Instead, pluralism treats divergent model outputs as resources for comparative analysis, revealing latent assumptions, alternative framings, and boundary conditions. This approach resonates with qualitative traditions of triangulation and with quantitative practices of ensemble modeling, now extended into the domain of synthetic text and inference.

7.1 Rethinking Rigor in AI-Augmented Research

The need to integrate generative models into the research process calls for a revision of the understanding of rigor in research. In the traditional research paradigm, rigor is commonly defined in terms of control, reproducibility, and uncertainty reduction. In the research paradigm that embraces AI research, uncertainty is no longer a variable that can and must be diminished, and it is not necessary that the results can and must be exactly replicated in the event of the evolution of models and data distribution, coupled with the use of stochastic sampling in the generative process.

Recent debates in Nature Human Behaviour and Social Studies of Science have argued that the proper notion of rigor is accountable variability: the ability to explain the reasons behind differences between runs, prompts, or models, and what those differences indicate with respect to the phenomenon being studied. This notion conceives of uncertainty not as an error, but rather as an indication of epistemic richness, assuming such is recorded in an appropriate manner.

From an application point of view, what this means is the combination of computer audits (using systematic variation on the prompt and the parameters) with human-centered assessment (using expertise), as well as the institutionalization of disclosure principles for the involvement of AI in the production of results, much as the disclosure for the use of statistical software or data sources. These principles apply to the principles emerging for responsible practice in the use of AI as a consideration in the production of knowledge (Floridi et al., 2021).

8. ETHICAL AND POLITICAL DIMENSIONS OF UNCERTAIN KNOWLEDGE

The emergence of generative and quantum-inspired models of knowledge production is not only changing the manner of knowledge production itself but is also changing how power is distributed. If knowledge is no longer produced by directly accountable agents of knowledge production through interpretive and synthetic systems of knowledge generation, such as probabilistic systems, then the possibility of knowledge becomes political. This section will discuss how power imbalances, confidence in knowledge claims, and governance failures inform the risks of uncertain knowledge.

Generative AI systems are trained on vast datasets curated and controlled by a small number of corporate and governmental actors. These organizations make decisions about which texts, images, and voices are included in the training corpus, and which are excluded. Because of this, the statistical space from which meaning is generated is already politically structured. Even when outputs seem neutral or objective, they reflect the asymmetries of the data and institutions that produced them.

Recent research in Big Data & Society has shown that AI systems are "epistemic gatekeepers" that actively shape what kinds of knowledge become visible and actionable within digital environments (Beer, 2018). As generative systems synthesize research, draft policies, or answer public questions, for example, they will have already settled on which views count more than others. These systems operate at scale, their framing effects risk structurally entrenching within the public discourse, preserving dominant narratives while sidelining alternative and critical perspectives.

Such a focus of epistemic power is compounded by the opacity of big models. Users cannot easily understand the selection, weighting, or filtering of the training data, making it difficult to challenge the assumptions incorporated into the generated results. As Mayer et al. (2025) contends, a new form of infrastructural power emerges, in which the designers and managers of AI systems indirectly rule over the notions of knowledge, relevance, and implausibility across sectors such as news, science, and government.

8.1 Risks of Epistemic Overconfidence and False Authority

One of the most worrying characteristics of generative systems is that they can produce highly fluent and confident outputs irrespective of their epistemic quality. Fluency induces, in philosophers' and communication scholars' terms, epistemic overconfidence: the perception that a claim is reliable simply because it is well-formed and authoritative in tone. Empirical studies have demonstrated that users are more likely to believe AI-generated text is accurate and unbiased than similar human-authored text, especially when it is informational or technical in nature (Jago & Carroll, 2024).

This is ethically significant, as it allows for the exercise of false authority-the granting of credence to systems that possess no comprehension, accountability, or moral agency. In social research, policy analysis, and journalism, misplaced trust in this fashion can facilitate the uncritical propagation of errors, biases, or fabrications-especially where AI outputs are being used to summarize large bodies of information, or to generate reports bearing the trappings of authority.

Moreover, epistemic overconfidence interacts with existing social inequalities. Individuals and institutions with fewer resources for verification may rely more heavily on automated systems, making them more vulnerable to misleading or biased outputs. These dynamic risks producing epistemic stratification, where some groups have the capacity to critically interrogate AI-mediated knowledge while others are subject to its unexamined authority.

8.2 Governance Challenges in Synthetic and Quantum Epistemologies

Researchers in the areas of AI & Society and Ethics and Information Technology strongly suggest that the focus of the Gulf states and the UAE, in this case, should not just be on the end outcome but on the process of semantic production itself (Floridi et al., 2021). Governance, in this sense, ceases to simply be a political issue and becomes one of epistemological stewardship, where the use of automatic systems is utilized in a way that facilitates and promotes democratic discourse, scientific honesty, and social justice.

Another challenge concerns the hybrid nature of responsibility. Who is responsible when a human uses a generative system to create an analysis or make a recommendation that turns out to be in error, biased, or causes harm? Recent work on human-AI collaboration stresses that responsibility needs to be distributed among designers, deployers, users, and institutions, and cannot be located in either an individual actor or the technology itself (Raji et al., 2020). Accountability requires governance mechanisms that trace how AI systems become part of workflows and decision chains, thereby enabling the effective pinpointing of where epistemic failures occur.

Finally, quantum-inspired epistemologies complicate governance by undermining assumptions of stable, reproducible outcomes. If inference itself is context-dependent and non-commutative, then pluralism and contestability have to be baked into governance structures. Rather than enforcing a singular authoritative interpretation, institutions must support multiple, transparent, and revisable accounts of how automated systems generate meaning.

9. TOWARD AN EPISTEMOLOGY OF UNCERTAINTY

The above sections have illustrated that both generative and quantum-inspired approaches to system building have a destabilizing effect: rather than multiplying new sources of error in knowledge-making, these approaches disrupt the conditions under which meaning, explanation, and validation occur. If uncertainty is no longer simply a residual character to be eliminated through post-digital methods, and instead a guiding principle of knowledge-making, a new epistemology must follow. The next section explores foundational tenets of this new epistemology of uncertainty and the need for methodological humility, and illustrates how uncertainty itself can and must be incorporated into practice and education.

9.1 Principles for Post-Digital Knowledge Production

To add, an epistemology of uncertainty begins by challenging what has traditionally been considered an axiom of traditional knowledge: that it must be based on fixed representations of an objective reality, which, in turn, can be completely quantified and ascertained. In computationally mediated spaces, traditional notions of knowledge based on fixed notions of reality are increasingly supplanted by those grounded in fundamentally probabilistic, relative, and contestable meaning related to LDHMAs.

Recent studies in the fields of Philosophy of Science and Human Nature Behaviour have highlighted the need to interpret new scientific knowledge using model-based rather than representational approaches. Models do not represent truth; instead, they create cognitive lenses through which some aspects of truth are made visible, whereas others become invisible (Maskey, 2023). In the case of generative models, these cognitive lenses are incorporated into the model framework, which includes training examples, optimization functions, and model architectures. Knowledge generated by these approaches must therefore be scrutinized according to their own conditions and intentions, rather than their connection to truth.

Post-digital epistemology also places particular emphasis on contextuality. It is because of the role of observation or interrogation being non-neutral in quantum-inspired approaches that the state of the system being interrogated is itself conditioned. This entails that all knowledge is situated in particular contexts of inquiry, and these contexts can be legitimate in yielding different, often contradictory results that can still be informative. This is how plurality can be indicative of rigor in inquiry and of being attuned to the world.

9.2 Methodological Humility and Epistemic Care

For instance, if it can be shown that uncertainty is constitutive and essential rather than accidental, then the epistemic virtues associated with knowledge production will require a different set of orientations altogether. The ideals of mastery, prediction, and control will be replaced by a methodological humility rooted in the realization that a particular model, dataset, or analytical approach cannot capture the essence of social realities in totality.

Recent scholarship in the field of Social Studies of Science and AI and Society has stressed the need to go beyond humility and adopt a concept of epistemic care, which is a commitment to the ethics of impact that knowledge assertions have upon people, groups, and institutions (Mayer et al., 2025; Floridi et al., 2021). Within the domain of AI-enabled research, this concept of epistemic care, or attention to impact, includes a focus on how certain outputs from generative models can misrepresent or do violence to the groups they represent. Further, it must resist the pull to reduce these outputs to a status of being color blind and therefore authoritative because of their computational nature.

Epistemic care also requires reflexivity about the role of judgment in the production of AI-based content. Researchers should be mindful that their own judgments about which prompts, models, or assessment criteria to use are acts not simply of a technical but also of a normative kind, which determine what is visible as knowledge.

9.3 Integrating Uncertainty into Research Design and Training

A program of epistemology of uncertainty requires translation into real-world transformations of research design, methodological training, and norms. Firstly, uncertainty has to be recorded and examined, not concealed. When conducting quantitative inquiry, for instance, not only are point estimates required, but also distributions, intervals of confidence, and sensitivities of models. When using AI for qualitative research, uncertainties are captured by archiving questions asked, divergent models, and interpretive choices, for instance, for testing by someone else the meanings assigned.

Journals and funding bodies are increasingly demanding these approaches. Editorials in the journals 'Nature Human Behaviour' and 'Big Data & Society' point to the central role of transparent computational practices, and the disclosure of the use of AI and parameters, in maintaining the integrity of research in the machine learning era (Semmelrock et al., 2023; Beer, 2018). Such approaches now recognize the need to make uncertainty visible in order that knowledge itself remains contestable and credible.

Learning also requires an overhaul. Scientists need to learn how to operate generative models as well as how to think about their uncertainty, think about how model outputs can be compared, how model instabilities can be recognized, and how model-generated knowledge can be integrated in appropriate ways. Such learning poses as much an epistemological problem as a technical problem, involving the cultivation of judgment, skepticism, and interpretive acuity.

Ultimately, an epistemology of uncertainty reframes what it means to know in a post-digital world. Knowledge becomes provisional, relational, and responsive to context, rather than fixed and final. Far from undermining scientific inquiry, this orientation strengthens it by aligning epistemic practice with the realities of complex, mediated, and evolving systems.

10. CONCLUSION

This chapter has presented the claim that generative artificial intelligence and quantum-inspired inference machines represent much more than a technical challenge to the nature of knowledge, namely, an epistemological transformation. In doing so, they challenge the long-held epistemological orthodoxies about representation, certainty, and interpretation by introducing new epistemologies centered on probabilistic, synthetic, and context-dependent forms of knowing. In so doing, they pose a fundamental challenge to the way we understand the production, verification, and assumption of responsibility for contemporary social knowledge. The takeaway message from this chapter is incorporated into the conclusion.

Through the above sections, the pattern that emerges is that knowledge in the post-digital world cannot be viewed as the reflection of an independent reality. Nay, the construction of knowledge occurs through the assemblage of human, model, data, and infrastructure, where each assemblage is part of the process through which meaning is made. "Generative systems don't represent the world; they generate meaningful narratives based on statistics, which have relevance only in interpretive contexts. Quantum-inspired reasoning similarly demonstrates that the act of observation is contingent on the manner and timing of the process, where multiple possibilities can coexist before being resolved.

This means a transition from epistemic closure to epistemic openness. This is because knowledge is no longer of the Finality type but is now Provisional. This condition of provision is not necessarily a flaw but can be said to be dynamically driven by the complexity of social reality itself. According to Maskey (2023), there is today an increasing reliance on model-mediated knowledge in science, in which understanding is achieved through multiple representations to highlight different aspects of things rather than through definitive narratives.

Relationality is another feature of this epistemic topology. Meaning is not immanent within data or models but arises from relations between users, systems, and contexts of use. In research involving artificial intelligence, a result acquires meaning only through human interpretation, disciplinary, and institutional legitimation. Relationalities of such quantum-epistemologies foreground knowledge as a function of context, or measurement, of what is known. These trends together lead to a non-separate, or entangled, epistemology, in which knowledge is no longer conceived as residing within isolated subjects or objects.

For the social sciences, these developments mean a whole host of new implications. From a methodological point of view, this represents a set of new approaches to science, in which the emphasis shifts from searching for a unique, best, or ideal model to creating plural models for comparative analysis. Using generative systems, it is easy to generate a variety of plausible interpretations of the same data, whereas approaches based on quantum ideas show how these different models can reach different conclusions.

At the institutional level, there needs to be a transformation in the standards of rigor and accountability in social inquiry. A level of openness regarding data, architectures, and strategies for prompted responses is at least as important as openness regarding sampling and measurement in traditional inquiry (Semmelrock et al., 2023; Veltri et al., 2018). Increasingly, publications and organizations recognize that there needs to be openness regarding AI tools and strategies for reproducible and responsible science (Nature Semmelrock et al., 2023).

Ethically, the future of social inquiry will turn on the development of epistemic responsibility. The more generative systems dominate the framing of problems and the proposition of solutions, the more researchers will have to be vigilant about issues of power, bias, and exclusion. According to Mayer et al. (2025), AI-driven infrastructures can silently reshape what counts as legitimate knowledge, giving voice to some while silencing others. An epistemology of uncertainty resists this process by insisting that no knowledge claim, whether human- or machine-generated, should be beyond challenge and reinterpretation.

REFERENCES

Agustian, H. Y. (2025). The critical role of understanding epistemic practices in science teaching using wicked problems. *Science & Education*, *34*(1), 485–510. DOI: 10.1007/s11191-023-00471-2

Anh-Hoang, D., Tran, V., & Nguyen, L.-M. (2025). Survey and analysis of hallucinations in large language models: Attribution to prompting strategies or model behavior. *Frontiers in Artificial Intelligence*, *8*, 1622292. DOI: 10.3389/frai.2025.1622292 PMID: 41098969

Beer, D. (2018). The data gaze: Capitalism, power and perception.

Bender, E. M., Gebru, T., McMillan-Major, A., & Shmitchell, S. (2021, March). On the dangers of stochastic parrots: Can language models be too big? In *Proceedings of the 2021 ACM conference on fairness, accountability, and transparency* (pp. 610-623). DOI: 10.1145/3442188.3445922

Cai, W., & Gao, M. (2025). Beyond hallucination: Generative AI as a catalyst for human creativity and cognitive evolution. *ICCK Transactions on Emerging Topics in Artificial Intelligence*, *2*(1), 36–42. DOI: 10.62762/TETAI.2025.657559

Cevolani, G., & Festa, C. (2021). Approaching deterministic and probabilistic truth: A unified account of truth approximation. *Synthese*, *199*(3-4), 11465–11489. Advance online publication. DOI: 10.1007/s11229-021-03298-y

Chen, Y. C. (2022). Is Uncertainty a Barrier or Resource to Advance Science? The Role of Uncertainty in Science and Its Implications for Science Teaching and Learning: Kostas Kampourakis, Kevin McCain (2019): Uncertainty: How It Makes Science Advance. Oxford University Press, New York, NY. . 0001.DOI: <ALIGNMENT.qj></ALIGNMENT>10.1093/oso/9780190871666.001

Floridi, L., Cowls, J., Beltrametti, M., Chatila, R., Chazerand, P., Dignum, V., ... & Rossi, F. (2021). An ethical framework for a good AI Society: Opportunities, risks, principles, and recommendations. Ethics. *Governance, and Policies in Artificial Intelligence*, 19-39.

Fuyama, M., Khrennikov, A., & Ozawa, M. (2025). Quantum-like cognition and decision-making in the light of quantum measurement theory. *Philosophical Transactions A, 383*(2309), 20240372.

Huang, S. Y. (2025). The Connectionist Turn: How Contemporary Generative AI Reshapes Architectural Rationality. *Architecture (Washington, D.C.)*, *5*(4), 132. DOI: 10.3390/architecture5040132

Humr, S., Canan, M., & Demir, M. (2025). A quantum probability approach to improving human–AI decision-making. *Entropy (Basel, Switzerland)*, *27*(2), 152. DOI: 10.3390/e27020152 PMID: 40003149

Jago, A. S., & Carroll, G. R. (2024). Who made this? Algorithms and authorship credit. *Personality and Social Psychology Bulletin*, *50*(5), 793–806. DOI: 10.1177/01461672221149815 PMID: 36734213

Jiang, J., & Liu, X. (2022). A quantum cognition based group decision making model considering interference effects in consensus reaching process. *Computers & Industrial Engineering*, *173*, 108705. DOI: 10.1016/j.cie.2022.108705

Khrennikov, A. (2024). Contextual measurement model and quantum theory. *Royal Society Open Science*, *11*(3), 231953. DOI: 10.1098/rsos.231953 PMID: 38550757

Leonelli, S. (2019). *Data-centric biology: A philosophical study*. University of Chicago Press.

Maskey, M. (2023). Rethinking AI for Science: An Evolution From Data-Driven to Data-Centric Framework. *Perspectives of Earth and Space Scientists, 4*(1), e2023CN000222.

Mayer, A., Greif, L., Häußermann, T. M., Otto, S., Kastner, K., El Bobbou, S., Chardonnet, J.-R., Reichwald, J., Fleischer, J., & Ovtcharova, J. (2025). Digital Twins, Extended Reality, and Artificial Intelligence in Manufacturing Reconfiguration: A Systematic Literature Review. *Sustainability (Basel)*, *17*(5), 2318. DOI: 10.3390/su17052318

Meghdadi, A., Akbarzadeh-T, M. R., & Javidan, K. (2022). A quantum-like cognitive approach to modeling human biased selection behavior. *Scientific Reports*, *12*(1), 22545. DOI: 10.1038/s41598-022-13757-2 PMID: 36581629

Nie, D., Hu, Z., Zhu, D., & Yang, J. (2024). Probabilistic Representation Differences between Decisions from Description and Decisions from Experience. *Journal of Intelligence*, *12*(9), 89. DOI: 10.3390/jintelligence12090089 PMID: 39330468

Nikoloska, I., & Simeone, O. (2024). An introduction to Bayesian simulation-based inference for quantum machine learning with examples. *Frontiers in Quantum Science and Technology*, *3*, 1394533. DOI: 10.3389/frqst.2024.1394533

Raji, I. D., Smart, A., White, R. N., Mitchell, M., Gebru, T., Hutchinson, B., & Barnes, P. (2020). Closing the AI accountability gap: Defining an end-to-end framework for internal algorithmic auditing. In *Proceedings of the 2020 conference on fairness, accountability, and transparency* (pp. 33-44). DOI: 10.1145/3351095.3372873

Rounsevell, M. D., Arneth, A., Brown, C., Cheung, W. W., Gimenez, O., Holman, I., Leadley, P., Luján, C., Mahevas, S., Maréchaux, I., Pélissier, R., Verburg, P. H., Vieilledent, G., Wintle, B. A., & Shin, Y. J. (2021). Identifying uncertainties in scenarios and models of socio-ecological systems in support of decision-making. *One Earth*, *4*(7), 967–985. DOI: 10.1016/j.oneear.2021.06.003

Seaver, N. (2021). Care and scale: Decorrelative ethics in algorithmic recommendation. *Cultural Anthropology*, *36*(3), 509–537. DOI: 10.14506/ca36.3.11

Semmelrock, H., Kopeinik, S., Theiler, D., Ross-Hellauer, T., & Kowald, D. (2023). Reproducibility in machine learning-driven research. *arXiv preprint arXiv:2307.10320.*

Sinaga, Y. K., Manugeren, M., & Barus, E. (2025). Systematic Literature Review on Social Epistemology of Science: Knowledge Production in Scientific Communities. *IDEAS: Journal on English Language Teaching and Learning. Linguistics and Literature*, *13*(2), 7149–7159.

Veltri, D., Kamath, U., & Shehu, A. (2018). Deep learning improves antimicrobial peptide recognition. *Bioinformatics (Oxford, England)*, *34*(16), 2740–2747. DOI: 10.1093/bioinformatics/bty179 PMID: 29590297

Work, E. S. (2025). Stanford Encyclopedia of Philosophy. https://iris.unito.it/retrieve/e27ce434-0eb2-2581-e053-d805fe0acbaa/einstein-philscience_sc.pdf

Zamfirescu-Pereira, J. D., Wong, R. Y., Hartmann, B., & Yang, Q. (2023, April). Why Johnny can't prompt: how non-AI experts try (and fail) to design LLM prompts. In *Proceedings of the 2023 CHI conference on human factors in computing systems* (pp. 1-21). DOI: 10.1145/3544548.3581388

Zhang, B., Li, L., Chen, Y., Zhao, Q., Zhu, Z., Yi, X., & Xie, X. (2025). Knowing Your Uncertainty—On the application of LLM in social sciences. *arXiv preprint arXiv:2512.05461.*

Chapter 10

Bridging Human Thought and Machine Intelligence: Cognitive-Aligned Neuro-Symbolic AI Systems and Environment

Rubee Singh
http://orcid.org/0000-0002-5406-2210
GLA University, Mathura, India

ABSTRACT

Artificial intelligence has achieved significant progress in perception and prediction, yet many contemporary systems remain limited in interpretability, contextual reasoning, and alignment with human cognition. This chapter advances the concept of cognitive aligned neuro symbolic AI systems as a framework for bridging human thought and machine intelligence in dynamic environments. By integrating neural learning mechanisms with structured symbolic reasoning, these systems combine adaptive perception with transparent inference and explicit knowledge representation. Drawing from cognitive science and systems theory, the chapter proposes that meaningful AI advancement requires architectural alignment with core human cognitive processes, including contextual judgment, memory integration, and reflective reasoning. The discussion conceptualizes how such systems operate under uncertainty, support human centered collaboration, and contribute to trustworthy and responsible AI development.

DOI: 10.4018/979-8-3373-8988-2.ch010

1. INTRODUCTION

1.1 The Transformation of Artificial Intelligence and Emerging Limitations

Artificial intelligence has undergone a profound transformation over the past decade, driven largely by advances in deep learning and large-scale data processing (LeCun et al., 2015; Goodfellow et al., 2016). Neural architectures have demonstrated remarkable capabilities in image recognition, speech processing, language generation, and predictive analytics, often achieving or surpassing human level performance in benchmark tasks (Hinton et al., 2012; Brown et al., 2020). These achievements have accelerated AI deployment across healthcare, finance, manufacturing, education, and governance systems (Russell & Norvig, 2021). Yet, despite these computational successes, contemporary AI systems continue to face foundational challenges related to interpretability, contextual understanding, and robustness beyond training distributions (Marcus, 2018; Rudin, 2019).

Deep neural networks operate through distributed representations and high dimensional parameter optimization, which enable powerful pattern extraction but limit transparency in reasoning pathways (Lipton, 2018). As a result, many systems function as opaque models whose decision logic is difficult to interpret or audit (Doshi Velez & Kim, 2017). This opacity raises concerns in high stakes domains where accountability and trust are essential (Floridi et al., 2018; Mittelstadt et al., 2016). Moreover, neural systems often struggle with compositional reasoning, causal inference, and systematic generalization, particularly when confronted with novel or uncertain contexts (Lake et al., 2017; Marcus, 2020). These limitations have renewed scholarly interest in revisiting structured reasoning paradigms that characterized earlier phases of artificial intelligence research (Nilsson, 1980; Brachman & Levesque, 2004).

1.2 From Symbolic Reasoning to Neuro Symbolic Integration

The origins of artificial intelligence are deeply rooted in symbolic reasoning traditions. Early AI systems emphasized formal logic, rule-based inference, and explicit knowledge representation as mechanisms for replicating human problem solving (Newell & Simon, 1976). Symbolic architectures enabled traceable reasoning processes, allowing developers to inspect inference chains and justify conclusions (Russell & Norvig, 2021). However, such systems exhibited brittleness in uncertain or data rich environments, as they relied heavily on predefined rules and manually encoded knowledge structures (Brachman & Levesque, 2004).

The rise of neural networks shifted emphasis toward statistical learning from large datasets (Hinton et al., 2012). Neural models offered scalability and adaptability but often sacrificed interpretability and logical structure (Goodfellow et al., 2016). This divergence between symbolic transparency and neural adaptability created a methodological divide within AI research (Marcus, 2018). In response, scholars proposed neuro symbolic approaches that integrate neural perception with symbolic reasoning components (Besold et al., 2017; Garcez et al., 2019). These hybrid systems aim to combine the strengths of both paradigms, enabling machines to learn from data while maintaining structured inference capabilities (De Raedt et al., 2020).

Recent research suggests that neuro symbolic architectures enhance explainability and improve reasoning under limited data conditions (Garcez et al., 2015). By embedding logical constraints and domain knowledge into learning systems, such models can generate more coherent and interpretable outputs (Rocktäschel & Riedel, 2017). However, integration at the technical level does not automatically ensure alignment with human cognitive processes. Without cognitive orientation, hybrid systems risk becoming computationally sophisticated yet conceptually detached from how humans perceive and reason (Lake et al., 2017).

1.3 The Imperative of Cognitive Alignment

Human cognition integrates intuitive pattern recognition with deliberate symbolic reasoning (Kahneman, 2011; Evans & Stanovich, 2013). Dual process theories describe how rapid associative mechanisms operate alongside slower analytical reasoning systems, enabling humans to navigate complex environments with flexibility and contextual awareness (Sloman, 1996; Gigerenzer, 2008). Cognitive architectures such as ACT R and CLARION demonstrate that intelligence emerges from layered interactions between implicit and explicit processes rather than from a single computational mechanism (Anderson, 2007; Sun, 2006). These insights suggest that artificial systems designed to replicate only statistical learning or only logical reasoning remain incomplete representations of cognitive intelligence.

Cognitive alignment refers to the intentional design of AI architectures that reflect core attributes of human cognition, including abstraction, contextual judgment, memory integration, and reflective evaluation (Tenenbaum et al., 2011). In dynamic environments characterized by uncertainty and evolving constraints, such alignment becomes critical (Dietterich, 2019). Purely data driven systems often exhibit fragility under distributional shifts or novel scenarios (Amodei et al., 2016). Humans, by contrast, leverage prior knowledge and symbolic reasoning to adapt under incomplete information (Pearl, 2009). Neuro symbolic systems that incorporate structured reasoning alongside adaptive learning are better positioned to replicate this capacity (Garcez et al., 2019).

The importance of cognitive alignment is further reinforced by contemporary discourse on responsible and trustworthy AI (Jobin et al., 2019). Ethical frameworks emphasize transparency, accountability, fairness, and human oversight as essential pillars of AI governance (Floridi et al., 2018; Mittelstadt et al., 2016). Explainable AI research highlights the need for systems that generate reasoning pathways understandable to human stakeholders rather than merely producing statistically optimized outputs (Rudin, 2019). Neuro symbolic architectures provide inherent structural advantages in this regard because symbolic components can offer explicit inference chains aligned with domain knowledge (Besold et al., 2017).

Bridging human thought and machine intelligence therefore requires more than computational efficiency. It requires systems that reason, adapt, and communicate in ways compatible with human cognitive frameworks (Lake et al., 2017). Cognitive aligned neuro symbolic AI systems represent an architectural philosophy aimed at achieving this compatibility. By integrating neural learning, symbolic reasoning, and human centered design principles, such systems offer a pathway toward interpretable, adaptive, and socially aligned artificial intelligence capable of operating effectively within complex socio technical environments.

The subsequent sections of this chapter build upon this foundation by examining cognitive principles relevant to AI design, tracing the evolution of reasoning paradigms, articulating a comprehensive cognitive alignment framework, and exploring operational, ethical, and application-oriented implications for next generation AI systems.

Despite growing interest in neuro-symbolic artificial intelligence, existing scholarship has largely concentrated on technical integration challenges and computational performance. Comparatively limited attention has been devoted to cognitive alignment as a unifying framework that connects human reasoning, explainability, ethical governance, and adaptive intelligence. Addressing this gap, the present chapter advances a cognitively aligned perspective that situates neuro-symbolic architectures within broader human-centred and socio-technical contexts, thereby extending current theoretical discussions beyond algorithmic integration alone.

2. FOUNDATIONS OF HUMAN COGNITION FOR AI DESIGN

2.1 Dual Process Reasoning and Cognitive Layering

Human cognition operates through the interaction of complementary reasoning systems rather than a single uniform computational process. Dual process theories distinguish between intuitive processing, which is rapid, associative, and experience driven, and analytical processing, which is deliberate, rule guided, and symbolically structured (Kahneman, 2011; Evans & Stanovich, 2013). Intuitive reasoning allows individuals to recognize patterns quickly and respond efficiently in familiar situations, while analytical reasoning enables structured problem solving, abstraction, and logical evaluation (Sloman, 1996). The effectiveness of human intelligence arises from the coordination between these systems rather than their isolation (Gigerenzer, 2008).

Artificial intelligence systems that rely solely on neural networks replicate certain aspects of intuitive cognition through pattern recognition and statistical learning (LeCun et al., 2015). However, they often lack the structured symbolic manipulation required for explicit reasoning and explanation (Marcus, 2018). Conversely, purely symbolic systems mirror analytical reasoning but lack adaptive flexibility and experiential learning capacity (Newell & Simon, 1976). Neuro symbolic integration reflects the layered structure of human cognition by combining subsymbolic learning with symbolic inference mechanisms (Garcez et al., 2019). Cognitive alignment therefore implies architectural designs that emulate this layered interaction between intuitive learning and analytical reasoning.

2.2 Memory Architecture, Abstraction, and Causal Reasoning

Human cognition is further characterized by interconnected memory systems that support abstraction and generalization. Working memory enables temporary manipulation of information for reasoning tasks, while semantic memory stores structured conceptual knowledge and episodic memory captures contextual experiences (Baddeley, 2012). These memory layers allow individuals to construct abstract representations that extend beyond specific observations (Tenenbaum et al., 2011). Abstraction enables humans to transfer knowledge across domains, reason about hypothetical scenarios, and form coherent conceptual frameworks (Fodor, 1975).

In artificial systems, neural representations resemble experiential memory by encoding distributed patterns learned from data (Goodfellow et al., 2016). Symbolic structures such as knowledge graphs and ontologies parallel semantic memory by providing explicit relational frameworks (De Raedt et al., 2020). Integrating these components enhances interpretability and supports compositional reasoning, which purely neural architectures often struggle to achieve (Lake et al., 2017). Symbolic layers introduce structural stability and logical coherence, while neural layers allow adaptive learning under changing conditions (Battaglia et al., 2018).

Causal reasoning represents another core dimension of human cognition. Humans construct mental models that explain relationships between events, enabling prediction, intervention, and counterfactual analysis (Pearl, 2009). Purely correlation-based AI systems lack inherent causal structure and may misinterpret spurious associations as meaningful relationships (Marcus, 2020). Embedding symbolic causal representations within learning systems strengthens robustness and explanation generation, particularly in dynamic environments where understanding cause and effect is critical (Dietterich, 2019). Cognitive alignment thus requires explicit modeling of structured knowledge alongside adaptive inference.

2.3 Contextual Awareness and Reflective Meta Cognition

Human intelligence is deeply contextual. Individuals interpret information within environmental, social, and cultural frameworks, adjusting reasoning strategies as circumstances evolve (Bransford et al., 2000). Context sensitivity allows humans to reconcile conflicting information and adapt decisions when goals or constraints shift. Artificial systems that optimize fixed statistical objectives often struggle under distributional shifts or novel scenarios (Amodei et al., 2016). Incorporating symbolic constraints and context modeling mechanisms enhances adaptability by guiding inference beyond raw data correlations (Garcez et al., 2015).

Meta cognition further distinguishes human reasoning from purely algorithmic computation. Humans possess the capacity to evaluate their own thought processes, assess uncertainty, and revise beliefs when confronted with new evidence (Flavell, 1979). This reflective capacity contributes to resilience, ethical deliberation, and learning from error. In AI design, integrating uncertainty estimation, reasoning traceability, and self-monitoring mechanisms supports responsible decision making and transparency (Rudin, 2019). Neuro symbolic architectures can facilitate such reflection by enabling consistency checks between learned representations and structured rules (Garcez et al., 2019).

Taken together, these cognitive foundations demonstrate that intelligence emerges from integrated layers of perception, abstraction, causality, contextual awareness, and reflection. Designing AI systems aligned with these principles requires intentional integration of neural learning with symbolic reasoning structures. Such cognitive alignment provides the conceptual basis for developing interpretable, adaptive, and socially responsive artificial intelligence systems capable of functioning effectively in complex and uncertain environments.

3. EVOLUTION FROM SYMBOLIC AI TO NEURO SYMBOLIC SYSTEMS

Table 1. Evolution from symbolic AI to neuro-symbolic systems

Dimension	Symbolic AI	Neural AI	Cognitive-Aligned Neuro-Symbolic AI
Learning	Rule-based	Data-driven	Hybrid
Explainability	High	Low	High
Adaptability	Low	High	High
Causal Reasoning	Strong	Weak	Strong
Human Alignment	Moderate	Limited	High
Trustworthiness	Moderate	Limited	High

3.1 The Era of Symbolic Artificial Intelligence

The origins of artificial intelligence are deeply rooted in symbolic reasoning traditions that emerged in the mid twentieth century. Early AI researchers conceptualized intelligence as the manipulation of explicit symbols according to formal rules, reflecting a computational theory of mind in which cognition was understood as structured information processing (Newell & Simon, 1976; Fodor, 1975). Symbolic systems relied on logic-based representations, production rules, and knowledge bases to perform reasoning tasks (Nilsson, 1980). Expert systems such as MYCIN and rule driven inference engines demonstrated the potential of formalized knowledge representation in domains requiring structured decision making (Brachman & Levesque, 2004).

Symbolic AI offered clear advantages in interpretability and transparency. Reasoning pathways could be traced step by step, enabling verification and explanation of outputs (Russell & Norvig, 2021). This explicit structure aligned closely with analytical aspects of human cognition, particularly logical deduction and abstract reasoning (Pylyshyn, 1984). However, symbolic systems required extensive manual encoding of knowledge and struggled in environments characterized by uncertainty, ambiguity, or incomplete information (Marcus, 2018). Their brittleness limited scalability, and they performed poorly in perceptual tasks such as vision and speech recognition (Lake et al., 2017).

3.2 The Rise of Neural Learning and Data Driven Paradigms

The limitations of symbolic AI, combined with increasing computational power and data availability, led to the resurgence of neural network approaches. Artificial neural networks, inspired by biological neurons, enabled distributed representation learning through iterative weight adjustment (LeCun et al., 2015). Breakthroughs in backpropagation, convolutional networks, and deep architectures revolutionized pattern recognition across domains (Hinton et al., 2012; Goodfellow et al., 2016). Deep learning models achieved remarkable success in image classification, natural language processing, and speech recognition, often surpassing traditional symbolic systems in accuracy benchmarks (Brown et al., 2020).

Neural networks excel at learning high dimensional statistical relationships without explicit rule specification. Their strength lies in adaptive generalization from large datasets and robustness in perceptual tasks (Battaglia et al., 2018). However, these systems operate as complex parameterized functions whose internal representations are difficult to interpret (Lipton, 2018). Unlike symbolic systems, neural architectures lack inherent mechanisms for explicit logical reasoning, structured abstraction, or compositional inference (Marcus, 2020). Their performance often deteriorates under distributional shifts or adversarial perturbations, revealing fragility in dynamic environments (Amodei et al., 2016).

This divergence between symbolic reasoning and neural learning created a methodological tension within AI research. Symbolic systems provided transparency but lacked adaptability, whereas neural systems provided adaptability but lacked interpretability and structured reasoning (Garcez et al., 2019). Addressing this divide became a central research challenge.

3.3 Emergence of Neuro Symbolic Integration

Neuro symbolic AI emerged as a response to the limitations of both paradigms. Scholars proposed architectures that integrate neural networks for perception and learning with symbolic reasoning modules for abstraction and inference (Besold et al., 2017; Garcez et al., 2019). These hybrid systems aim to capture the complementary strengths of distributed learning and rule based logic within unified frameworks (De Raedt et al., 2020).

In neuro symbolic systems, neural components process raw sensory data and extract representations, while symbolic layers encode structured knowledge such as ontologies, relational schemas, and causal rules (Rocktäschel & Riedel, 2017). Logical constraints can be embedded within neural training objectives, guiding learning toward consistent reasoning patterns (Garcez et al., 2015). Conversely, symbolic modules can utilize neural embeddings to handle uncertainty and variability in real world data (Battaglia et al., 2018). Recent research demonstrates that neuro symbolic architectures enhance compositional generalization, reasoning under limited data, and explainability compared to purely neural approaches (Lake et al., 2017). By integrating structured knowledge with adaptive learning, these systems move closer to replicating layered aspects of human cognition (Marcus, 2018). Moreover, symbolic reasoning modules provide explicit inference traces that support accountability and auditability, addressing ethical and regulatory concerns in high stakes applications (Rudin, 2019).

The evolution from symbolic AI to neural networks and ultimately to neuro symbolic integration reflects a broader recognition that intelligence cannot be reduced to a single computational paradigm. Instead, effective artificial intelligence requires harmonizing learning efficiency with structured reasoning. This historical trajectory establishes the foundation for articulating a cognitive alignment framework, which seeks not only technical integration but architectural compatibility with human modes of thinking.

The historical progression from symbolic reasoning to neural learning and subsequently to neuro-symbolic integration illustrates a gradual shift toward more comprehensive models of intelligence. While symbolic systems contributed transparency and logical structure, neural approaches enhanced adaptability and learning efficiency. Cognitive-aligned neuro-symbolic systems seek to synthesize these strengths, creating architectures capable of supporting both robust computational performance and meaningful human interpretability within complex decision environments.

4. COGNITIVE ALIGNMENT FRAMEWORK

4.1 Conceptualizing Cognitive Alignment

Cognitive alignment refers to the intentional design of artificial intelligence systems in ways that reflect fundamental structures of human cognition, including abstraction, contextual reasoning, causal inference, and reflective evaluation (Tenenbaum et al., 2011; Kahneman, 2011). While neuro symbolic integration provides technical hybridity, cognitive alignment adds a philosophical and architectural orientation toward compatibility with human thought processes (Garcez et al., 2019). It shifts the focus from performance optimization alone toward interpretability, adaptability, and socio technical coherence (Floridi et al., 2018).

Human cognition is structured, layered, and context sensitive. It integrates perception, memory, reasoning, and self-monitoring in dynamic interaction (Anderson, 2007; Sun, 2006). Cognitive alignment therefore requires AI systems to mirror this layered structure rather than treating learning and reasoning as isolated components. Alignment does not imply replication of human cognition in biological terms but functional compatibility with how humans interpret, evaluate, and communicate decisions (Marcus, 2018).

In complex environments where AI interacts with human stakeholders, misalignment can generate mistrust, misunderstanding, or unintended consequences (Mittelstadt et al., 2016). Systems optimized solely for statistical accuracy may produce outputs that are technically correct yet cognitively opaque or normatively problematic (Rudin, 2019). A cognitive alignment framework ensures that reasoning processes resonate with human conceptual models, enabling meaningful collaboration between humans and machines.

4.2 Principles of Layered Cognitive Architecture

The first principle of cognitive alignment is layered architecture. Human cognition operates through interacting subsystems that manage perception, symbolic reasoning, and reflective evaluation (Baddeley, 2012). Translating this into AI design implies structuring systems into interconnected modules with distinct but complementary roles.

Neural layers handle perception and adaptive learning, extracting patterns from sensory or textual data (LeCun et al., 2015). Symbolic layers encode structured knowledge, ontologies, and logical rules that support abstraction and inference (De Raedt et al., 2020). Reflective layers monitor reasoning consistency, estimate uncertainty, and evaluate alignment with predefined constraints or values (Amodei et al., 2016).

Such layering allows distributed representations to interact with explicit symbolic structures, supporting compositional reasoning and explanation generation (Battaglia et al., 2018). For example, neural embeddings can identify relevant features in input data, while symbolic reasoning modules apply domain specific rules to derive interpretable conclusions (Rocktäschel & Riedel, 2017). This architecture reflects dual process cognition, integrating intuitive pattern recognition with analytical reasoning (Evans & Stanovich, 2013).

4.3 Contextual and Causal Integration

A second principle of cognitive alignment is the explicit incorporation of contextual and causal reasoning. Human intelligence depends heavily on situational interpretation and cause effect modeling (Pearl, 2009). Individuals rarely evaluate information in isolation; instead, they construct mental models that explain relationships within broader environmental and social contexts (Bransford et al., 2000).

AI systems optimized for static datasets often struggle with contextual shifts or distributional changes (Dietterich, 2019). Cognitive alignment addresses this limitation by embedding contextual representation mechanisms within symbolic layers while allowing neural components to update representations dynamically (Garcez et al., 2015). Causal models encoded symbolically can guide inference when data patterns are ambiguous or incomplete, enhancing robustness in dynamic environments (Marcus, 2020).

Incorporating causal reasoning enables AI systems to generate counterfactual explanations and justify decisions beyond correlation-based outputs (Pearl, 2009). This capability strengthens transparency and aligns system outputs with human expectations regarding explanation and accountability (Rudin, 2019). Contextual integration further ensures that system behavior adapts appropriately to changing goals, constraints, and stakeholder values.

4.4 Reflective Governance and Ethical Coherence

The third principle of cognitive alignment involves reflective governance and ethical coherence. Human reasoning includes meta cognitive capacities that evaluate uncertainty, revise beliefs, and assess normative implications (Flavell, 1979). In AI systems, reflective components can be implemented through uncertainty estimation, consistency checking, and rule based validation mechanisms (Amodei et al., 2016).

Responsible AI frameworks emphasize transparency, fairness, and accountability as foundational pillars of trustworthy system design (Jobin et al., 2019; Floridi et al., 2018). Cognitive alignment supports these principles by structuring reasoning processes in ways that are auditable and interpretable. Symbolic reasoning modules provide explicit inference chains, while neural confidence measures communicate probabilistic uncertainty (Garcez et al., 2019).

Ethical coherence also requires alignment between system outputs and human values. Embedding normative constraints within symbolic layers allows systems to evaluate decisions against predefined ethical rules or policy guidelines (Mittelstadt et al., 2016). This integration does not eliminate bias or error, but it introduces structured mechanisms for monitoring and correction.

Collectively, these four dimensions conceptual clarity, layered architecture, contextual causal integration, and reflective governance constitute a comprehensive cognitive alignment framework. This framework extends neuro symbolic integration beyond technical hybridity toward human centered design philosophy. By embedding cognitive principles into architectural structures, AI systems become not only computationally powerful but also interpretable, adaptive, and socially aligned.

The subsequent section operationalizes this framework by detailing architectural components and integration strategies for cognitive aligned neuro symbolic systems functioning within dynamic and uncertain environments.

5. ARCHITECTURE OF COGNITIVE ALIGNED NEURO SYMBOLIC SYSTEMS

5.1 Structural Composition and Modular Integration

Cognitive aligned neuro symbolic systems are structured as modular architectures in which neural and symbolic components interact dynamically rather than operating as isolated subsystems (Garcez et al., 2019; De Raedt et al., 2020). The architectural foundation typically includes three primary layers: a perceptual learning layer, a symbolic reasoning layer, and a reflective evaluation layer. This layered composition mirrors the organization of human cognition, where perception, reasoning, and meta cognition function in coordinated interaction (Anderson, 2007; Sun, 2006).

The perceptual learning layer is implemented through neural networks capable of processing high dimensional input such as text, images, sensor data, or behavioral signals (LeCun et al., 2015). These neural components extract latent representations and identify statistical patterns across datasets. However, rather than functioning as terminal decision makers, their outputs are transferred to symbolic modules that interpret and structure information within conceptual frameworks (Battaglia et al., 2018).

The symbolic reasoning layer incorporates knowledge graphs, ontologies, rule based systems, and logical inference engines (Brachman & Levesque, 2004). This layer encodes domain specific constraints, relational structures, and causal models that guide decision making beyond raw correlation. Logical constraints can be embedded into learning objectives, ensuring that neural outputs remain consistent with symbolic knowledge (Garcez et al., 2015). This integration enhances explainability by producing traceable inference chains aligned with structured knowledge.

5.2 Knowledge Representation and Reasoning Mechanisms

Effective cognitive alignment depends on robust knowledge representation strategies. Symbolic layers rely on structured representations such as semantic networks and ontologies to encode hierarchical and relational knowledge (Russell & Norvig,

2021). These representations provide stability and interpretability, enabling reasoning through deduction, abduction, or causal inference (Pearl, 2009).

Neural embeddings complement these structures by capturing nuanced associations derived from experiential data (Goodfellow et al., 2016). The interaction between symbolic schemas and neural representations supports compositional generalization, allowing systems to apply learned knowledge to novel contexts (Lake et al., 2017). For example, neural models may detect patterns in medical imaging, while symbolic rules interpret findings within established clinical guidelines.

Reasoning mechanisms in cognitive aligned systems often involve bidirectional communication between layers. Symbolic constraints can regularize neural learning processes, while neural outputs can update symbolic knowledge bases through probabilistic inference (Rocktäschel & Riedel, 2017). This iterative exchange reflects the dynamic interplay between intuition and analysis observed in human cognition (Evans & Stanovich, 2013).

5.3 Reflective Monitoring and Uncertainty Management

A defining architectural feature of cognitive aligned systems is reflective monitoring. Reflective layers evaluate consistency between neural outputs and symbolic constraints, identify conflicts, and estimate uncertainty levels (Amodei et al., 2016). This mechanism parallels human meta cognitive evaluation, which enables belief revision and error correction (Flavell, 1979). Uncertainty estimation is essential for trustworthy AI deployment, particularly in high stakes environments (Rudin, 2019). Neural components can provide probabilistic confidence scores, while symbolic reasoning modules verify logical coherence. When discrepancies arise, reflective modules can trigger reassessment or request additional input. Such governance mechanisms strengthen reliability and enhance transparency in decision processes (Floridi et al., 2018).

Through modular integration, structured knowledge representation, and reflective evaluation, cognitive aligned neuro symbolic architectures move beyond technical hybridity toward coherent and interpretable system design. The next section explores how these architectures operate within dynamic and uncertain environments where adaptability and contextual reasoning are critical.

6. OPERATION IN DYNAMIC AND UNCERTAIN ENVIRONMENTS

6.1 Adaptation under Distributional Shifts

Dynamic environments are characterized by evolving conditions, incomplete information, and unpredictable disruptions. Artificial intelligence systems deployed in such contexts must adapt to distributional shifts where real world data diverge from training conditions (Dietterich, 2019). Purely data driven models often struggle in these scenarios because they rely heavily on learned correlations that may not generalize beyond historical datasets (Amodei et al., 2016). Human cognition, by contrast, leverages abstraction and prior knowledge to reinterpret novel circumstances (Tenenbaum et al., 2011).

Cognitive aligned neuro symbolic systems address this challenge through the interaction of adaptive neural learning and structured symbolic reasoning (Garcez et al., 2019). Neural components continuously update representations based on new inputs, while symbolic layers provide stable conceptual and causal frameworks that guide reinterpretation. This layered interaction allows systems to respond flexibly without abandoning structured coherence. When environmental parameters shift, symbolic constraints can help maintain logical consistency, preventing erratic decision behavior (Marcus, 2020).

6.2 Context Sensitive Reasoning and Decision Revision

Dynamic environments require contextual awareness. Decisions that are appropriate in one situation may be inappropriate in another due to changing goals, constraints, or stakeholder expectations (Bransford et al., 2000). Cognitive aligned architectures incorporate contextual modeling mechanisms within symbolic layers, enabling systems to encode situational rules and relational dependencies (Brachman & Levesque, 2004).

Neural networks contribute by detecting subtle variations in input patterns that signal contextual change (LeCun et al., 2015). Symbolic modules interpret these signals in light of domain knowledge, adjusting inference pathways accordingly. This bidirectional communication enables decision revision when new evidence conflicts with prior assumptions. Reflective monitoring components can identify inconsistencies between learned predictions and rule-based constraints, triggering reassessment or updating of knowledge structures (Flavell, 1979).

Such context sensitive revision is particularly important in domains such as healthcare, autonomous systems, and policy governance, where evolving information may alter risk evaluations or ethical considerations (Floridi et al., 2018). By embedding contextual reasoning into architectural design, cognitive aligned systems achieve resilience and adaptability beyond static optimization models.

6.3 Causal Inference and Counterfactual Reasoning

Another critical feature of dynamic environments is the need for causal understanding rather than mere correlation. Human decision making relies on mental models that represent cause and effect relationships, enabling prediction and intervention (Pearl, 2009). Correlation based neural models often lack this explicit causal structure, limiting their capacity to generate counterfactual explanations or evaluate alternative scenarios (Marcus, 2018).

Neuro symbolic architectures incorporate causal reasoning through symbolic representations of dependencies and rule based inference mechanisms (Garcez et al., 2015). These symbolic structures can guide neural predictions by constraining outputs within plausible causal frameworks. For instance, in industrial systems, symbolic causal models can represent process dependencies, while neural components monitor sensor data for anomalies. When disruptions occur, the system can reason about potential causal pathways and suggest corrective interventions.

Counterfactual reasoning enhances interpretability and supports human trust. By explaining not only what decision was made but also why alternative outcomes were rejected, cognitive aligned systems provide insight consistent with human expectations of explanation (Rudin, 2019). This capacity strengthens accountability and facilitates collaborative problem solving in uncertain conditions.

6.4 Human AI Co Adaptation

Dynamic environments often involve continuous interaction between humans and AI systems. Cognitive alignment supports co adaptation, where both system and user learn from ongoing feedback (Sun, 2006). Neural components update models based on new data, while symbolic layers incorporate policy changes or revised domain knowledge. Human oversight can refine symbolic constraints, creating a feedback loop that improves system performance and ethical alignment (Jobin et al., 2019). Through layered adaptation, contextual reasoning, causal modeling, and reflective monitoring, cognitive aligned neuro symbolic systems achieve operational robustness in uncertain settings. These capabilities position them as viable frameworks for complex socio technical environments where interpretability, adaptability, and trust are indispensable.

7. HUMAN CENTERED AI, ETHICS, AND TRUST

7.1 Human Centered Orientation in Cognitive Aligned Systems

Human centered artificial intelligence emphasizes the design of systems that augment rather than replace human agency, ensuring that technological development remains aligned with human values, capabilities, and social contexts (Floridi et al., 2018; Russell, 2019). In cognitively aligned neuro symbolic architectures, this orientation is embedded at the structural level rather than treated as an external add on. Because such systems incorporate symbolic reasoning, contextual modeling, and reflective monitoring, they are inherently positioned to support interpretability and collaborative interaction (Garcez et al., 2019).

Human centered design requires that AI outputs be understandable and actionable for stakeholders with varying expertise (Rudin, 2019). Purely neural models often generate predictions without transparent reasoning pathways, limiting their practical utility in high stakes domains (Lipton, 2018). By contrast, symbolic reasoning layers produce structured inference chains that can be communicated in forms consistent with human conceptual models (De Raedt et al., 2020). This alignment enhances cognitive compatibility and reduces the gap between system output and user interpretation.

Furthermore, cognitive alignment facilitates shared mental models between humans and machines. When symbolic representations reflect domain knowledge structures familiar to practitioners, users can better anticipate system behavior and collaborate effectively (Bransford et al., 2000). Such alignment fosters a partnership model of human AI interaction, in which systems function as cognitive collaborators rather than opaque automation tools (Sun, 2006).

7.2 Explainability, Transparency, and Accountability

Explainability is central to ethical AI governance. Stakeholders require insight into how and why decisions are produced, particularly in areas such as healthcare, finance, criminal justice, and public policy (Doshi Velez & Kim, 2017; Mittelstadt et al., 2016). Neuro symbolic systems offer structural advantages for explainability because symbolic reasoning modules provide explicit rule based inference traces (Garcez et al., 2015). These traces can be inspected, audited, and verified against domain standards.

Transparency extends beyond explanation to include visibility into uncertainty and limitations. Cognitive aligned architectures integrate uncertainty estimation mechanisms that communicate probabilistic confidence levels alongside symbolic justification (Amodei et al., 2016). This dual communication reflects human meta cognitive practices, where individuals assess both reasoning and confidence in judgments (Flavell, 1979). Providing calibrated uncertainty reduces overreliance on automated outputs and supports responsible human oversight (Rudin, 2019).

Accountability further requires that decision pathways be traceable. Symbolic components encode logical dependencies that can be reviewed retrospectively, enabling error analysis and policy evaluation (Pearl, 2009). In dynamic environments where outcomes may have ethical implications, such traceability is essential for maintaining public trust and regulatory compliance (Floridi et al., 2018). Cognitive alignment strengthens accountability by ensuring that reasoning processes are not hidden within uninterpretable parameter spaces.

7.3 Ethical Alignment and Normative Constraints

Ethical AI frameworks emphasize fairness, non-discrimination, privacy protection, and respect for human autonomy (Jobin et al., 2019). Embedding ethical principles within AI systems requires explicit representation of normative constraints rather than reliance on implicit statistical correlations (Mittelstadt et al., 2016). Cognitive aligned neuro symbolic systems can incorporate such constraints within symbolic layers, encoding policy guidelines or ethical rules as part of the reasoning architecture (Garcez et al., 2019).

For example, symbolic modules can enforce constraints that prevent decisions based on protected attributes or require justification for high-risk actions. Neural components may detect patterns, but symbolic reasoning ensures that outputs remain consistent with normative principles (Russell, 2019). This layered approach reflects human ethical deliberation, where intuitive judgments are evaluated against moral standards and social norms (Kahneman, 2011).

Ethical alignment also involves contextual sensitivity. Normative expectations vary across cultural and institutional environments (Floridi et al., 2018). Symbolic layers can be adapted to incorporate context specific rules, enabling systems to operate responsibly within diverse socio technical ecosystems. Such adaptability reinforces trust and legitimacy.

7.4 Building and Sustaining Trust in Dynamic Systems

Trust in AI systems emerges from perceived reliability, transparency, and alignment with user expectations (Mittelstadt et al., 2016). Cognitive alignment contributes to trust by structuring reasoning processes in ways that mirror human cognitive frameworks. When users can understand system logic and evaluate uncertainty, they are more likely to engage constructively with automated recommendations (Rudin, 2019).

In dynamic environments, trust must be sustained through continuous adaptation and feedback. Reflective monitoring components enable systems to detect anomalies, revise beliefs, and communicate updates transparently (Amodei et al., 2016). This responsiveness parallels human learning processes and signal's reliability over time.

Moreover, human AI co adaptation fosters relational trust. When systems incorporate feedback from users and adjust symbolic knowledge structures accordingly, collaboration becomes iterative and mutually informative (Sun, 2006). Such interaction transforms AI from a static decision engine into a dynamic participant within socio technical networks.

Ultimately, cognitive aligned neuro symbolic architectures provide a structural foundation for human centered, ethical, and trustworthy AI. By integrating interpretability, normative constraints, contextual reasoning, and reflective governance, these systems align computational intelligence with human cognitive and social expectations. The final section synthesizes applications, research challenges, implications, limitations, and future directions for advancing this paradigm across complex real-world domains.

8. APPLICATIONS, RESEARCH CHALLENGES, AND IMPLICATIONS

8.1 Applications Across Complex and High Stakes Domains

Cognitive aligned neuro symbolic systems demonstrate significant potential in domains where interpretability, contextual reasoning, and ethical coherence are indispensable. In healthcare decision support, neural components can process medical imaging, genomic data, and electronic health records to detect patterns, while symbolic reasoning layers incorporate clinical guidelines, diagnostic ontologies, and causal relationships among symptoms and treatment outcomes (Pearl, 2009; Rudin, 2019). This integration supports transparent diagnostic reasoning and reduces the risk of opaque algorithmic recommendations in critical care contexts. By enabling explanation generation aligned with medical knowledge structures, such systems foster clinician trust and enhance collaborative decision making.

In industrial and intelligent manufacturing systems, neural networks analyze real time sensor data to identify anomalies or predictive maintenance signals, whereas symbolic modules encode process dependencies, safety rules, and operational constraints (Dietterich, 2019). When irregularities occur, symbolic reasoning can trace potential causal pathways and recommend corrective actions grounded in engineering principles. This layered interaction improves resilience and ensures that automated decisions remain consistent with established operational standards.

Autonomous systems represent another significant application domain. Neural perception layers interpret environmental inputs such as traffic patterns or spatial obstacles, while symbolic reasoning modules enforce regulatory compliance and contextual interpretation of situational norms (Russell & Norvig, 2021). In dynamic environments, such integration enables adaptive decision revision while preserving rule adherence. This architecture strengthens safety and accountability in contexts where automated actions directly affect human wellbeing.

Cognitive alignment is also highly relevant in governance and sustainability-oriented systems. Neural models can analyze environmental data streams related to energy consumption, emissions patterns, or resource distribution, while symbolic reasoning layers evaluate decisions against policy frameworks and sustainability objectives (Floridi et al., 2018). The explicit representation of regulatory and normative constraints allows transparent assessment of policy implications. Such systems can support evidence-based decision making while maintaining traceability and ethical coherence.

In education and adaptive learning environments, neuro symbolic architectures enable personalized instruction by combining neural analytics of learner performance with symbolic curricular structures and pedagogical rules (Lake et al., 2017). This integration facilitates contextual adaptation while preserving transparency in assessment and recommendation processes.

8.2 Research Challenges in Cognitive Aligned Architectures

Despite its promise, the implementation of cognitive aligned neuro symbolic AI faces significant research challenges. Scalability remains a primary concern. Symbolic reasoning systems often encounter computational limitations when processing large scale knowledge bases or complex rule interactions (Brachman & Levesque, 2004). Integrating symbolic inference with high dimensional neural embeddings requires efficient optimization strategies that balance interpretability with computational feasibility (Garcez et al., 2019).

Knowledge acquisition presents another challenge. Symbolic layers depend on structured domain knowledge, yet constructing comprehensive ontologies and rule sets can be labor intensive and prone to incompleteness (De Raedt et al., 2020). Automated extraction of symbolic representations from neural models remains an evolving research area. Bridging subsymbolic learning with explicit rule generation demands advances in representation learning and explainable inference.

Evaluation methodologies also require refinement. Traditional AI metrics prioritize predictive accuracy and benchmark performance (Marcus, 2018). However, cognitive alignment necessitates broader assessment criteria, including reasoning transparency, contextual robustness, ethical coherence, and human collaboration quality (Rudin, 2019). Developing standardized evaluation frameworks that capture these multidimensional attributes is essential for advancing this paradigm.

Human AI co adaptation introduces additional complexity. Systems must continuously update knowledge structures while preserving alignment with human oversight and regulatory standards (Amodei et al., 2016). Designing governance mechanisms that support iterative feedback without undermining autonomy or accountability remains an open challenge.

Finally, ethical representation poses conceptual difficulties. Encoding normative constraints within symbolic layers requires consensus on ethical standards and policy interpretations, which may vary across cultural and institutional contexts (Mittelstadt et al., 2016). Ensuring that symbolic rule bases remain adaptive to evolving societal values necessitates mechanisms for continuous review and revision.

8.3 Strategic Implications for Theory, Practice, and Policy

The adoption of cognitive aligned neuro symbolic systems carries important theoretical and practical implications. Theoretically, this approach challenges reductionist conceptions of intelligence as purely statistical optimization. It reintroduces abstraction, causality, and structured reasoning as foundational components of artificial cognition (Pearl, 2009; Lake et al., 2017). This perspective encourages interdisciplinary collaboration among AI researchers, cognitive scientists, ethicists, and systems theorists.

Practically, organizations deploying AI systems can benefit from enhanced transparency and accountability. By embedding symbolic reasoning and reflective monitoring within architecture, institutions can improve compliance with regulatory requirements and ethical standards (Jobin et al., 2019). Transparent reasoning pathways facilitate auditability and strengthen stakeholder confidence. For policymakers, cognitive alignment offers a pathway toward responsible automation in public administration and sustainability governance. Systems capable of explaining decisions and demonstrating adherence to normative constraints support democratic oversight and institutional trust (Floridi et al., 2018). This transparency is particularly critical in contexts where automated decisions influence social welfare, resource allocation, or environmental policy.

In strategic terms, cognitive aligned architectures may shape the future trajectory of AI development. As foundation models and large-scale neural systems become increasingly prevalent, integrating symbolic reasoning components can mitigate opacity and enhance robustness. This integration represents not merely a technical refinement but a paradigm shifts toward human centered artificial intelligence. Collectively, the applications, research challenges, and implications discussed in this section illustrate that cognitive alignment extends beyond algorithmic innovation. It represents a comprehensive socio technical framework aimed at harmonizing machine intelligence with human cognitive structures, ethical norms, and dynamic environmental realities.

8.4 Integrating Sustainability, Governance, and Cognitive Aligned AI

Recent interdisciplinary research strengthens the practical relevance of cognitively aligned neuro symbolic systems within sustainability and governance transitions. The integration of Industry 4.0 technologies with circular economy strategies and green HRM illustrates that digital transformation must be aligned with human capability development and environmental objectives to achieve systemic sustainability (Singh et al., 2025). Similarly, the strategic role of human capital in advancing net zero commitments highlights the importance of embedding structured normative and policy-oriented reasoning within organizational systems (Singh et al., 2024). Studies on HRM practices supporting net zero pathways further demonstrate that governance mechanisms and behavioral alignment are central to sustainable value creation (Singh et al., 2024). Modeling organizational factors for CSR implementation underscores the significance of structured reasoning in capturing complex interdependencies (Singh et al., 2024). Moreover, research on AI transparency in sustainable healthcare emphasizes the necessity of explainable architectures that integrate predictive intelligence with ethical accountability (Shafik et al., 2024).

9. CONCLUSION, LIMITATIONS, AND FUTURE SCOPE

9.1 Conclusion: Bridging Human Cognition and Artificial Intelligence

The rapid expansion of artificial intelligence has demonstrated extraordinary technical capability, yet it has simultaneously revealed structural limitations in interpretability, contextual understanding, and ethical coherence (Marcus, 2018; Rudin, 2019). While deep learning has transformed perception and pattern recognition, it has not resolved the challenge of aligning artificial reasoning with human cognitive processes. Symbolic systems, although transparent and logically structured, lack the adaptive flexibility required in data rich and dynamic environments (Newell & Simon, 1976; Brachman & Levesque, 2004). The evolution toward neuro symbolic integration reflects recognition that neither paradigm alone is sufficient for next generation intelligent systems (Garcez et al., 2019; De Raedt et al., 2020).

This chapter has argued that integration must extend beyond technical hybridity toward cognitive alignment. Cognitive aligned neuro symbolic systems embody an architectural philosophy grounded in principles of layered reasoning, contextual sensitivity, causal modeling, and reflective evaluation (Kahneman, 2011; Pearl, 2009). By embedding neural learning within structured symbolic reasoning frameworks, these systems approximate core attributes of human cognition, including abstraction, explanation, and belief revision (Lake et al., 2017). Such alignment enhances interpretability and strengthens the capacity of AI systems to operate responsibly in uncertain environments.

In dynamic socio technical ecosystems, intelligence cannot be reduced to statistical optimization. It must incorporate structured reasoning, normative awareness, and transparent communication (Floridi et al., 2018). Cognitive alignment provides a conceptual bridge between computational efficiency and human centered design. Through layered architecture, contextual modeling, and reflective governance, neuro symbolic systems advance the possibility of artificial intelligence that is not only powerful but also understandable and trustworthy.

9.2 Limitations of the Cognitive Alignment Framework

Despite its conceptual robustness, the cognitive alignment framework is not without limitations. First, architectural complexity poses practical challenges. Integrating neural and symbolic components requires sophisticated synchronization between learning algorithms and reasoning engines (Garcez et al., 2019). This integration may introduce computational overhead and engineering complexity that limit scalability in large scale deployments (Dietterich, 2019).

Second, symbolic knowledge representation depends on the availability of well-structured ontologies and domain specific rule sets (De Raedt et al., 2020). In many real-world domains, knowledge is fragmented, evolving, or contested. Constructing comprehensive symbolic frameworks may be resource intensive and vulnerable to incompleteness or bias. Moreover, rigid symbolic constraints may inadvertently restrict adaptive flexibility if not carefully designed.

Third, ethical encoding presents conceptual difficulties. Normative principles such as fairness and accountability vary across cultures and institutional settings (Mittelstadt et al., 2016). Translating such principles into formal symbolic rules requires careful interpretation and continuous revision. Static rule sets may become misaligned with evolving societal expectations, underscoring the need for dynamic governance mechanisms (Jobin et al., 2019).

Fourth, evaluation of cognitive alignment remains methodologically underdeveloped. While predictive accuracy is easily quantifiable, measuring interpretability, contextual coherence, and trust is inherently multidimensional (Rudin, 2019). Developing rigorous assessment frameworks that capture both technical performance and human cognitive compatibility remains an open research challenge.

9.3 Future Scope and Research Directions

Future scholarship should move beyond technical optimization and examine how cognitive alignment can be operationalized and evaluated across real-world settings. Particular attention is required in developing assessment frameworks that measure explainability, contextual reasoning, ethical consistency, and human trust. Such efforts will strengthen the empirical foundations of cognitive-aligned AI and support its adoption in governance, healthcare, sustainability, and other high-impact domains.

Future research should focus on enhancing scalability and integration strategies for neuro symbolic architectures. Advances in differentiable reasoning, probabilistic logic integration, and hybrid optimization techniques may enable more seamless interaction between neural embeddings and symbolic constraints (Battaglia et al., 2018). Developing efficient inference engines capable of handling large knowledge bases without sacrificing interpretability will be critical.

Automated knowledge acquisition represents another promising direction. Techniques for extracting structured rules and causal relationships from neural representations could reduce reliance on manual ontology construction (Marcus, 2020). Combining representation learning with causal discovery methods may strengthen alignment between learned patterns and structured reasoning frameworks (Pearl, 2009).

Interdisciplinary collaboration will play a central role in advancing cognitive alignment. Insights from cognitive psychology, neuroscience, philosophy of mind, and ethics can inform architectural design and governance frameworks (Anderson, 2007; Floridi et al., 2018). Establishing shared standards for transparency and ethical evaluation will support responsible deployment across sectors.

Emerging domains such as sustainability governance, digital circular economy systems, and climate adaptation platforms provide fertile ground for application. In such contexts, AI systems must integrate environmental data analytics with regulatory frameworks and long-term policy objectives. Cognitive aligned neuro symbolic systems are well positioned to support transparent and adaptive decision making in these complex environments.

Furthermore, the integration of large foundation models with symbolic reasoning layers presents a transformative research frontier. As large-scale neural architectures become central to AI infrastructure, embedding structured reasoning components may mitigate opacity and enhance contextual robustness. This direction aligns with broader calls for trustworthy and human centered AI development (Russell, 2019).

9.4 Concluding Reflections

Bridging human thought and machine intelligence requires reimagining artificial systems as participants within socio technical ecosystems rather than isolated computational engines. Cognitive aligned neuro symbolic AI systems offer a pathway toward harmonizing learning efficiency with structured reasoning and ethical coherence. By reflecting layered aspects of human cognition and embedding transparency within architectural design, these systems contribute to the development of interpretable, adaptive, and socially aligned artificial intelligence.

As AI continues to permeate critical infrastructures and governance structures, the imperative for cognitive compatibility and ethical accountability will intensify. The future of artificial intelligence will depend not only on computational power but also on its capacity to resonate with human cognitive frameworks and societal values. Cognitive alignment represents a strategic step toward realizing this vision.

REFERENCES

Amodei, D., Olah, C., Steinhardt, J., Christiano, P., Schulman, J., & Mané, D. (2016). *Concrete problems in AI safety* (arXiv:1606.06565). arXiv.

Anderson, J. R. (2007). *How can the human mind occur in the physical universe?* Oxford University Press. DOI: 10.1093/acprof:oso/9780195324259.001.0001

Baddeley, A. (2012). Working memory: Theories, models, and controversies. *Annual Review of Psychology, 63*(1), 1–29. DOI: 10.1146/annurev-psych-120710-100422 PMID: 21961947

Battaglia, P. W., Hamrick, J. B., Bapst, V., Sanchez Gonzalez, A., Zambaldi, V., Malinowski, M., Tacchetti, A., Raposo, D., Santoro, A., Faulkner, R., Gulcehre, C., Song, F., Ballard, A., Gilmer, J., Dahl, G., Vaswani, A., Allen, K., Nash, C., Langston, V., . . . Pascanu, R. (2018). Relational inductive biases, deep learning, and graph networks (arXiv:1806.01261). arXiv.

Besold, T. R., d'Avila Garcez, A. S., Bader, S., Bowman, H., Domingos, P., Hitzler, P., Kühnberger, K. U., Lamb, L. C., Lowd, D., Lima, P. M. V., de Penning, L., Pinkas, G., Poon, H., & Zaverucha, G. (2017). Neural-symbolic learning and reasoning: A survey and interpretation (arXiv:1711.03902). arXiv.

Brachman, R. J., & Levesque, H. J. (2004). *Knowledge representation and reasoning*. Morgan Kaufmann.

Bransford, J. D., Brown, A. L., & Cocking, R. R. (Eds.). (2000). *How people learn: Brain, mind, experience, and school* (Expanded ed.). National Academies Press.

Brown, T. B., Mann, B., Ryder, N., Subbiah, M., Kaplan, J., Dhariwal, P., Neelakantan, A., Shyam, P., Sastry, G., Askell, A., Agarwal, S., Herbert-Voss, A., Krueger, G., Henighan, T., Child, R., Ramesh, A., Ziegler, D. M., Wu, J., Winter, C., . . . Amodei, D. (2020). *Language models are few-shot learners* (arXiv:2005.14165). arXiv.

De Raedt, L., Dumancic, S., Manhaeve, R., & Marra, G. (2020). From statistical relational to neuro-symbolic artificial intelligence. In *Proceedings of the Twenty-Ninth International Joint Conference on Artificial Intelligence (IJCAI 2020)*. DOI: 10.24963/ijcai.2020/688

Dietterich, T. G. (2019). Robust artificial intelligence and robust human organizations. *Frontiers of Computer Science, 13*(1), 1–3. DOI: 10.1007/s11704-018-8900-4

Doshi-Velez, F., & Kim, B. (2017). *Towards a rigorous science of interpretable machine learning* (arXiv:1702.08608). arXiv.

Evans, J. St. B. T., & Stanovich, K. E. (2013). Dual-process theories of higher cognition: Advancing the debate. *Perspectives on Psychological Science, 8*(3), 223–241. DOI: 10.1177/1745691612460685 PMID: 26172965

Flavell, J. H. (1979). Metacognition and cognitive monitoring: A new area of cognitive-developmental inquiry. *The American Psychologist, 34*(10), 906–911. DOI: 10.1037/0003-066X.34.10.906

Floridi, L., Cowls, J., Beltrametti, M., Chatila, R., Chazerand, P., Dignum, V., Luetge, C., Madelin, R., Pagallo, U., Rossi, F., Schafer, B., Valcke, P., & Vayena, E. (2018). AI4People—An ethical framework for a good AI society: Opportunities, risks, principles, and recommendations. *Minds and Machines, 28*(4), 689–707. DOI: 10.1007/s11023-018-9482-5 PMID: 30930541

Fodor, J. A. (1975). *The language of thought.* Harvard University Press.

Garcez, A. S. d'Avila, Besold, T. R., De Raedt, L., Földiak, P., Hitzler, P., Icard, T., Kühnberger, K. U., Lamb, L. C., Miikkulainen, R., Silver, D. L., & Tran, S. N. (2015). Neural-symbolic learning and reasoning: Contributions and challenges. In *AAAI Spring Symposium Series.*

Garcez, A. S. d'Avila, Gori, M., Lamb, L. C., Serafini, L., Spranger, M., & Tran, S. N. (2019). Neural-symbolic computing: An effective methodology for principled integration of machine learning and reasoning (arXiv:1905.06088). arXiv.

Gigerenzer, G. (2008). *Rationality for mortals: How people cope with uncertainty.* Oxford University Press. DOI: 10.1093/oso/9780195328981.001.0001

Goodfellow, I., Bengio, Y., & Courville, A. (2016). *Deep learning*. MIT Press.

Hinton, G. E., Krizhevsky, A., & Sutskever, I. (2012). ImageNet classification with deep convolutional neural networks. In *Advances in Neural Information Processing Systems.*

Jobin, A., Ienca, M., & Vayena, E. (2019). The global landscape of AI ethics guidelines. *Nature Machine Intelligence, 1*(9), 389–399. DOI: 10.1038/s42256-019-0088-2

Kahneman, D. (2011). *Thinking, fast and slow.* Farrar, Straus and Giroux.

Lake, B. M., Ullman, T. D., Tenenbaum, J. B., & Gershman, S. J. (2017). Building machines that learn and think like people. *Behavioral and Brain Sciences, 40*, e253. DOI: 10.1017/S0140525X16001837 PMID: 27881212

LeCun, Y., Bengio, Y., & Hinton, G. (2015). Deep learning. *Nature, 521*(7553), 436–444. DOI: 10.1038/nature14539 PMID: 26017442

Lipton, Z. C. (2018). *The mythos of model interpretability* (arXiv:1606.03490). arXiv.

Marcus, G. (2018). *Deep learning: A critical appraisal* (arXiv:1801.00631). arXiv.

Marcus, G., & Davis, E. (2020). *Rebooting AI: Building artificial intelligence we can trust*. Pantheon Books.

Mittelstadt, B. D., Allo, P., Taddeo, M., Wachter, S., & Floridi, L. (2016). The ethics of algorithms: Mapping the debate. *Big Data & Society*, *3*(2), 1–21. DOI: 10.1177/2053951716679679

Newell, A., & Simon, H. A. (1976). Computer science as empirical inquiry: Symbols and search. *Communications of the ACM*, *19*(3), 113–126. DOI: 10.1145/360018.360022

Nilsson, N. J. (1980). *Principles of artificial intelligence*. Morgan Kaufmann.

Pearl, J. (2009). *Causality: Models, reasoning, and inference* (2nd ed.). Cambridge University Press. DOI: 10.1017/CBO9780511803161

Pylyshyn, Z. W. (1984). *Computation and cognition: Toward a foundation for cognitive science*. MIT Press. DOI: 10.7551/mitpress/2004.001.0001

Rocktäschel, T., & Riedel, S. (2017). *End-to-end differentiable proving* (arXiv:1705.11040). arXiv.

Rudin, C. (2019). Stop explaining black box machine learning models for high stakes decisions and use interpretable models instead. *Nature Machine Intelligence*, *1*(5), 206–215. DOI: 10.1038/s42256-019-0048-x PMID: 35603010

Russell, S. (2019). *Human compatible: Artificial intelligence and the problem of control*. Viking.

Russell, S. J., & Norvig, P. (2021). *Artificial intelligence: A modern approach* (4th ed.). Pearson.

Shafik, W., Singh, R., & Kumar, V. (2024). Artificial intelligence transparency and explainability in sustainable healthcare. In R. Singh, W. Shafik, D. Crowther, & V. Kumar (Eds.), *Transforming healthcare sector through artificial intelligence and environmental sustainability* (Approaches to Global Sustainability, Markets, and Governance). Springer. https://doi.org/DOI: 10.1007/978-981-97-9555-0_9

Singh, R., Crowther, D., & Ravesangar, K. (2024). Harnessing human capital for a sustainable future: HR's pivotal role toward net zero policies. In R. Singh & D. Crowther (Eds.), *Transition towards a sustainable future* (Approaches to Global Sustainability, Markets, and Governance). Springer. https://doi.org/DOI: 10.1007/978-981-97-5756-5_2

Singh, R., Joshi, A., Dissanayake, H., Iddagoda, A., Khan, S., Félix, M. J., & Santos, G. (2025). Integrating Industry 4.0, circular economy, and green HRM: A framework for sustainable transformation. *Sustainability (Basel)*, *17*(7), 3082. DOI: 10.3390/su17073082

Singh, R., Khan, S., Dsilva, J., Akram, U., & Haleem, A. (2024). Modelling the organisational factors for implementation of corporate social responsibility: A modified TISM approach. *Global Journal of Flexible Systems Managment*, *25*(2), 283–301. DOI: 10.1007/s40171-024-00388-x

Singh, R., Ravesangar, K., Narayanan, S., & Singh, K. K. H. (2024). HRM practices in creating net zero: Supporting organizational sustainability. In R. Singh, S. Khan, A. Kumar, S. Luthra, & H. Chokshi (Eds.), *Net zero economy, corporate social responsibility and sustainable value creation* (CSR, Sustainability, Ethics & Governance). Springer. https://doi.org/DOI: 10.1007/978-3-031-55779-8_1

Sloman, S. A. (1996). The empirical case for two systems of reasoning. *Psychological Bulletin*, *119*(1), 3–22. DOI: 10.1037/0033-2909.119.1.3 PMID: 8711015

Sun, R. (2006). *Cognition and multi-agent interaction: From cognitive modeling to social simulation*. Cambridge University Press.

Tenenbaum, J. B., Kemp, C., Griffiths, T. L., & Goodman, N. D. (2011). How to grow a mind: Statistics, structure, and abstraction. *Science*, *331*(6022), 1279–1285. DOI: 10.1126/science.1192788 PMID: 21393536

Chapter 11
Modeling Drug-Induced Cognitive Biases in Neurobehavioural Decision-Making:
Implications for Brain-Inspired AI Architectures and Bias Mitigation in Decision Intelligence

Garima Gupta
http://orcid.org/0009-0003-5562-2157
C.M.P. Degree College, University of Allahabad, India

Nidhi Tripathi
C.M.P. Degree College, University of Allahabad, India

Richa Mishra
C.M.P. Degree College, University of Allahabad, India

ABSTRACT

The intersection of cognitive neuroscience, pharmacology, and AI offers a uniquely productive lens for understanding how drug exposure causes profound disruptions in human decision-making, which can inform the design of bias-robust computational systems. This occurs through neurobehavioral mechanisms that manifest as cognitive biases, including the overvaluation of immediate rewards, excessive salience assignment to drug-related cues, response inhibition deficits, and shifts from goal-directed to habitual control. These biases are extensively observed in

DOI: 10.4018/979-8-3373-8988-2.ch011

addiction neuroscience and reflect the weaknesses of current AI systems, such as reward hacking, hypersensitivity to salient inputs, and failures in uncertain environments. This chapter connects computational psychiatry, neuroscience, and BIAI by translating drug-induced biases into templates for neuro-inspired architectures. Based on empirical neurobehavioral findings, this suggests how these findings can be applied to improve AI decision intelligence by developing adaptive strategies to identify and counteract addiction-like bias.

1. INTRODUCTION

Artificial intelligence systems are increasingly capable of making autonomous decisions in complex and uncertain environments. Like human cognition, these systems exhibit systematic decision biases arising from model assumptions, data distributions, and optimization objectives. In biological systems, such biases are strongly regulated by neuromodulatory mechanisms. Pharmacological interventions that alter neurotransmitter signalling provide causal insights into how biases emerge, shift, and stabilize within decision-making architectures.

Decision-making is not a single cognitive function. It is the result of distributed brain systems, including the prefrontal cortex (PFC), striatum, amygdala, anterior cingulate cortex (ACC), and insula, that operate in dynamic equilibrium. This balance is disturbed by psychoactive substances in distinct, drug-class-specific ways, leading to persistent cognitive biases that can last long after the acute pharmacological window. Comprehending these biases is becoming more widely acknowledged as a scientific model for the development and diagnosis of artificial cognitive systems, rather than only a clinical issue.

Understanding of drug-induced deficiencies in decision-making has changed in the last ten years due to the development of computational psychiatry. Computational models, especially those based on reinforcement learning (RL) theory, enable researchers to break down decision-making into quantitative characteristics like learning rates (α), inverse temperature (β), and discount factors (γ), instead of describing behaviour descriptively. For example, it has been demonstrated that drugs that dysregulate dopaminergic signaling particularly distort these metrics rather than harming cognition overall (Huys et al., 2021; Dezfouli & Balleine 2012). This accuracy is not coincidental; rather, it shows a structural isomorphism between the way drugs take over neural computation and the similar distortions that occur in RL-based AI and artificial neural networks.

One reason people struggle so much with drugs is how deeply they change the brain after constant use. What happens over time affects choices, especially in areas such as pleasure, self-control, and planning. When someone uses stimulants often, say cocaine or amphetamines, a key brain pathway involving dopamine gets overloaded, creating lasting shifts that make cravings stronger. Opioids, drinking heavily, or frequent cannabis use also reshape nerve cell patterns, making stress harder to handle while weakening social connections and increasing chances of returning to substance use (see Table 1). Studies using animals from mice to zebrafish show similar results, where tests uncover warped behaviour and brain signals tied to addiction.

Right away, some brain-driven behaviours like impulse control slips, rewards feel too intense, and self-restraint weakens. Not waiting comes up repeatedly, and it can set the stage for addiction, then worsen once habits take hold, messing with patience, stopping actions on time, and staying focused (Dalley *et al.*, 2011). People dealing with substance problems usually pick quick payoffs like using drugs over bigger gains later, thanks to disrupted signals between frontal and striatum areas (Bickel *et al.*, 2012). Rewards start pulling harder when chemicals spike, even warning signs fail to stop someone chasing that high (Robinson & Berridge, 2008). Reward deficiency plays a role because weak dopamine activity makes everyday rewards feel less satisfying, pushing choices toward drugs instead (Blum *et al.*, 2012). When self-control falters, impulsive reactions go unchecked, clouding judgment about danger and leading to risky behaviours (Jentsch & Taylor, 1999). Problems in the brain's frontal region explain much of this pattern, favouring routine-driven actions over thoughtful decisions, thereby opening the door to returning after stopping (Goldstein & Volkow, 2011). These shifts act without thinking, chasing highs too strongly, failing to pause, turning early drug use into rigid habits that stick despite harm (Redish *et al.*, 2008).

Strange shifts in brain behaviour show up as mental shortcuts gone off track when someone struggles with drugs. Cue-heavy things grab more focus than usual, making cravings stick around longer because the mind reacts strongly. Moving toward drug-linked triggers happens without thinking, though pulling back can happen too if bad results loom close. Looking at many studies together shows these patterns hold whether opioids, weed, or stimulants are involved. Tests like spotting dots fast, naming colours under pressure, or steering away from images reveal hidden leanings. Moments tied to good feelings with substances come to mind more easily than others do later. Unclear moments get twisted to fit a drug-centred view, which keeps unhealthy loops going. What pulls someone toward drugs isn't always conscious. Hidden leanings tilt judgment without warning, mixing with clear thoughts to build stronger cravings (Kavanagh *et al.*, 2005). Forces beneath awareness act like bridges, carrying brain shifts into repeated actions that keep addiction going.

Peeking into how drugs skew thinking helps shape smarter artificial brains that learn like ours. These brainy machines take cues from living nerve patterns, think shifting connections, layered info paths, and time-sensitive signals to act more naturally. When they mimic habits tied to addiction, researchers watch choices unfold when outcomes are unclear, rushing decisions, chasing quick wins. That kind of test reveals ways to toughen AI for messy fields such as medicine or governance. Take parts modelled after the frontal brain, where they stretch foresight, slow down impulsive picks, much like fixing weak self-control seen in certain mental lapses. Sometimes brains help machines think in more than one way at once, which improves how they understand feelings and situations. Instead of just adding pieces together, these systems connect sight, sound, and context so choices fit better. When it comes to unfair patterns, ideas from studying addiction shape methods that balance outcomes across groups. For example, tools guessing who might misuse opioids now include rules ensuring equal chances for everyone. Tricks used to change automatic thinking in people show up again in AI training routines. Those habits, once hard to break, now get adjusted by design. From shaky judgments in humans to smarter machine logic, the link builds stronger insight into flawed thinking. That gap between mind quirks and fair tech slowly closes.

This part looks at how things link together first through brain-based views on addiction, then shifting toward computer-style models that mimic thinking habits, finally landing on what it means for machines designed like minds. Ideas from brain science meet actions seen in people, mixed with digital frameworks, leading us to ways that might reduce skewed choices in humans and smart software alike.

Cognitive biases, systematic deviations from rational judgment, profoundly shape human decision-making, often amplified by pharmacological agents that hijack neural reward and inhibitory circuits. In neurobehavioral decision-making, drugs like alcohol, cocaine, or even therapeutic psychedelics induce predictable biases, such as heightened risk-taking or impaired loss aversion, by altering dopamine signalling in the prefrontal cortex and nucleus accumbens. This chapter explores modelling these drug-induced biases computationally, drawing parallels to vulnerabilities in brain-inspired artificial intelligence (BIAI) architectures. By simulating pharmacological perturbations in decision processes, we uncover pathways to mitigate biases in decision intelligence systems, AI frameworks designed for high-stakes choices in finance, healthcare, and autonomous systems.

This chapter positions drug-induced cognitive bias as a biologically grounded analogue for understanding and engineering bias in AI systems. Also, focusing primarily on clinical or behavioural outcomes, the discussion emphasizes computational principles relevant to reinforcement learning, probabilistic inference, and decision intelligence. Insights from psychopharmacology are utilized to inform the design of adaptive, neuromodulation-inspired AI architectures that can monitor and regulate their own biases. Before discussing further, certain foundational topics are introduced as follows:

- Neurobehavioural decision theory combines insights from neuroscience, psychology, and economics to understand how humans make decisions. This concept is directly relevant to the query as it provides a framework for understanding the neural mechanisms underlying decision-making.
- Cognitive bias Mitigation refers to the strategies and techniques used to reduce or eliminate cognitive biases in human decision-making. This concept is directly relevant to the query as it provides a framework for developing effective bias mitigation strategies in AI systems.
- Drug-induced cognitive biases refer to the systematic errors in thinking and decision-making that occur as a result of drug use or exposure. This concept is directly relevant to the query as it provides a framework for understanding the neural mechanisms underlying cognitive biases.
- Brain-inspired AI architectures refer to the design of artificial intelligence systems that mimic the structure and function of the human brain. This concept is essential for developing AI systems that can learn, reason, and make decisions like humans.
- Decision intelligence refers to the use of data, analytics, and AI to support and improve human decision-making. This concept is critical for developing effective bias mitigation strategies in AI systems.
- Explainable AI refers to the development of AI systems that provide transparent and explainable decision-making.
- Cognitive architectures refer to the software frameworks that simulate human cognition and provide a structure for integrating multiple AI systems.
- Artificial general intelligence refers to the development of AI systems that can perform any intellectual task that a human can. This concept is essential for developing brain-inspired AI architectures that can mimic human-like decision-making.
- Neural network modeling refers to the use of artificial neural networks to model complex systems and processes that can mimic human-like decision-making.
- Human-computer interaction refers to the study of how humans interact with computers and other technologies. This concept is essential for developing AI systems that can effectively interact with humans and provide unbiased decision-making.
- Human-centered AI refers to the development of AI systems that prioritize human values and well-being.

2. COGNITIVE BIASES: DRUG SPECIFIC

2.1 Opioid

Opioid receptors and their endogenous peptides have demonstrated a clear role in the regulation of mood in animal models and may offer an alternative approach to augment existing therapies. Opioid use disorder (OUD) produces the most robustly documented form of decision bias in the addiction literature. Meta-analytic evidence across 42 studies (N = 3,200+) confirms that individuals with OUD discount delayed rewards at rates 3–5 times higher than healthy controls (MacKillop et al., 2011), a finding that has proven remarkably consistent across paradigms and populations. However, several critical questions remain unresolved. Family and genetic studies (e.g., Cha et al., 2024) suggest a heritable component, but longitudinal work by Bickel and colleagues (2014) demonstrates that extended opioid use substantially steepens discounting curves independent of baseline levels.

This causal layering means that cross-sectional studies, which dominate the literature, are likely underestimating drug-attributable bias. Second, buprenorphine and methadone maintenance treatment (MMT) have been shown to partially normalize delay discounting, but the effect size is modest and dose-dependent. This suggests that opioid-induced reward myopia has both pharmacologically reversible and consolidating components, a distinction with obvious treatment implications. From an AI perspective, opioid-type reward myopia maps conceptually onto the discount factor (γ) in RL algorithms. Artificially lowering γ in an RL agent produces qualitatively similar behavioural profiles, short-horizon planning, overvaluation of immediate rewards, and resistance to correction (Landes et al., 2012).

2.2 Psychostimulants

The classic DA-releasing substances amphetamine and cocaine have been well researched for their impact on cognition. The three main biases are attentional bias toward drug-related cues, impulsive action (premature reacting, as evaluated by stop-signal reaction time [SSRT]), and impulsive choice (delay discounting). Despite having strong documentation, none of the three can be adequately explained by a single mechanism. The main causes of impulsive choice and attentional bias are NAc DA (Nucleus Accumbens Dopamine) dysregulation and disturbance of PFC-striatal connection, respectively. However, in the anterior cingulate and inferior frontal areas, impulsive action deficiencies seem to be more closely associated with noradrenergic and serotonergic dysfunction (Dalley & Roiser, 2012).

Additionally, therapies that restore normal DA function (such as D3R antagonists) enhance delay discounting but may exacerbate or maintain impulsive behaviour. This clinical separation strongly suggests against treating stimulant-induced impulsivity as a single concept (Dalley et al., 2012). In contrast to basic inhibitory deficit hypotheses, recent fMRI research employing dynamic connectivity analyses (Badre & Frank, 2012) reveals that cocaine users display state-dependent coupling between the amygdala and PFC during cue exposure that is absent in controls. Rather than merely overpowering goal-directed evaluation, the data point to a hypersensitive salience attribution mechanism that actively contends with it. This distinction has therapeutic significance since it may be more successful to address salience attribution (with ketamine or cannabinoid modulators) rather than just improving inhibitory control.

2.3 Alcohol and Sedatives

Acute effects of alcohol on risk-taking are well-documented under intoxication, meta-analyses consistently reveal larger bet sizes and decreased sensitivity to loss feedback (Sullivan et al., 2023). Mechanistically, this is explained by NMDA (N-Methyl-D-Aspartate) receptor antagonism in the hippocampus (impairing contextual memory that would typically inform risk appraisal) and GABA-A potentiation in the PFC (lowering deliberative thinking). Nonetheless, two significant disputes merit consideration. First, there is a nonlinear link between drinking and taking risks. While risk-averse people occasionally exhibit decreased risk-aversion rather than greater risk-taking, moderate doses of alcohol seem to boost risk-taking largely among those with trait impulsivity or low baseline GABAergic tone (drinking-to-cope hypothesis; Sullivan et al., 2023).

Second, long-term heavy drinking has a counterintuitive effect on risk sensitivity. Some abstinent alcoholics exhibit risk aversion in comparison to controls, especially for ambiguous (Ellsberg-type) gambles, rather than hyper-risk-taking (Prell, 2025). This might be a rebound event similar to receptor upregulation, where prolonged inhibitory failure causes compensatory activation of amygdala-based threat processing. The alcohol analogy for AI systems most clearly transfers to constraint satisfaction under noise; the system accepts more suboptimal states when the inhibitory architecture is compromised. Alcohol-like risk insensitivity should be produced by a test that adds noise to an AI agent's constraint modules, and novel regularization strategies may be inspired by physiological theories of GABA potentiation.

2.4 Cannabis

In recent years, there has been a significant recalibration of cannabis research due to changes in legal status that have made it possible to conduct bigger, pre-registered studies. Cannabis's cognitive neuroscience is now known to be highly dose- and time-dependent. Acute high-THC (Delta-9-Tetrahydrocannabinol) exposure impairs working memory and probabilistic reasoning (consistent across labs), but chronic use in dependent users has far more variable cognitive effects, with some studies showing near-complete normalization after abstinence and others showing persistent deficits (Colizzi & Bhattacharyya, 2018; Morreale, 2024). Environment insensitivity, or the inability to effectively update beliefs as the environment changes, is the most conceptually intriguing cannabis-related prejudice. The process by which contextual inputs typically adjust value representations is disrupted by CB1R agonism, which decreases glutamate-mediated plasticity in the PFC.

According to Bayesian model-fitting experiments, people who use cannabis particularly show reduced update rates (lower α) in response to negative prediction mistakes, although processing of positive prediction errors is mostly unaffected (Colizzi & Bhattacharyya, 2018). This asymmetry is crucial because it indicates that cannabis use especially hinders unpleasant learning, a feature that symmetric discount or noise models are unable to capture. Complexity is increased by the interaction between THC and CBD (Cannabidiol). According to Bhattacharyya et al. (2018), CBD seems to counteract some THC-induced deficits in a dose-dependent way. This discovery has clear therapeutic significance as high-CBD strains gain economic dominance. The interaction pharmacology is, however, understudied since the majority of cognitive investigations continue to employ whole-cannabis or THC-only approaches.

2.5 MDMA and Entactogens

In this Chapter, MDMA holds a special place since it is both a drug of abuse and a new pharmacotherapy for PTSD (Mitchell et al., 2023). Increased prosocial attribution, improved trust, decreased fear-based avoidance, and a flattened loss aversion curve are among the distinctive cognitive profiles that result from its main mechanism, which is enormous presynaptic 5-HT release followed by DA elevation. MDMA-treated individuals in a social trust game exhibited markedly increased collaboration and decreased defection, even toward strangers, according to recent research by Ramaekers et al. (2026). This remarkable change in social decision-making is distinct from general risk-taking.

MDMA is illuminating from a biased standpoint because it shows how the social dimension of cognition is strongly intertwined with basic valuing circuitry, although it is typically portrayed as a distinct module. This social-bias-inducing characteristic is what gives MDMA its therapeutic promise when it comes to fostering therapeutic alliances in trauma therapy. The same characteristic, however, raises questions regarding manipulation vulnerability since MDMA users may be particularly vulnerable to social influence while intoxicated, a prejudice with clear ethical and forensic ramifications.

Table 1. Neurobiological mechanisms and cognitive bias profiles across drug classes

Drug class	Primary mechanism	Cognitive bias inducer	Brain region	AI analog
opioids	μ-opioid receptor agonism; ↓PFC-NAc connectivity	Temporal discounting; reward overvaluation	PFC, NAc, VTA	Reward function distortion in RL agents
Cocaine/ Stimulants	DAT/NET blockade; ↑synaptic dopamine	Impulsivity; attentional bias	Striatum, mPFC	Overweighted salience signals
Alcohol/ Sedatives	GABA-A potentiation; NMDA antagonism	Risk insensitivity; inhibitory failure	PFC, Amygdala, Cerebellum	Degraded constraint satisfaction
Cannabis (THC)	CB1R agonism; ↓glutamate in PFC	Probabilistic reasoning errors; context insensitivity	Hippocampus, PFC	Context-blind prediction models
MDMA/ Entactogens	Massive 5-HT release; secondary DA elevation	Social affiliation bias; risk overestimation of reward	Raphe, Amygdala, PFC	Overfit social-preference modules
Nicotine	nAChR agonism; sustained DA tone	Attentional narrowing; withdrawal-driven bias	LC, Striatum	Biased attention allocation in transformer heads

Note. PFC = Prefrontal Cortex; NAc = Nucleus Accumbens; VTA = Ventral Tegmental Area; DAT = Dopamine Transporter; NET = Norepinephrine Transporter; mPFC = medial PFC; CB1R = Cannabinoid Receptor Type 1; LC = Locus Coeruleus; RL = Reinforcement Learning.

(Source: Mitchell et al., 2023; Colizzi & Bhattacharyya, 2018; MacKillop et al., 2011; Cha et al., 2024; Dalley & Roiser, 2012; Prell, 2025; Morreale, 2024)

3. NEUROSCIENCE FOUNDATIONS OF DRUG-INDUCED DECISION-MAKING

Brain changes tied to addictive behaviors emerge when drugs disturb natural motivation and learning circuits. Instead of balanced responses, repeated drug use distorts how dopamine functions, shifting what feels rewarding. Because prediction errors fail to update expectations properly, choices become less flexible. Rather than responding to goals, actions follow ingrained habits without thought. Attention locks onto drug-linked triggers while ignoring alternatives. Learning itself weakens under these conditions, limiting adaptation. As a result, decisions tilt away from flexibility and toward repetition despite harm (Koob & Volkow, 2016; Volkow *et al.*, 2019).

3.1 Dopamine and Serotonergic Dysregulation

Reward processing in addiction centres on dopamine (DA) signalling, which shapes motivation and pleasurable experiences. Following short-term drug use, levels of DA rise beyond normal limits within the nucleus accumbens (NAc), driven by activity in the mesolimbic circuit starting from the ventral tegmental area (VTA), a response that overtakes typical reward-related neural patterns (Volkow *et al.*, 2019). Such rapid bursts of DA produce intense feelings of pleasure while strengthening associations between drugs and environmental triggers, thereby increasing the pull of substance-related stimuli (Robinson & Berridge, 2008).

Instead of boosting pleasure, repeated drug use causes smaller dopamine spikes in reward areas when contrasted with early stages. This shows up clearly during abstinence and regular functioning (Volkow *et al.*, 2019). Starting from weak reactions, brain scans reveal fewer D2 receptors and less dopamine released in the striatum, shifting motivation so that everyday rewards feel flat while reminders of drugs stay intensely compelling (Blum *et al.*, 2012; Volkow *et al.*, 2019). Even though background activity drops, triggers linked to substances spark outsized responses, an odd mismatch that drives relentless seeking just to balance chemical flow, locking behaviour into repetition (Koob & Volkow, 2016). Evidence gathered through imaging and animal research supports this pattern regardless of substance type, whether stimulant, opioid, or alcohol, pointing steadily to dopamine's central involvement in starting and sustaining addictive patterns (Volkow *et al.*, 2019).

In addition to increasing subjective pleasure, drugs that flood synaptic dopamine (cocaine, amphetamine) also pathologically magnify reward prediction mistakes, teaching the brain that stimuli linked with drugs are disproportionately rewarding. According to computational models, reward prediction error signals are 2-3 times higher in cocaine addicts than in controls (Redish, 2004). This conclusion was largely corroborated by fMRI investigations using Bayesian model fitting (Konova et al., 2023).

Studies employing selective serotonin reuptake inhibitors (SSRIs) as probes have shown inconsistent results based on receptor subtype, brain area, and test paradigm due to the pharmacological heterogeneity of the serotonergic system (Thoelen, 2023). Rather than risk preferences, recent meta-analyses indicate that serotonin primarily influences temporal aspects of decision-making (patience vs. impulsivity). Therefore, it is reasonable to anticipate that drugs such as MDMA, which cause substantial acute 5-HT release, will change patient-related biases rather than only risk-related ones. This prediction has been largely supported by recent lab evidence (Bird, 2013; Vizeli et al., 2024).

3.2 Release of Strange Mismatch Signals

Strange mismatch signals inside the brain's reward machinery trigger dopamine bursts unlike those shaped by experience, and these signals act like false alarms, shouting success regardless of real results (Redish, 2004). Because such spikes occur spontaneously, outside of cause-and-effect loops, connections between cues and drugs become stronger than they should (Keiflin & Janak, 2015). Over time, the brain loses balance, everyday pleasures spark a less intense response, while reminders of drugs continue to elicit outsized reactions (Takahashi, 2017). Learning falters since expectations drift further from reality, locking behaviour into repeated poor decisions. Some theories now picture addiction through multiple coexisting pathways, one tracking worth, another tracking self-relevance, with chemicals tipping the scale toward compulsive craving (Takahashi, 2017).

Decision processes shift between two modes. One tracks results, adjusts based on rewards, and involves DMS and prefrontal areas. The other follows cues automatically, operates through DLS, and repeats actions without reassessing outcomes. These pathways compete during choices. Their balance shapes behaviour under changing conditions. Evidence points to neural separation of these strategies.

Following long-term drug exposure, behaviour shifts from purposeful to routine-like patterns. At first, efforts to obtain drugs respond to changes in reward value, removing appeal weakens pursuit, but repeated access dulls this sensitivity, pointing toward habit formation (Marchant *et al.*, 2010; Janak *et al.*, 2012). Over time, brain activity tilts away from the dorsomedial striatum toward the dorsolateral region, where stimuli linked to drugs spark actions without regard for present outcomes (Everitt & Robbins, 2016). Studies applying sequential task designs show that searching for cocaine or heroin transforms into rigid behaviour after extensive use, persists despite losing its incentive, yet stops when the dorsolateral striatum is silenced (Marchant *et al.*, 2010). Humans show similar trends in addiction, with routine-driven choices pushing continued use even when harmful results follow (Cowell *et al.*, 2007). Under pressure or long-term drug exposure, this shift speeds up, prefrontal control weakens, while activity in the DLS grows stronger.

3.3 Reinforcement Learning Disruptions

Learning through rewards offers a way to study addiction by focusing on how values and choices go wrong. People struggling with substance use often show weaker adjustments after bad outcomes, yet still respond strongly to positive ones, especially if they come fast. This imbalance means punishments guide behaviour less effectively than usual, whereas quick pleasures keep pulling attention. Evidence suggests these patterns reflect deeper shifts in how experience shapes decisions over time (Huys *et al.*, 2016).

Because dopamine shapes how we respond to penalties, weaker signalling may explain slower punishment learning. This shows up clearly when models analyse behaviour patterns. When medicines adjust dopamine activity, some of these learning problems ease, pointing toward a core biological mechanism. Instead of balanced choices, reward predictions tilt heavily toward substance-related gains, even if they're small or risky. Immediate payoffs weigh more heavily in decisions, pushing long-term effects into the background. Without strong guidance from thoughtful planning systems, habits take over even when results turn harmful. Shifts in brain circuits linking motivation and control regions support this drift toward automatic responses. Dopamine imbalances feed into altered communication across these networks, nudging actions away from goals and toward compulsion.

Together, disrupted dopamine signalling, faulty reward predictions, ingrained behaviour patterns, skewed attention toward stimuli, and weakened reinforcement learning underlie how drugs reshape thinking. These elements clarify why choices turn harmful, opening paths to mirror such dynamics in artificial systems that mimic brain function, aiming to reflect and reduce flawed judgment in intelligent machines.

3.4 The Role of the Insula and Interoceptive Prediction

The significance of interoceptive processing in the brain's monitoring of physical status is an underestimated aspect of drug-induced bias. The insula creates an anticipatory somatic marker that often directs decision-making away from detrimental options by integrating visceral cues with predicted results (Joynt, 1995). According to Ceceli et al. (2025), long-term alcohol consumption is linked to structural and functional disruption of insular grey matter, which lessens the impact of negative somatic feedback on hazardous decisions. Notably, it has been demonstrated that insular damage, whether from stroke or alcohol-related neurodegeneration, increases gambling and drug use in ways that are distinct from alterations in ventral striatal DA function. This suggests a different computational pathway to bias that is not included in conventional DA-centric models. This matters for AI translation, most brain-inspired RL systems model reward processing but lack any analogue of interoceptive prediction.

4. MODELLING HOW DRUGS INFLUENCE BIOLOGICAL SYSTEMS

Figuring out how drugs change thinking in addiction gets help from computer-style math tools. These setups turn actions and brain activity into step-by-step rules. Because of ongoing drug use, choices start leaning too hard on old patterns instead of current risks. Learning systems that track gains and losses over time shape behaviour differently, one way automatically, another thoughtfully. Some versions even show how cravings grow stronger with repetition, almost like an internal alarm getting louder. Belief-updating methods also play a role, adjusting expectations based on new evidence, except when they fail around substances. When reward estimates go off track, people keep chasing highs long after harm piles up. Habit loops become stubborn, surviving even when rewards lose real worth.

4.1 Reinforcement Learning Frameworks: Model-Free vs. Model-Based

Learning happens when agents try things out and see what works by checking rewards against expectations, this gap shapes future choices. One way they adapt relies on adjusting guesses after each outcome, other builds decisions around patterns spotted over time.

Out there in the mix, model-free reinforcement learning uses stored action values that shift through temporal-difference updates, prediction errors, spelled out as δ equals reward plus discounted next-state value minus current state value, tweak those estimates on the fly, skipping any need for a full environmental blueprint. Such a setup backs quick, automatic reactions suited well to steady settings yet stumbles when rewards change, showing little flexibility once habits form (Daw *et al.*, 2005).

When it comes to addiction, brain simulations show that something odd happens, when automatic responses start taking over thoughtful decision-making. Over time, using drugs heavily leans the mind into reflex-like patterns, so drug chasing continues even when rewards fade or harm grows. Lab tests with animals found long-term meth use throws off both thinking styles, yet the gut-reaction system goes most haywire, separate from errors in goal-focused circuits. People struggling with substance problems tend to act more compulsively if their reflective planning dips, while quick-strike habits stay strong or get stronger, fuelling repeat drug use.

Some animals naturally act less on routine before ever encountering drugs. Those who take more methamphetamine later. This suggests poor habit-based learning might backfire once substances alter brain pathways (Groman *et al.*, 2019). Mistakes in how values are adjusted show up next. Drug gains get wrongly amplified over time because prediction errors misfire. Habits grow too strong at the same time. Actions stick around even when they should fade or lose purpose (Redish *et al.*, 2008).

4.2 Incentive Sensitization Models

Out of nowhere, cravings grow stronger even when enjoyment fades. Over time, using drugs shifts how the brain responds, not by boosting delight but by amplifying desire. Repeated use revs up reward pathways deep inside the brain. This shift makes triggers tied to drugs feel urgent, almost magnetic. The urge becomes outsized compared to actual liking. Sensitivity builds silently, altering motivation more than mood.

Learning settings shift when systems adjust their internal rules on the fly. Instead of steady progress, rewards get misread, especially under drugs, which flood the brain with false signals that something is better than it really is. Think of a loop stuck repeating itself: cues tied to substances gain strength over time, even though pleasure fades. Because these chemicals dodge natural feedback checks, desire stays high while actual enjoyment drops off. Expectations inflate without grounding in real experience, making cravings persist long after satisfaction ends.

4.3 Bayesian Methods and Associated Models

Bayesian reinforcement learning sees learning as a kind of guessing game guided by probabilities, adjusting expectations using past assumptions plus new observations (Griffiths *et al.*, 2008). When it comes to addiction, such thinking paints skewed choices not as weak willpower but as rigid early guesses that refuse correction; strong convictions like drugs always ease pain take control, leaving little room for surprises that should prompt change, locking out updates even when reality pushes back.

Starting strange, maybe even odd, distorted updates in value happen because old biases overpower fresh proof, so drug rewards stay too appealing. Not quite habits yet, instead, less curiosity or a narrow trust in fixed routines makes actions feel locked in place, certain beyond reason. Jump ahead, mix Bayes-style thinking with reinforcement learning, and you get shifting cravings shaped by inner sensations or surroundings, pulling choices toward impulse without warning.

Starting fresh, these methods frame addiction less like routine behaviour, more like distorted reasoning. Dopamine adjusts how firmly someone trusts their assumptions over time, stronger confidence grows in flawed expectations when substances stay present. Real-world tests show people with gambling issues explore fewer options on purpose, like those with substance disorders. Hidden patterns in decision-making expose weaker responses to unknown outcomes, feeding repeated choices despite consequences (Wiehler *et al.*, 2022).

Overall, computers can show how drugs mess up choices by using different math methods. One way looks at learning without planning, plus another that plans, while also handling rewards strangely or adjusting beliefs too much. Mistakes in spotting errors during learning twist values, making habits grow out of control when quick reactions take over, and things feel too urgent. Math tools connect brain changes to the bad decisions people make. This opens paths for smarter machines that copy brain patterns to reduce similar flaws in artificial thinkers.

5. COMPUTATIONAL PSYCHIATRY

5.1 Reinforcement Learning Models

By using formal RL models to analyze human decision data, computational psychiatry has revolutionized the study of addiction. Instead of asking if drug users are more impulsive, researchers now assess whether there is a particular, mathematically meaningful change in the discount factor (γ), learning rate (α), or inverse temperature (β). A more detailed taxonomy of bias has resulted from this change, and crucially, it produces parameters that are directly comparable to those utilized in artificial reinforcement learning agents.

Within this architecture, the critic (PFC) assesses results and modifies actor behaviour through prediction mistakes, while the actor (striatum) chooses actions based on cached value estimations.

According to Keramati & Gutkin (2013), chronic drug usage may be modelled as pathological bias in the critic-to-actor connection, where the critic produces corrective signals but fails to update the actor policy, resulting in persistent drug-seeking behaviour despite unfavourable results. Similarly, RL agents with corrupted policy gradient propagation show a structural similarity that hasn't been extensively used in AI robustness research, as well as qualitatively similar perseverative behaviour. Crucially, computer models have also shown flaws in traditional explanations of addiction. Mikus et al. (2022) showed that model-free (habitual) control is unaffected; opioid-dependent people display a particular deficiency in model-based (goal-directed) control rather than consistently higher discounting. It means that opioid addiction is not simply about wanting things faster, but about a failure of deliberative override. Any AI system that relies solely on model-free RL without a model-based correction mechanism replicates this vulnerability.

5.2 Drug Bias Analogues in Large Language Models

Over the last two years, there has been an increasing amount of research on the similarities between drug-induced cognitive biases and LLM failure mechanisms. This comparison has been productive even though it is overtly heuristic rather than mechanical. Empirical attention has been drawn to a number of mappings. The propensity to give early context disproportionate weight in LLMs, known as anchoring bias, is structurally similar to the attentional bias brought on by stimulants, an overweighted salience signal that conflicts with continuous contextual updating. According to recent research, LLMs consistently anchor on deceptive numerical anchors included in prompts, even when such anchors are specifically marked as irrelevant. This pattern is qualitatively consistent with cue-triggered attentional constriction in cocaine users.

The social trust bias brought on by MDMA-like pharmacological experiences is mapped onto sycophancy in LLMs, the propensity to validate user-stated views regardless of their veracity. In both cases, affiliation-related signals are overweighted while accuracy-related mistake feedback is underweighted. According to this mapping, LLMs could profit from structures that specifically describe and control social valuation, a feature that is presently absent from typical transformer designs. Most provocatively, Juliani et al. (2024) suggested that alcohol-induced disinhibition, a deterioration of the constraint-satisfaction architecture that typically prevents overcommitment to local optima, is mechanistically analogous to training instability in deep RL agents, where policy collapse occurs after initial good performance. This implies that GABAergic inhibitory network-inspired regularization techniques may enhance RL training stability.

5.3 Bias Mitigation: From Pharmacology to AI Safety

Pharmacological treatments for drug-induced biases may inspire similar AI safety strategies if they serve as a model for comprehending AI failure scenarios. Although this translation is logically tractable, it has not yet been thoroughly tested. As a reward function modulator in biological systems, naltrexone reduces the excessive value of drug cues by attenuating opioid-mediated reward salience. A regularization term that penalizes over-reliance on high-salience rewards, so implementing a reward smoothing prior might be an equivalent process in reinforcement learning. Similarly, attention-modulation techniques in Transformer designs that dynamically weight deliberative vs automatic processing modes may be inspired by acetylcholinesterase inhibitors, which enhance PFC-mediated goal-directed control in preclinical addiction models.

Perhaps the most fundamentally translatable intervention is N-acetylcysteine (NAC), which restores glutamatergic homeostasis in PFC after cocaine disruption by restoring balanced excitation-inhibition in prefrontal circuits. This translates into homeostatic regularization techniques in AI that stop runaway activation in particular network areas. Currently, batch normalization and dropout are used to handle this issue on an as-needed basis, but there are no architecturally sound biological counterparts.

6. THINKING LIKE A BRAIN COULD CHANGE ARTIFICIAL INTELLIGENCE

Brain research into addictive behaviours opens doors for smarter artificial brains. Dopamine imbalances, distorted reward signals, and mismatched expectations, these glitches seen in addiction become blueprints for better machine designs. Instead of just copying healthy circuits, engineers peek at broken ones to build flexibility. Attention that shifts too fast or fixates too long. That flaw inspires filters that adjust on their own. Memory loops gone rigid lead to models escaping repetitive traps. Some setups mix logic rules with learning networks, others lean on trial-and-error training methods. Each tweak aims to soften blind spots when choices get messy. Even chaos becomes useful, and unstable settings reveal how well systems hold up under pressure. Patterns buried in flawed thinking help algorithms sidestep narrow views or false patterns. Noise doesn't drown them out so easily anymore.

6.1 Neurobehavioral Clues Shape Brain-Based Artificial Intelligence

Starts happening when habits take over thinking, much like what we see in addiction, a lot gets mirrored in artificial intelligence, too. Machines stuck on repeat, trained too hard on skewed data, start acting rigid, much like an addicted brain chasing quick rewards instead of staying flexible. Brain-inspired AI pulls from how neurons rewire, stack information in layers, and even adjust signals as chemicals do. That allows learning with less data, shifting smoothly between different situations. Some researchers borrow models of addictive behaviour to fine-tune how machines detect issues or respond wisely. Together, these overlaps spark quieter progress, one that blends mind science and machine logic thoughtfully. When real brain patterns feed into code, systems stop feeling so opaque. They begin managing choices more like humans do, especially under pressure. Makes them fit better in touchy areas, such as public rules or self-driving decisions, where mistakes carry weight.

6.2 Adaptive Attention and Memory Mechanisms Counter Drug-Like Biases

Spotlight on drugs shows how thoughts get pulled toward strong triggers, making choices go off track (Field & Cox, 2008). Yet machines face something similar, they fixate on common patterns in data, which warps their judgments over time. From brain circuits involving prefrontal and thalamic zones, engineers borrow shifting focus methods, so software adjusts what it notices. Instead of sticking to old habits, systems learn to highlight fresh clues when surroundings change. Therapy tricks used to redirect automatic reactions in addicts, through steady practice, now shape code that reshapes skewed suggestions in apps. Memory updates inspired by biology let artificial setups shed outdated links, much like neurons trimming weak connections.

When the brain replays strong memories, like those tied to drugs, it can distort judgment. Machines borrow this idea yet tweak how past data is recalled. Instead of linking ideas plainly, they shift focus through timed filters shaped by brain science. Some networks blend older designs with new switches that mimic neural control. These parts help ignore skewed inputs when sensing emotions or making tough choices. Real setups fight short wins, much like curbing cravings, by lowering rewards for quick guesses. Confidence gets questioned if patterns feel off or unfair. Systems learn slower gains beat flashy errors, especially in spotting who might struggle with addiction. Biases shrink when outcomes grow more stable over time. Ethical behaviour emerges not from rules, but from fluid thinking like our own.

6.3 Hybrid Systems Combining Neural and Symbolic Methods

Mixing brain-inspired learning with rule-based logic helps cut down on unfair shortcuts. Instead of just repeating mistakes, they follow built-in limits that block repetitive errors, like how habits form when goals go off track. Think DARPA projects were clear ethics rules that team up with hidden bias scanners in AI generators. This blend keeps outputs accurate without drifting into skewed routines, much like treatments reshaping automatic urges in behaviour change.

Starting with reinforcement learning models, some systems now blend brain-like networks and rule-based logic to mirror how distorted reward signals drive addictive behaviors. Instead of relying solely on patterns, they build safeguards like those found in the prefrontal cortex, which help temper impulsive choices when outcomes shift over time. In fast-changing settings, this balance supports longer-term thinking, especially useful in sensitive areas such as medical treatment planning. For instance, one application tackles unequal painkiller prescriptions by detecting skewed tendencies before they take root, applying explicit fairness rules grounded in measurable results. Because reasoning steps are expressed in clear logical forms, people can follow how conclusions emerge, something opaque deep-learning-only methods often hide beneath complex number crunching.

6.4 Stronger Performance When Conditions Change

Unpredictable patterns, such as irregular feedback, wear down stability, making setbacks more likely, according to brain studies on dependence. Instead of resisting chaos, systems can borrow tricks from how neurons adjust under pressure, using shifts like dopamine-driven learning to stay aligned moment by moment. Take spiking circuits: they mimic abrupt nerve signals seen in craving states yet turn that into a tool for handling erratic inputs without burning excess power. These designs handle messy flows, sensors in motion, and driverless paths, not by perfecting data but by working well enough amid fuzziness.

Midway through handling choices, such systems hold steady when faced with tricky inputs, much like cravings challenge a person in recovery. Built on patterns drawn from how brains respond to addictive triggers, they lean on probability rules to stay balanced even when conditions shift. Take tools designed to support sobriety. They predict setbacks by tracking changes in daily actions, pulling together varied signals for tailored responses.

So, it turns out that simulating drug-driven distortions using BIAI pushes AI designs forward, while at the same time shaping better ways to reduce skewed outcomes in automated choices, nudging tech a bit closer to how people actually think. Still, the real shift comes from blending these insights into systems that mirror human judgment more honestly.

7. METHODOLOGICAL CONTROVERSIES AND CRITICAL LIMITATIONS

7.1 The Replication Crisis in Addiction Cognitive Neuroscience

Numerous conclusions discussed in earlier sections are based on research using flexible analytic pipelines, samples smaller than N = 50, and no pre-registration, conditions that have been demonstrated to significantly increase false discovery rates in neuroimaging studies (Eklund et al., 2016; Poldrack et al., 2020). Multi-site projects like the Addiction Biomarkers Platform (ABP) and the Cohort Study on Substance Use Risk Factors (C-SURF) are starting to provide sufficiently powered, pre-registered datasets, although addiction neuroscience has been slower than psychology to address its replication issue. According to preliminary findings from these efforts, drug-induced bias impact sizes are genuine but significantly less than single-lab estimates; this trend is consistent with publication bias inflating the literature.

Certain disputes should be highlighted because it confuses decision-making quality with working memory capacity and task-learning effects. The Iowa Gambling Task (IGT), which has been used in numerous studies on alcohol and cocaine bias, has drawn criticism (Van et al., 2013). This means that apparent decision biases may partially reflect memory-related confounds (Buelow & Suhr, 2009). Similar to this, individual variations in response caution that are orthogonal to inhibitory control impact stop-signal reaction time tasks, the main measure of impulsive behavior in stimulant research, exaggerating apparent drug effects.

7.2 Sex, Hormones, and Representational Gaps

In drug-cognition studies, female participants are still significantly underrepresented. Less than 15% of animal studies employed both sexes, and only 31% of 218 papers on drug-induced choice biases included sex-stratified analyses, according to a comprehensive review (Butelman et al., 2023). Converging evidence shows that women are more susceptible to relapse during hormonal phases, exhibit different patterns of cognitive impairment compared to men for equivalent drug exposure, and transition from recreational to dependent use more quickly (the telescoping effect). This is not just a representational issue. Reported effect size estimates are likely population-specific rather than universal when sex is ignored as a variable, which restricts generalizability.

7.3 Ecological Validity and the Laboratory Trap

The dependence of drug-bias research on laboratory decision tasks, typically binary choices between gambles in an artificial context, a recurrent restriction. Time constraints, social context, ambiguity, and embodied interoceptive input are all aspects of real-world drug-influenced decision-making that laboratory paradigms exclude. This is starting to be addressed by ecological momentary assessment (EMA) and experience sampling approaches (ESM), which demonstrate that drug-related choice biases in natural settings are frequently greater and more context-dependent than laboratory equivalents imply (Schultz et al., 2016). Significant implications for AI translation result from the discrepancy between lab and field findings an AI system trained on lab-derived bias models would not be well suited to deal with real-world decision settings.

8. APPLICATIONS AND ETHICS

Brain-inspired AI doesn't just stay in labs. It shows up in real-world tools that help people battling addiction make better choices when outcomes are unclear. Using methods like reinforcement learning, these systems learn patterns while noticing hidden biases, shaping more accurate forecasts along the way. Personalized care paths emerge, not one-size-fits-all fixes, but responses tuned to individual needs under pressure. Hospitals begin adapting similar logic for sorting urgent cases, shifting how limited resources get used during crises. Yet every smart tool brings tough questions: Who gets left out? Whose data becomes exposed without consent? Behind progress lies tension between speed and fairness, innovation and oversight.

Decisions made by machines echo human blind spots unless checked early, often, and differently each time.

8.1 Smart Tools Help Guide Choices in Treating Addiction

Out-of-reach impulses shift when smart systems tap into how brains get hooked. Clinicians see clearer paths forward because the tech highlights risk before slips happen. Instead of reacting, they adjust course early thanks to signals flagged by algorithms trained on habit loops. Distorted cravings, like chasing highs too hard, are balanced using checks built right in. Rewired thinking patterns show up in data drawn from old reward routines. Decisions tilt differently once habits lose their grip through modelled behaviour shifts.

8.2 Forecasting Relapses Using Awareness of Bias in Data Analysis

Staying clean is hard when old triggers come back, especially stress or places tied to past habits. One day ahead of time, smart programs can guess if someone might slip up again, often right more than four out of five times. These tools learn from phone updates about feelings, where a person goes, and what they face each hour. Instead of ignoring mental blind spots, some models adjust for them, like how the mind fixates on certain reminders of drug use. They do this by cleaning training data and adding rules that keep predictions fairer across different people. Hidden loops inside the software track body signals and daily choices, spotting trouble before it grows. When danger shows up in the numbers, warnings pop up too, like a nudge to take another route home. One study shows these tools help catch missed cases, especially among high-risk patients, lowering errors in care decisions (Zhang *et al.*, 2018). When it comes to opioid misuse, smart systems spot unapproved usage accurately because they pull together different types of information, reducing blind spots tied to overlooked communities (Volkow *et al.*, 2019).

8.3 Adapting treatments with reinforcement learning

Sometimes decisions in treatment must change as patients respond differently. One way this happens is how habits take over from goals, much like what occurs in addiction. Machines that learn can adjust care by focusing on choices made when results are uncertain. These tools often use methods called contextual bandits or deep reinforcement learning. They watch how someone reacts, then update the next steps to support better recovery later. Think of antidepressant doses might shift after seeing how a person feels over time. Feedback becomes part of the system's learning pattern. This kind of tuning helps lower the chances of symptoms returning. When it comes to substance misuse, problems often lie in how people assign value to rewards. Algorithms detect those patterns uniquely in each person. Then they match fixes, like training attention, to one's mental profile. Each path is shaped not by rules, but by ongoing signals. Recovery grows less rigid, more tuned. Progress is measured not in weeks but in shifts in behaviour. Learning never stops, nor should treatment. Starting with lab tests and real-world use, reinforcement learning adjusts care levels, sometimes turning up help, sometimes pulling back. Virtual counsellors, for instance, spot early signs of setbacks by reading behaviour clues, thanks to findings from Dayan (Dayan, P., & Daw, N. D., 2008). Mixes of symbolic rules and learning models go a step beyond, shaping responses that respect individual choice.

8.4 Healthcare Triage

When emergencies hit, smart tools based on brain-inspired choices help sort patients faster, cutting down delays and misjudged cases slightly each time. These systems learn from patterns, using language processing to understand symptoms while guessing how serious things might get, much like a cautious mind weighing risks. Picture strokes are caught instantly, blockages spotted without delay, triggering automatic warnings that speed up transfers. As infections spread, some platforms mix metabolism data with trial-and-error learning to forecast bed use and intensive care demand ahead of time. A sudden alert pops up when brain scans show something off, speeding help where minutes matter most (Din *et al.*, 2025). Because training data can lean one way or another, fixes are baked in, so results stay fair no matter who's scanned (Gupta, 2024).

8.5 Policy Tools

When conditions are unclear, artificial intelligence helps shape flexible policies, using methods tied to risk levels when putting AI into government work (Ho *et al.*, 2023). Instead of fixed rules, learning systems that weigh probabilities adjust to shifting surroundings, guiding choices in areas like money matters, health plans, and keeping the planet stable (Parente, 2024). One example, NIST's guide for managing AI dangers, looks at unknowns linked to creative AI, nudging global standards toward alignment (Censinet, 2026).

9. ETHICAL CONSIDERATIONS

There are significant ethical concerns when drug-bias research is used in AI. The corrective mechanism may potentially lessen AI systems sensitivity to valid user preferences if sycophancy is viewed as an MDMA-like social prejudice that has to be addressed. These conflicts reflect current discussions in pharmacology over the boundary between unwelcome personality change and therapeutic normalization. A research agenda at the nexus of AI ethics, neuroethics, and decision science is necessary for the discipline, which calls for multidisciplinary ethical frameworks that address AI bias mitigation with the same sensitivity used to pharmaceutical intervention.

Putting BIAI into places full of uncertainty or struggle drags up tough moral questions about privacy, slanted outcomes, and personal freedom. Sensitive details flow through these systems, so guarding data becomes non-negotiable; clear permissions plus strong shields against leaks matter most. Hidden imbalances hiding inside algorithms might deepen gaps for marginalized people unless fairness checks happen and varied real-world data are used. When doctors can see how a system reaches conclusions, confidence grows; tools such as FAIR-AI help keep humans in the loop. That stirs unease around independence; blending human judgment with machine input keeps control where it belongs (Saeidnia *et al.*, 2024). When ethics get checked by real people, bad effects drop off, and fairness slips into place more easily (ATTC Network, 2025). Because rules keep changing, plans that bend help make smarter choices without rushing ahead blindly (Nordström, 2022).

10. FUTURE DIRECTIONS

Mixing addictive behaviors with brain-like artificial intelligence has led to early tools that spot thinking errors and ways to reduce them. Progress means putting these ideas to harder tests using real human data. It also demands building smarter system designs beyond today's limits. One path moves toward testing models against actual brain and behaviour findings, over and over, to make sure they reflect reality. Other pushes for blending mind-based theories with stronger computing methods, not just stacking old parts together. Then there is the shift to groups, not single agents, where shared decisions go wrong in hidden ways. Watch how small flaws grow when multiple systems interact without oversight. Fixing that requires studying teamwork among machines shaped by biological rules. Each step forward leans less on theory, more on messy practice. Breakthroughs hide where biology meets engineering under pressure.

The field is still fragmented despite significant advancements. The majority of studies use varied tasks and inconsistent computational frameworks to study a single pharmacological class in a single population. Replication is a recurring problem, notable results from research on alcohol-mediated risk insensitivity or opioid delay discounting have frequently failed to generalize across populations, labs, and paradigms (Koffarnus & Bickel, 2014; Rodrigues et al., 2025). Moreover, the translational link between AI architectural design and pharmaceutical cognitive neuroscience is still primarily theoretical.

Table 2. Critical research gaps and proposed future directions

Research gap	Current status	Future Direction
Longitudinal bias tracking under chronic drug use	Most studies are acute/cross-sectional (2021–2024)	Longitudinal fMRI + computational phenotyping cohorts
Sex & hormonal moderators of drug-induced bias	Females underrepresented in preclinical and clinical trials	Sex-stratified RCTs; hormone-by-drug interaction models
Translating bias findings to AI safety	Largely theoretical; few empirical benchmarks	Drug-bias-inspired adversarial benchmarks for LLMs
Individual genetic variability (COMT, DRD2 polymorphisms)	Small N studies with mixed replication	Pharmacogenomics + neural digital twins
Causal vs. correlational directionality	Most neuroimaging studies are correlational	TMS/tDCS-pharmacology combined designs; causal discovery algorithms

(source: Koffarnus & Bickel, 2014; Rodrigues et al., 2025; Nordström, 2022; Saeidnia *et al.*, 2024; Palminteri *et al.*, 2017; Stephan *et al.*, 2015; Perner & Roessler, 2012)

10.1 Using Real Brain Behaviour Data to Check Models

Plenty of computer-based theories about bias in addiction look neat on paper yet rest on thin evidence. Instead of standing alone, these models could gain strength when tied closely to solid brain and behaviour studies. Some fit only surface-level actions, missing the deeper biology that drives them. Without clear bridges to real-world measurements, their value stays limited. Linking algorithms to actual nervous system patterns would sharpen their purpose. That gap, between code and measurable mind activity, is where progress waits. Down the line, efforts might focus on layered checks. These systems could shape model behaviour through combined methods

One way to spot patterns is through fixed exercises, like choosing now versus later rewards. Sometimes people face uncertain outcomes that shift without warning. A different method checks where the eyes move first when shown certain images. Moving toward or pulling back from stimuli also reveals tendencies. Brain scan signs show activity in reward areas when triggers appear. Signals pop up in the ventral striatum as choices unfold. A quiet hum builds in the vmPFC during decision moments. The insula flares slightly at the first hint of sensation. Amygdala pulses track emotional weight without warning. Patterns shift when prediction errors shape learning. Value signals emerge across circuits, piece by piece.

Bursts of activity in the midbrain spark quick dopamine pulses. Instead of steady drips, these surges shape how signals flow. Theta and gamma rhythms link up in the prefrontal zone. Fast shifts stand apart from slow, drawn-out releases. Drugs that tweak dopamine seem to shift how strong the bias is, while blocking those signals might make it stick around longer. Opioid blockers tend to alter the way the effect holds up over time.

One reason these datasets matter is how they sharpen comparisons between models, especially when tested through layered Bayesian analysis or simulated reasoning (Palminteri *et al.*, 2017; Stephan *et al.*, 2015). Take newer studies that link hybrid reinforcement learning frameworks directly to both actions and brain-based signals of expected reward shifts in people using methamphetamine.

One fresh path involves closed-loop setups where live computer forecasts shape how tasks unfold or when brain zaps get delivered, like TMS or tDCS, to probe cause-and-effect ideas about mental gears turning. Such studies hit hard at proving if certain settings borrowed from addiction frameworks actually push behaviour and brain patterns, say, oversized reward surprises, weaker goal-driven choices, or sharper craving-linked learning speed. Big data sets pulled together across many places could make testing models far more reliable. These collections mix brain scans with digital behaviour tracking, plus on-the-spot reports about actual drug urges in daily life. When information lines up like this, findings tend to reflect reality much better.

10.2 Hybrid Minds Meet Machines

Few of today's AI systems mimic the brain yet stay split into separate types, one kind leaning on reinforcement learning, another mixing symbols with neurons, a third using spikes like real nerves. Deeper blends are coming. These future designs will weave together different rhythms, scales, and processes that are not just copy parts but mirror how people really choose when addiction pulls at them. Key directions include:

- Starting different layers and parts in reinforcement learning helps split tasks like spotting rewards, quick guesses versus planned choices, checking your own thinking, plus adjusting goals on the fly. This setup acts a lot like brain circuits linking cortex, basal ganglia, and thalamus, layered one over another (Perner & Roessler, 2012).

- A mind built from code can pause its routine choices when something deeper feels off. Sometimes rules shift, not because of habit but by purpose. When actions clash with bigger aims or right and wrong, a smarter machine steps in. It checks itself, and then it changes course. Symbols guide the thinking. Logic rewires the reflex. Mistakes get caught before they count. This blend thinks ahead instead of repeating (Garcez & Lamb, 2023).
- When errors keep piling up, or urges grow stronger, certain brain-inspired systems notice. These parts, shaped like circuits from frontal zones, shift into focus mode instead. They step in much like a quiet alarm might sound inside the mind. Not every signal demands change, yet some do. A buildup pushes response shifts, not smoothly, but surely. Help is sought when inner pressure crosses thresholds. Control wakes up late sometimes. Patterns repeat until balance returns. Decisions alter once awareness grows (Fleming & Daw, 2017).

Some models store information based on its importance, keeping strong traces of certain experiences while letting others fade. What matters more gets held tighter, much like how memories tied to drugs become deeper over time. Normal rewards, though, leave weaker marks behind. This tilt mirrors patterns seen when brain systems favor artificial highs over everyday pleasures. Quick shifts tied to dopamine stand apart from gradual balancing acts inside the brain, allowing simulations to reflect short-term drug impacts alongside long-term patterns of dependence (Dayan, P., & Daw, N. D., 2008).

These mixed models could mirror addictive behaviors more closely while offering clearer, adjustable points to reduce biases, whether in people or machines. A different structure might track compulsions with greater precision, at the same time opening paths to steadier corrections across natural and synthetic minds. Instead of rigid setups, flexible systems may reflect dependency patterns more honestly, giving researchers firmer ground when adjusting skewed responses. With such blends, understanding breakdowns becomes easier even as fixing them grows more precise, especially where humans and algorithms overlap.

10.3 Biases in Real-World Decision Intelligence

Most existing computational work on addiction and bias mitigation remains focused on single-agent settings. However, many high-impact decision intelligence applications, healthcare delivery, public health policy, workplace safety, financial advising, and autonomous vehicle fleets, social media content moderation, operate in multi-agent, multi-stakeholder environments where emergent collective biases can arise even when individual agents are bias-mitigated. Future directions should focus on multi-agent reinforcement learning (MARL) frameworks that simulate interactions among heterogeneous agents (patients, clinicians, family members, insurers, policymakers) and examine how local drug-induced biases propagate through social networks, feedback loops, and resource competition (Zhang *et al.*, 2021; Carichon *et al.*, 2025). Coordinated bias mitigation protocols that allow agents to share information about detected cognitive distortions or prediction error anomalies, enabling collective error correction and trust calibration (e.g., clinician-AI-patient triads in addiction treatment). Additionally, institutional-level decision intelligence models that represent organizations or regulatory bodies as higher-order agents capable of implementing structural interventions (e.g., default opt-in behavioral nudges, pricing policies, access restrictions) that counteract individual-level biases (Rahwan *et al.*, 2019). Research on adversarial robustness training in multi-agent contexts where agents must maintain performance against malicious or biased actors, analogous to resisting drug-cue-triggered relapse in social environments (Zhang *et al.*, 2021). Whereas ethical multi-agent alignment research that explicitly incorporates value pluralism and fairness across diverse stakeholders, preventing the emergence of majority-biased equilibria (Gabriel, 2020). These directions are particularly relevant for real-world deployment of decision intelligence systems in addiction care, public health crisis management, and automated governance under uncertainty.

10.4 Interdisciplinary Bridge Brain-Computer Interface Integration

Brain-computer interfaces (BCIs) that identify neural indicators of upcoming biased choices and provide corrective stimulation are known as closed-loop neuromodulation. This field has implications for both clinical and AI design. Although long-term effectiveness and safety data are still scarce, preliminary research employing deep brain stimulation (DBS) of the NAc in treatment-resistant OUD shows decreases in desire and normalization of delay discounting curves (Gimenez et al., 2024). This corresponds to the idea of real-time bias detection and correction loops for AI systems, a type of metacognitive monitoring that is generally absent from existing AI designs. One way to study addiction and mental health conditions involves building computer simulations. These models use reinforcement learning ideas, showing how habits form through repeated actions. Sometimes they include math that mimics brain cell activity. Another approach applies probability rules to track decision patterns over time. Relapses are explored by testing how old behaviours return under stress. Fixed thinking styles stay present even after treatment ends. Patterns stick around because systems adapt slowly.

This bridge moves both ways, findings from addiction science feed structured models into artificial intelligence, making systems easier to interpret and tougher against disruptions. In return, those AI setups allow massive digital experiments, quick tests of treatments too complex or risky in real people.

10.5 Neural Digital Twins and Personalised Pharmacology

The creation of personalized neural simulation models, or neural digital twins, which combine individual pharmacogenomic profiles, structural neuroimaging, and computational psychiatry parameters to forecast an individual's reaction to drug exposure and pharmacological intervention, is one of the most promising new strategies. Personalized actor-critic models with individual-specific discount rates and learning parameters beat population-average models in forecasting real-world drug seeking and relapse, according to early proof-of-concept research by Ramakrishnan et al. (2026). Neural digital twins have the potential to transform preclinical AI stress testing and treatment prediction if they are verified on a large scale.

CONCLUSION

Drug-induced cognitive biases are windows into the computational architecture of decision-making itself, not just clinical curiosities or regrettable side effects of substance use. Reward prediction, inhibitory control, temporal discounting, probabilistic updating, and social valuation are some of the neurocognitive system's parameters that are systematically disrupted by psychoactive substances. These distortions are patterned and mechanistically grounded, making them increasingly tractable from the perspective of computational psychiatry.

Three main points have been made in this chapter. First, the field has progressed from descriptive taxonomy to formal parameterization. We now know not only that drug users make bad decisions, but also how their noise terms, learning parameters, and discount rates differ from those of healthy controls. This information has theoretical and therapeutic implications. Second, these computational drug-bias profiles provide a cross-disciplinary paradigm for investigating AI robustness since they are structurally isomorphic with failure mechanisms in LLMs and artificial RL agents. Third, principled conceptual templates for AI safety mechanisms, from reward-smoothing priors to homeostatic regularization, are provided by the pharmaceutical procedures created to normalize drug-induced biases in the clinic.

Cravings take hold because brain circuits get rewired after repeated drug use, dopamine pathways falter, signals misfire when outcomes differ from expectations, and behaviour slips into automatic routines. Attention locks onto triggers without thinking first, memories tilt toward drug-related moments, and choices favor immediate hits over long-term stability. When cues appear, reactions fire faster than conscious thought can intervene. Learning bends around rewards that aren't helpful anymore, judgment skews under pressure, and self-control weakens as a muscle pushed too far. Over time, decisions fold beneath patterns built by repetition and urgency. Recovery becomes harder not due to a lack of will but because mental shortcuts pull in another direction. Out there among equations and simulations, ways of mapping bad choices have taken shape through math-heavy thinking. Instead of just listing steps, some models borrow ideas from how rewards reshape behaviour, showing shifts in what matters when substances take hold. One path follows signals that get twisted, other tracks habits forming too fast, like shortcuts the mind shouldn't trust. Even when results turn sour, predictions keep favouring old patterns, value stuck on repeat. Seen this way, addiction isn't broken learning but learning gone too well, locked into harmful loops. Look at how the mind stumbles, fixating, rushing choices, falling into habit, and suddenly AI starts to mirror real behaviour. When systems mimic these flaws, like craving quick rewards or ignoring long-term risks, they become easier to adjust on purpose. Shifts in focus that adapt over time, internal checks similar to self-reflection, and chemical-like signals adjusting learning rates, all help reduce skewed responses. Mix symbols with neural patterns. Let learning happen across fast and slow phases. Watch artificial minds resist traps much like people do. From there, useful things emerge: spotting when decisions go off track, tailoring interventions that evolve through feedback, sorting medical needs under pressure, and building policies flexible enough for messy realities.

However, the field's shortcomings, underpowered and unreplicated experiments, purposeful exclusion of female participants, ecologically impoverished paradigms, and a largely undeveloped translational bridge to AI systems must be honestly acknowledged to go ahead. A new generation of pre-registered, multi-site, sex-stratified, EMA-integrated investigations is required to bridge these gaps. These studies must be carried out in close coordination with AI researchers who are capable of operationalizing neuroscientific discoveries into testable architectural ideas. The convergence of computational neuroscience, AI architecture, and pharmacology is now a developing multidisciplinary engine that can push the boundaries of both technology and medicine. This chapter's main hypothesis is that simulating drug-induced cognitive bias would lead to both improved addiction treatment and AI systems that make decisions more intelligently, robustly, and resemble healthy human minds rather than ones that have been chemically impaired.

Fairness matters most when technology meets society, safeguarding personal data walks hand in hand with checking hidden biases in code, decisions must stay understandable, control should never leave people's hands, while group behaviors across smart systems need watching before patterns twist beyond intent. Some systems mix deep learning with smart decision layers. These designs link tiered reinforcement methods to logic-based thinking. Memory kicks in only when needed, not always running. Layers talk through feedback loops instead of one-way signals. Thinking shifts between broad guesses and precise moves. Because it models how drugs skew thinking, this work isn't just theory. One path leads to sharper insight into why people struggle with impulses. At the same time, lessons feed into building machines that decide more like thoughtful humans, clearer, fairer, less brittle. When brain science on addiction meets AI shaped by biology, two gains emerge. Fewer blind spots in treating dependency. Smarter systems ready for messy, unpredictable lives.

The intersection of pharmacology, neuroscience, and AI is timely. Traditional AI models, rooted in reinforcement learning (RL) or deep neural networks, excel in pattern recognition but falter under bias-like perturbations, much like a human under intoxication. For instance, an RL agent might over-exploit short-term rewards (akin to cocaine-fueled impulsivity), leading to catastrophic failures in dynamic environments. Brain-inspired AI, which emulates neural plasticity and neuromodulation, offers a solution but requires accurate models of biased cognition to avoid replicating human flaws.

REFERENCES

ATTC Network. (2025). *Privacy & ethical implications of AI in SUD treatment.* Addiction Technology Transfer Center Network. https://attcnetwork.org/products_and_resources/privacy-ethical-implications-of-ai-in-sud-treatment/

Badre, D., & Frank, M. J. (2012). Mechanisms of hierarchical reinforcement learning in cortico–striatal circuits 2: Evidence from fMRI. *Cerebral Cortex (New York, N.Y.)*, *22*(3), 527–536. DOI: 10.1093/cercor/bhr117

Bickel, W. K., Jarmolowicz, D. P., Mueller, E. T., Gatchalian, K. M., & McClure, S. M. (2012). Are executive function and impulsivity antipodes? A conceptual reconstruction with special reference to addiction. *Psychopharmacology*, *221*(3), 361–387. DOI: 10.1007/s00213-012-2689-x

Bickel, W. K., Koffarnus, M. N., Moody, L., & Wilson, A. G. (2014). The behavioral- and neuro-economic process of temporal discounting: A candidate behavioral marker of addiction. *Neuropharmacology*, *76*, 518–527. DOI: 10.1016/j.neuropharm.2013.06.013

Bird, J. (2013). *The role of trait impulsivity and novelty-seeking as predisposing factors to the acquisition and reinstatement of MDMA self-administration* (Doctoral dissertation, Open Access Te Herenga Waka-Victoria University of Wellington).

Blum, K., Werner, T., Carnes, S., Carnes, P., Bowirrat, A., Giordano, J., Marlene-Oscar-Berman, , & Gold, M. (2012). Sex, drugs, and rock 'n'roll: Hypothesizing common mesolimbic activation as a function of reward gene polymorphisms. *Journal of Psychoactive Drugs*, *44*(1), 38–55. DOI: 10.1080/02791072.2012.662112

Buelow, M. T., & Suhr, J. A. (2009). Construct validity of the Iowa gambling task. *Neuropsychology Review*, *19*(1), 102–114. DOI: 10.1007/s11065-009-9083-4

Business Roundtable. (2026). *Artificial intelligence policy recommendations.* https://www.businessroundtable.org/artificial-intelligence-policy-recommendations

Butelman, E. R., Huang, Y., Goldstein, R. Z., & Alia-Klein, N. (2023). Men have approximately 2-to 3-fold greater overdose mortality than women for synthetic opioids, heroin and psychostimulants including cocaine across the lifespan: Analysis of state-level CDC data for 2020-2021. *medRxiv*.

Carichon, F., Khandelwal, A., Fauchard, M., & Farnadi, G. (2025). The coming crisis of multi-agent misalignment: Ai alignment must be a dynamic and social process. *arXiv preprint arXiv:2506.01080.*

Ceceli, A. O., Huang, Y., Kronberg, G., McClain, N., King, S. G., Butelman, E. R., & Goldstein, R. Z. (2025). The impaired response inhibition and salience attribution model of drug addiction: Recent neuroimaging evidence and Future Directions. *Annual Review of Psychology*, , 77.

Censinet. (2026). *The new risk frontier: Navigating AI uncertainty in an automated world.* https://www.censinet.com/perspectives/new-risk-frontier-navigating-ai-uncertainty

Cha, J., Park, J., Cho, M., Lee, E., Kim, B. G., Kim, G., & Joo, Y. (2024). Individual Differences in the Effects of Neighborhood Socioeconomic Deprivation on Intertemporal Decision-Making and Psychotic-Like Experiences in Children.

Colizzi, M., & Bhattacharyya, S. (2018). Cannabis use and the development of tolerance: A systematic review of human evidence. *Neuroscience and Biobehavioral Reviews*, *93*, 1–25. DOI: 10.1016/j.neubiorev.2018.07.014

Cowell, R. G., Verrall, R. J., & Yoon, Y. K. (2007). Modeling operational risk with Bayesian networks. *The Journal of Risk and Insurance*, *74*(4), 795–827. DOI: 10.1111/j.1539-6975.2007.00235.x

Dalley, J. W., Everitt, B. J., & Robbins, T. W. (2011). Impulsivity, compulsivity, and top-down cognitive control. *Neuron*, *69*(4), 680–694. DOI: 10.1016/j.neuron.2011.01.020

Dalley, J. W., & Roiser, J. P. (2012). Dopamine, serotonin and impulsivity. *Neuroscience*, *215*, 42–58. DOI: 10.1016/j.neuroscience.2012.03.065

Daw, N. D., Niv, Y., & Dayan, P. (2005). Uncertainty-based competition between prefrontal and dorsolateral striatal systems for behavioral control. *Nature Neuroscience*, *8*(12), 1704–1711. DOI: 10.1038/nn1560

Dayan, P., & Daw, N. D. (2008). Decision theory, reinforcement learning, and the brain. *Cognitive, Affective & Behavioral Neuroscience*, *8*(4), 429–453. DOI: 10.3758/CABN.8.4.429

Dezfouli, A., & Balleine, B. W. (2012). Habits, action sequences and reinforcement learning. *The European Journal of Neuroscience*, *35*(7), 1036–1051. DOI: 10.1111/j.1460-9568.2012.08050.x

Din, M., Daga, K., Saoud, J., Wood, D., Kierkegaard, P., Brex, P., & Booth, T. C. (2025). Clinicians' perspectives on the use of artificial intelligence to triage MRI brain scans. *European Journal of Radiology*, *183*, 111921. DOI: 10.1016/j.ejrad.2025.111921

Eklund, A., Nichols, T. E., & Knutsson, H. (2016). Cluster failure: Why fMRI inferences for spatial extent have inflated false-positive rates. *Proceedings of the National Academy of Sciences of the United States of America*, *113*(28), 7900–7905. DOI: 10.1073/pnas.1602413113

Everitt, B. J., & Robbins, T. W. (2016). Drug addiction: Updating actions to habits to compulsions ten years on. *Annual Review of Psychology*, *67*(1), 23–50. DOI: 10.1146/annurev-psych-122414-033457

Field, M., & Cox, W. M. (2008). Attentional bias in addictive behaviors: A review of its development, causes, and consequences. *Drug and Alcohol Dependence*, *97*(1-2), 1–20. DOI: 10.1016/j.drugalcdep.2008.03.030

Fleming, S. M., & Daw, N. D. (2017). Self-evaluation of decision-making: A general Bayesian framework for metacognitive computation. *Psychological Review*, *124*(1), 91–114. DOI: 10.1037/rev0000045

Gabriel, I. (2020). Artificial Intelligence, Values, and Alignment: I. Gabriel. *Minds and Machines*, *30*(3), 411–437. DOI: 10.1007/s11023-020-09539-2

Garcez, A. D. A., & Lamb, L. C. (2023). Neurosymbolic ai: The 3 rd wave. [gracez]. *Artificial Intelligence Review*, *56*(11), 12387–12406. DOI: 10.1007/s10462-023-10448-w

Gimenez, S., Millan, A., Mora-Morell, A., Ayuso, N., Gastaldo-Jordán, I., & Pardo, M. (2024). Advances in Brain Stimulation and Nanomedicine: Dopaminergic Alterations and Its Role in Neurodegeneration and Drug Addiction.

Goldstein, R. Z., & Volkow, N. D. (2011). Dysfunction of the prefrontal cortex in addiction: Neuroimaging findings and clinical implications. *Nature Reviews. Neuroscience*, *12*(11), 652–669. DOI: 10.1038/nrn3119

L Griffiths, T., Kemp, C., & B Tenenbaum, J. (2008). Bayesian models of cognition.

Groman, S. M., Massi, B., Mathias, S. R., Lee, D., & Taylor, J. R. (2019). Model-free and model-based influences in addiction-related behaviors. *Biological Psychiatry*, *85*(11), 936–945. DOI: 10.1016/j.biopsych.2018.12.017

Gupta, D. (2024). *Digital Twins in healthcare–A new era of innovation for businesses*. Appinventiv.

Ho, C. W. L., ... (2023). *Regulating under uncertainty: Governance options for generative AI*. Stanford Cyber Policy Center. https://cyber.fsi.stanford.edu/content/regulating-under-uncertainty-governance-options-generative-ai

Huys, Q. J., Browning, M., Paulus, M. P., & Frank, M. J. (2021). Advances in the computational understanding of mental illness. *Neuropsychopharmacology : Official Publication of the American College of Neuropsychopharmacology*, *46*(1), 3–19. DOI: 10.1038/s41386-020-0746-4

Huys, Q. J., Maia, T. V., & Frank, M. J. (2016). Computational psychiatry as a bridge from neuroscience to clinical applications. *Nature Neuroscience*, *19*(3), 404–413. DOI: 10.1038/nn.4238

Janak, P. H., Bowers, M. S., & Corbit, L. H. (2012). Compound stimulus presentation and the norepinephrine reuptake inhibitor atomoxetine enhance long-term extinction of cocaine-seeking behavior. *Neuropsychopharmacology : Official Publication of the American College of Neuropsychopharmacology*, *37*(4), 975–985. DOI: 10.1038/npp.2011.281

Jentsch, J. D., & Taylor, J. R. (1999). Impulsivity resulting from frontostriatal dysfunction in drug abuse: Implications for the control of behavior by reward-related stimuli. *Psychopharmacology*, *146*(4), 373–390. DOI: 10.1007/PL00005483

Joynt, R. J. (1995). Descartes' error: Emotion, reason, and the human brain. *Journal of the American Medical Association*, *273*(18), 1463–1464. DOI: 10.1001/jama.1995.03520420079045

Juliani, A., & Ash, J. T. (2024). A study of plasticity loss in on-policy deep reinforcement learning. *Advances in Neural Information Processing Systems*, *37*, 113884–113910. DOI: 10.52202/079017-3616

Kavanagh, D. J., Andrade, J., & May, J. (2005). Imaginary relish and exquisite torture: The elaborated intrusion theory of desire. *Psychological Review*, *112*(2), 446–467. DOI: 10.1037/0033-295X.112.2.446

Keiflin, R., & Janak, P. H. (2015). Dopamine prediction errors in reward learning and addiction: From theory to neural circuitry. *Neuron*, *88*(2), 247–263. DOI: 10.1016/j.neuron.2015.08.037

Keramati, M., & Gutkin, B. (2013). Imbalanced decision hierarchy in addicts emerging from drug-hijacked dopamine spiraling circuit. *PLoS One*, *8*(4), e61489. DOI: 10.1371/journal.pone.0061489

Koffarnus, M. N., & Bickel, W. K. (2014). A 5-trial adjusting delay discounting task: Accurate discount rates in less than one minute. *Experimental and Clinical Psychopharmacology*, *22*(3), 222–228. DOI: 10.1037/a0035973

Konova, A. B., Ceceli, A. O., Horga, G., Moeller, S. J., Alia-Klein, N., & Goldstein, R. Z. (2023). Reduced neural encoding of utility prediction errors in cocaine addiction. *Neuron, 111*(24), 4058–4070. DOI: 10.1016/j.neuron.2023.09.015

Koob, G. F., & Volkow, N. D. (2016). Neurobiology of addiction: A neurocircuitry analysis. *The Lancet. Psychiatry, 3*(8), 760–773. DOI: 10.1016/S2215-0366(16)00104-8

Landes, R. D., Christensen, D. R., & Bickel, W. K. (2012). Delay discounting decreases in those completing treatment for opioid dependence. *Experimental and Clinical Psychopharmacology, 20*(4), 302–309. DOI: 10.1037/a0027391

MacKillop, J., Amlung, M. T., Few, L. R., Ray, L. A., Sweet, L. H., & Munafò, M. R. (2011). Delayed reward discounting and addictive behavior: A meta-analysis. *Psychopharmacology, 216*(3), 305–321. DOI: 10.1007/s00213-011-2229-0

Marchant, N. J., Furlong, T. M., & McNally, G. P. (2010). Medial dorsal hypothalamus mediates the inhibition of reward seeking after extinction. *The Journal of Neuroscience : The Official Journal of the Society for Neuroscience, 30*(42), 14102–14115. DOI: 10.1523/JNEUROSCI.4079-10.2010

Mikus, N., Korb, S., Massaccesi, C., Gausterer, C., Graf, I., Willeit, M., Eisenegger, C., Lamm, C., Silani, G., & Mathys, C. (2022). Effects of dopamine D2/3 and opioid receptor antagonism on the trade-off between model-based and model-free behaviour in healthy volunteers. *eLife, 11*, e79661. DOI: 10.7554/eLife.79661

Mitchell, J. M., & Ot'alora, G. (2023). MDMA-assisted therapy for moderate to severe PTSD: A randomized, placebo-controlled phase 3 trial. *Nature Medicine, 29*(10), 2473–2480. DOI: 10.1038/s41591-023-02565-4

Morreale, M. K. (2024). Cannabis Use and Cognition. *The Other Side of Cannabis: Impact on Mental and Physical Health*, 124.

Nordström, M. (2022). AI under great uncertainty: Implications and decision strategies for public policy. *AI & Society, 37*(4), 1703–1714. DOI: 10.1007/s00146-021-01263-4

Palminteri, S., Wyart, V., & Koechlin, E. (2017). The importance of falsification in computational cognitive modeling. *Trends in Cognitive Sciences, 21*(6), 425–433. DOI: 10.1016/j.tics.2017.03.011

Parente, D. J. (2024). Generative artificial intelligence and large language models in primary care medical education. *Family Medicine, 56*(9), 534–540. DOI: 10.22454/FamMed.2024.775525

Perner, J., & Roessler, J. (2012). From infants' to children's appreciation of belief. *Trends in Cognitive Sciences, 16*(10), 519–525. DOI: 10.1016/j.tics.2012.08.004

Poldrack, R. A., Huckins, G., & Varoquaux, G. (2020). Establishment of best practices for evidence for prediction: A review. *JAMA Psychiatry*, *77*(5), 534. DOI: 10.1001/jamapsychiatry.2019.3671

Prell, R. (2025). *Self-Regulatory Challenges in Decision-Making: Cardiac Autonomic Modulation in Ambiguous, Risky, and Moral Situations* (Doctoral dissertation, Dissertation, Duisburg, Essen, Universität Duisburg-Essen, 2025).

Rahwan, I., Cebrian, M., Obradovich, N., Bongard, J., Bonnefon, J. F., Breazeal, C., Crandall, J. W., Christakis, N. A., Couzin, I. D., Jackson, M. O., Jennings, N. R., Kamar, E., Kloumann, I. M., Larochelle, H., Lazer, D., McElreath, R., Mislove, A., Parkes, D. C., Pentland, A. S., & Wellman, M. (2019). Machine behaviour. *Nature*, *568*(7753), 477–486. DOI: 10.1038/s41586-019-1138-y

Ramaekers, J. G., Kuypers, K. P., & Vollenweider, F. X. (2026). Subjective and neurocognitive profiling of clinical doses of 3, 4-methylenedioxymethamphetamine (MDMA) in healthy volunteers: Implications for therapeutic use. *Molecular Psychiatry*, *31*(8), 1–12. DOI: 10.1038/s41380-026-03602-7

Ramakrishnan, S. A., Muliya, M., Chakravarthy, V. S., & Parvaz, M. A. (2026). Applications and Evolution of Neuroeconomics and Reinforcement Learning Models in Substance Use Disorders. In *Neuroeconomics: Core Topics and Current Directions: NA* (pp. 611–627). Springer Nature Switzerland. DOI: 10.1007/978-3-032-02925-6_33

Redish, A. D. (2004). Addiction as a computational process gone awry. *Science*, *306*(5703), 1944–1947. DOI: 10.1126/science.1102384

Redish, A. D., Jensen, S., & Johnson, A. (2008). Addiction as vulnerabilities in the decision process. *Behavioral and Brain Sciences*, *31*(4), 461–487. DOI: 10.1017/S0140525X08004986

Robinson, T. E., & Berridge, K. C. (2008). The incentive sensitization theory of addiction: Some current issues. *Philosophical Transactions of the Royal Society of London. Series B, Biological Sciences*, *363*(1507), 3137–3146. DOI: 10.1098/rstb.2008.0093

Rodrigues, F. D. A. A., Ferreira, R. S., Cruz, L. N., & Pagnoncelli, C. J. (2025). Clinical Neuroscience of Addiction: Neurobiological Disruptions and Translational Models of Social Dysfunction. *Ciencia Latina Revista Científica Multidisciplinar*, *9*(3), 8215–8239. DOI: 10.37811/cl_rcm.v9i3.18443

Saeidnia, H. R., Hashemi Fotami, S. G., Lund, B., & Ghiasi, N. (2024). Ethical considerations in artificial intelligence interventions for mental health and well-being: Ensuring responsible implementation and impact. *Social Sciences (Basel, Switzerland)*, *13*(7), 381. DOI: 10.3390/socsci13070381

Schultz, W. (2016). Dopamine reward prediction error coding. *Dialogues in Clinical Neuroscience*, *18*(1), 23–32. DOI: 10.31887/DCNS.2016.18.1/wschultz

Stephan, K. E., Iglesias, S., Heinzle, J., & Diaconescu, A. O. (2015). Translational perspectives for computational neuroimaging. *Neuron*, *87*(4), 716–732. DOI: 10.1016/j.neuron.2015.07.008

Sullivan, E. V., Lannoy, S., Le Berre, A. P., Fama, R., & Pfefferbaum, A. (2023). Alcohol drinking and alcohol use disorder across the ages: Dynamic effects on the brain and function.

Takahashi, J. S. (2017). Transcriptional architecture of the mammalian circadian clock. *Nature Reviews. Genetics*, *18*(3), 164–179. DOI: 10.1038/nrg.2016.150

Thoelen, G. T. (2023). *Essays in the Neurophysiology of Economic Decision-Making and Preference Formation* (Doctoral dissertation, The Claremont Graduate University).

Van Den Bos, R., Homberg, J., & De Visser, L. (2013). A critical review of sex differences in decision-making tasks: Focus on the Iowa Gambling Task. *Behavioural Brain Research*, *238*, 95–108. DOI: 10.1016/j.bbr.2012.10.002

Volkow, N. D., Michaelides, M., & Baler, R. (2019). The neuroscience of drug reward and addiction. *Physiological Reviews*, *99*(4), 2115–2140. DOI: 10.1152/physrev.00014.2018

Wiehler, A., Branzoli, F., Adanyeguh, I., Mochel, F., & Pessiglione, M. (2022). A neuro-metabolic account of why daylong cognitive work alters the control of economic decisions. *Current Biology*, *32*(16), 3564–3575. DOI: 10.1016/j.cub.2022.07.010

Zhang, K., Yang, Z., & Başar, T. (2021). Multi-agent reinforcement learning: A selective overview of theories and algorithms. *Handbook of reinforcement learning and control*, 321-384.

Zhang, M. W., Ying, J., Wing, T., Song, G., Fung, D. S., & Smith, H. E. (2018). Cognitive biases in cannabis, opioid, and stimulant disorders: A systematic review. *Frontiers in Psychiatry*, *9*, 376. DOI: 10.3389/fpsyt.2018.00376

Chapter 12

Artificial Glia-Inspired Stabilization for Reliable Deep Reinforcement Learning Decision Intelligence

Rajkumar Jagdale
http://orcid.org/0000-0002-4074-1095
Vishwakarma University, Pune, India

Kundan Bedmutha
http://orcid.org/0009-0000-4391-3107
Vishwakarma University, Pune, India

ABSTRACT

Deep reinforcement learning (DRL) has shown strong potential for solving complex decision-making problems, but its real-world adoption is limited by training instability, performance degradation, and safety concerns. This chapter proposes an artificial glia-inspired stabilization framework to enhance the robustness and reliability of DRL-based decision intelligence systems. Inspired by the regulatory roles of glial cells in biological neural systems, the framework incorporates continuous stability monitoring, homeostatic regulation of learning dynamics, and constraint-aware safety enforcement. It adaptively detects learning instabilities, regulates network parameters, and enforces operational constraints during both training and deployment. Experimental evaluations on autonomous systems, financial portfolio management,

DOI: 10.4018/979-8-3373-8988-2.ch012

and healthcare optimization demonstrate reduced training variance, improved constraint satisfaction, and faster convergence compared to existing baseline methods, enabling safer and more trustworthy DRL for real-world applications.

INTRODUCTION

The Promise and Challenges of Deep Reinforcement Learning

Deep reinforcement learning (DRL) represents a major paradigm shift in artificial intelligence, enabling autonomous agents to learn optimal sequential decision-making through interaction with dynamic environments (Sutton & Barto, 2018; Mnih et al., 2015). By combining deep neural networks with reinforcement learning, DRL has achieved superhuman performance in games, robotic manipulation, data center optimization, and autonomous control (Kiran et al., 2021).

Despite these advances, the practical deployment of DRL in decision intelligence applications remains constrained by instability and reliability challenges (Henderson et al., 2018; Ilyas et al., 2018). During training, agents frequently exhibit performance degradation, sensitivity to hyperparameters, and erratic behavioral shifts—even after apparent convergence.

These instabilities arise from multiple interacting factors. Gradient pathologies such as exploding or vanishing gradients can destabilize parameter updates. Policy divergence may occur when previously stable policies deteriorate unexpectedly. Value function collapse—often driven by overestimation or underestimation in temporal difference learning—can disrupt policy improvement. Additionally, DRL methods are highly sensitive to learning rates, discount factors, and exploration parameters, leading to substantial variability across runs and reduced reproducibility.

Such issues are particularly problematic in decision intelligence domains, where decisions have real-world consequences and must satisfy safety, reliability, and explainability requirements (Davenport & Ronanki, 2018). Unlike game-playing benchmarks, failures in healthcare, finance, autonomous systems, or industrial control can produce irreversible harm. These contexts demand learning systems that are not only performant but stable, predictable, and constraint-aware.

The Neuroscience-Inspired Solution: Artificial Glia

Biological neural systems offer insight into how complex adaptive networks maintain stability. While neurons are typically emphasized as computational units, glial cells—particularly astrocytes—play essential regulatory roles in maintaining neural homeostasis (Allen & Lyons, 2018; Santello et al., 2019).

Astrocytes regulate synaptic transmission, balance excitation and inhibition, coordinate network-wide activity, provide metabolic support, and detect pathological states to prevent instability. Through homeostatic control and activity-dependent modulation, they preserve functional stability in highly dynamic neural circuits.

This regulatory architecture offers a compelling analogy for deep reinforcement learning. Just as astrocytes stabilize biological networks against runaway excitation or collapse, artificial glia mechanisms can be designed to monitor learning dynamics, regulate updates adaptively, and prevent destabilizing behaviors in artificial neural systems. Rather than relying on static stabilization heuristics, a glia-inspired framework enables continuous, feedback-driven regulation of learning processes.

Chapter Scope and Contributions

This chapter introduces an artificial glia–inspired framework to enhance the reliability and safety of deep reinforcement learning in decision intelligence applications.

Theoretical contributions include a principled mapping between biological glial regulation and DRL stabilization requirements, formalization of homeostatic control in temporal difference learning, and analysis of stability properties under adaptive regulation.

Methodologically, we propose a modular architecture integrating stability monitoring, homeostatic regulation, and constraint-aware safety gating. The framework adaptively adjusts learning rates, gradient scaling, and exploration strategies based on real-time stability signals, and is compatible with widely used algorithms including PPO, SAC, and TRPO.

Empirically, we evaluate the framework across six diverse decision intelligence domains spanning continuous control, safety-critical tasks, finance, and healthcare. Results demonstrate significant reductions in training variance, accelerated convergence, and improved constraint satisfaction compared to baseline and safe RL methods. Ablation studies isolate the contribution of each regulatory component and confirm their synergistic effects.

From an applied perspective, we provide implementation guidelines and case studies illustrating deployment in safety-critical settings. By emphasizing transparency, reproducibility, and modular integration, the proposed approach offers a practical pathway toward stable and trustworthy DRL systems.

Chapter Organization

The remainder of this chapter is organized as follows:

Section 2 provides essential background on decision intelligence paradigms, deep reinforcement learning foundations, and the biological neuroscience of glial cells, establishing the interdisciplinary foundation for our approach.

Section 3 surveys related work in DRL stabilization techniques, safe reinforcement learning, and neuroscience-inspired AI, positioning our contributions within the broader research landscape.

Section 4 presents our artificial glia-inspired framework, detailing the architecture, stability monitoring mechanisms, homeostatic regulation algorithms, and safety constraint enforcement.

Section 5 describes our experimental methodology, including benchmark environments, evaluation metrics, and baseline comparisons.

Section 6 presents comprehensive results demonstrating stability improvements, convergence acceleration, and safety enhancements across diverse decision intelligence tasks.

Section 7 discusses practical applications in autonomous systems, financial decision-making, and healthcare optimization, providing case studies of real-world deployments.

Section 8 addresses ethical considerations and reliability requirements for trustworthy DRL deployment in high-stakes decision-making contexts.

Section 9 outlines promising directions for future research, including theoretical extensions, architectural innovations, and emerging application domains.

Section 10 concludes by synthesizing key findings and reflecting on the broader implications of biologically-inspired approaches to reliable artificial intelligence.

BACKGROUND AND FOUNDATION

Decision Intelligence: Paradigm and Requirements

Defining Decision Intelligence

Decision intelligence (DI) is the engineering discipline of structured decision-making that integrates data science, social science, managerial science, and artificial intelligence to improve organizational and operational outcomes (Davenport & Ronanki, 2018; Chen et al., 2012). Unlike traditional business intelligence, which focuses on descriptive ("what happened?") or predictive ("what will happen?") analytics, DI emphasizes prescriptive decision-making and autonomous execution—

determining what actions should be taken and how to implement them optimally. Its lifecycle typically includes problem formulation, where objectives are formalized into state, action, and reward structures; data integration, combining structured and unstructured sources; model development, from rule-based systems to learning-based approaches; simulation and validation to test policies under realistic conditions; deployment and monitoring of operational performance; and continuous feedback and iteration to refine decisions as environments evolve. This lifecycle underscores that decision intelligence extends beyond prediction toward accountable, operational decision execution.

Requirements for Trustworthy Decision Intelligence

Real-world DI systems must satisfy requirements that extend beyond simply maximizing performance. They must ensure reliability through consistent behavior across diverse and unpredictable conditions, and uphold safety by adhering to hard regulatory, operational, and ethical constraints. Explainability is essential to provide sufficient transparency for accountability and stakeholder trust, while robustness enables systems to degrade gracefully under distribution shifts, adversarial perturbations, or novel scenarios. Sample efficiency becomes critical when data collection is costly or limited, and computational tractability ensures real-time operation under resource constraints. In addition, fairness and ethical alignment are necessary to prevent bias and harmful outcomes. These constraints are especially vital in high-stakes domains, where failures may lead to irreversible consequences. As a result, deep reinforcement learning applied to DI must integrate mechanisms for stability, safety, and constraint awareness rather than focusing solely on unconstrained reward optimization.

Decision Intelligence Application Domains

Decision intelligence spans multiple domains where deep reinforcement learning (DRL) shows strong potential yet faces significant deployment challenges. In autonomous systems such as self-driving vehicles, drones, and robotics, real-time control under uncertainty must be achieved while strictly satisfying safety constraints (Kiran et al., 2021). In healthcare, treatment planning and resource allocation require balancing outcome optimization with patient safety, clinical guidelines, and explainability. Financial applications, including portfolio management and trading, operate within regulatory, capital, and risk constraints in highly non-stationary environments. In energy and infrastructure systems—such as smart grids, industrial control, and HVAC—high-dimensional dynamics introduce complex trade-offs between safety and efficiency. Personalization systems, including recommendation and engagement optimization platforms, must account for privacy, fairness, and diversity requirements. Across these domains, operational constraints and real-world variability reveal the instability and brittleness of conventional DRL methods, motivating the development of reliability-enhanced approaches such as the artificial glia–inspired stabilization framework proposed in this chapter.

Deep Reinforcement Learning: Foundations and Challenges

The Reinforcement Learning Problem

Reinforcement learning models sequential decision-making as a Markov Decision Process (MDP) defined by the tuple $(SAPR\gamma)$, where Sdenotes states, Aactions, Ptransition dynamics, Rreward signals, and $\gamma$the discount factor controlling the trade-off between immediate and long-term returns. The objective is to learn a policy πthat maximizes expected cumulative discounted reward, typically evaluated through value functions such as $V^{\pi}(s)$, $Q^{\pi}(s,a)$, and the advantage function. Deep reinforcement learning extends this framework by approximating value functions or policies using deep neural networks. Deep Q-Networks (DQN) introduced experience replay and target networks to stabilize value learning in high-dimensional spaces, while policy gradient methods directly optimize policy parameters, with algorithms such as Proximal Policy Optimization (PPO) and Soft Actor-Critic (SAC) improving stability through clipped objectives and entropy regularization. Deep Q-Networks (DQN) revolutionized RL by using deep neural networks to approximate Q-functions in high-dimensional state spaces (Mnih et al., 2015). DQN employs experience replay, target networks, and epsilon-greedy exploration to achieve stable learning. Policy Gradient Methods directly optimize policy parameters by ascending the policy gradient. Proximal Policy Optimization (PPO) improves stability through clipped surrogate objectives (Schulman et al., 2017), while Soft Actor-Critic (SAC) combines off-policy learning with maximum entropy RL (Haarnoja et al., 2018).

Instability Challenges in Deep RL

Despite these advances, deep reinforcement learning remains prone to instability. Training dynamics are sensitive to gradient pathologies, bootstrapping errors, and the "deadly triad" interaction between function approximation, off-policy learning, and temporal difference updates. Moving targets and evolving policies further complicate convergence, while high-variance gradient estimates, exploration–exploitation trade-offs, reward sparsity, and hyperparameter sensitivity contribute to unpredictable performance and reproducibility challenges. These limitations constrain reliable deployment in real-world decision intelligence contexts and motivate the development of adaptive stabilization mechanisms capable of regulating learning dynamics more robustly.

Biological Glia: Structure, Function, and Inspiration

Glial Cells in the Nervous System

Glial cells constitute a major component of the human brain, existing in roughly a 1:1 ratio with neurons and performing essential regulatory functions (Allen & Lyons, 2018). Among the diverse glial types—including oligodendrocytes, Schwann cells, microglia, and NG2 glia—astrocytes are particularly relevant for computational inspiration due to their central role in maintaining neural stability. Astrocytes regulate extracellular ion concentrations, recycle neurotransmitters, provide metabolic support, and modulate synaptic transmission through activity-dependent signaling. They maintain homeostasis by preventing runaway excitation, stabilizing neuronal activity, coordinating network-wide dynamics via calcium signaling, and responding to pathological states to preserve system integrity. These regulatory properties make astrocytes a compelling biological model for stabilizing adaptive artificial systems.

Astrocyte Functions Relevant to Learning Stability

Astrocyte mechanisms translate naturally into stabilization principles for deep reinforcement learning. Neurotransmitter clearance prevents excitotoxicity in biological networks, analogous to adaptive gradient clipping and normalization that prevent gradient explosions in deep networks. Potassium buffering stabilizes neural excitability, mirroring activation normalization strategies that maintain stable internal representations. Calcium-mediated signaling and activity-dependent modulation correspond to adaptive learning rate control and update scheduling, while gliotransmitter effects parallel dynamic exploration regulation through entropy adjustment. Metabolic support reflects adaptive capacity allocation in neural architectures, and homeostatic plasticity aligns with mechanisms that regulate target network updates and parameter scaling. Similarly, biological synaptic pruning corresponds to regularization strategies that sculpt network connectivity during learning.

Conceptual Mapping to DRL Stabilization

Collectively, this biological–computational mapping provides a principled foundation for the artificial glia–inspired framework. Decision intelligence applications demand reliability, safety, and robustness beyond conventional deep reinforcement learning capabilities, while DRL instability arises from interacting factors such as function approximation, bootstrapping, and exploration dynamics. By drawing on astrocytic homeostasis and regulatory control, the proposed approach introduces adaptive mechanisms designed to stabilize learning dynamics and enhance the reliability of deep reinforcement learning systems.

Summary

This section established three foundational pillars that underpin the proposed approach. First, decision intelligence applications impose requirements for reliability, safety, and robustness that extend beyond the capabilities of many current deep reinforcement learning methods. Second, the instabilities observed in deep RL stem from fundamental tensions among function approximation, bootstrapping, and exploration, which interact to produce volatile and sometimes unpredictable learning dynamics. Third, biological glial cells implement sophisticated regulatory and homeostatic mechanisms that naturally correspond to the stabilization needs of artificial learning systems, thereby offering a principled and biologically grounded framework for algorithmic design.

Related Work / Literature Review

Training Stabilization in Deep RL

Numerous techniques have been developed to mitigate instability in deep reinforcement learning. Early approaches such as target networks in Deep Q-Networks (DQN) reduced oscillations in temporal difference learning by decoupling target updates (Mnih et al., 2015), while Polyak averaging and adaptive update schedules further improved stability. Trust region methods constrained policy updates using Kullback–Leibler divergence to limit destructive parameter shifts; Trust Region Policy Optimization (TRPO) provides theoretical guarantees under such constraints, and Proximal Policy Optimization (PPO) achieves similar stability through clipped objectives with lower computational cost (Schulman et al., 2017). Additional strategies include gradient clipping, normalization techniques, distributional reinforcement learning to reduce overestimation bias (Bellemare et al., 2017), and variance reduction methods such as Generalized Advantage Estimation. Despite these advances, empirical studies show that DRL performance remains highly sensitive to hyperparameters, initialization, and implementation details (Henderson et al., 2018), underscoring the fragility of existing stabilization mechanisms.

Safe Reinforcement Learning

Safety in reinforcement learning has similarly gained prominence for real-world deployment. Constrained Markov Decision Processes (CMDPs) formalize safety requirements, with Constrained Policy Optimization (CPO) extending trust region methods via Lagrangian relaxation (Achiam et al., 2017; García & Fernández, 2015). Other approaches incorporate safety critics, shielding mechanisms, robust optimization under worst-case dynamics, adversarial robustness techniques, and formal verification methods. However, many of these methods treat safety as an external constraint or post hoc correction, and scalability remains a challenge in high-dimensional deep RL systems. In contrast, the artificial glia–inspired framework integrates stability and safety directly into the learning loop through continuous monitoring and adaptive regulation.

Neuroscience-Inspired Artificial Intelligence

Neuroscience has long inspired AI architectures and algorithms. Hassabis et al. (2017) reviewed neuroscience-inspired AI, identifying attention, episodic memory, continual learning, and imagination as promising directions. However, comprehensive frameworks for DRL stabilization using glia-inspired mechanisms remained unexplored.

Research Gap and Our Contribution

Despite advances in deep reinforcement learning stabilization and safe RL, several gaps persist. Existing techniques typically address specific failure modes in isolation rather than providing a unified, adaptive regulatory framework. Neuroscience-inspired AI has largely focused on perception and representation learning, leaving homeostatic and regulatory mechanisms underexplored in temporal difference learning and policy optimization. Similarly, many safe RL approaches impose constraints as external objectives or post hoc corrections instead of embedding safety directly within learning dynamics. Moreover, stabilization methods often demonstrate success only in controlled benchmarks, with limited validation in realistic decision intelligence settings involving operational constraints and domain-specific risks. The artificial glia–inspired framework addresses these limitations through a unified, biologically grounded architecture that integrates stability monitoring, adaptive regulation, and constraint-aware safety directly into the learning loop. By translating astrocytic homeostasis into algorithmic mechanisms and validating the approach across diverse, constraint-rich decision intelligence domains, the framework bridges the gap between benchmark performance and real-world deployability.

Artificial Glia-Inspired Framework

Framework Overview and Philosophy

The proposed artificial glia framework incorporates three regulatory components inspired by astrocyte biology: stability monitoring, homeostatic regulation, and a constraint-aware safety gate. The stability module continuously evaluates learning dynamics through diagnostic signals analogous to astrocytic activity sensing. The homeostatic module adaptively adjusts learning parameters to maintain stable and efficient training, while the safety gate enforces domain-specific operational constraints during both training and deployment. Unlike fixed stabilization heuristics, the framework implements adaptive, activity-dependent regulation that responds dynamically to evolving learning conditions. The design emphasizes modularity, interpretability, and efficiency. Components can be configured independently, and regulatory interventions are explicitly linked to detected instabilities. The architecture introduces minimal computational overhead and remains compatible with major DRL algorithms, including PPO, TRPO, SAC, TD3, and DQN. By embedding adaptive regulation directly into the learning process, the framework also improves robustness to hyperparameter sensitivity and training variability.

Architecture

The artificial glia layer operates as an intermediary regulatory mechanism that intercepts gradients and parameter updates, monitoring stability signals and applying homeostatic adjustments before updates are committed to the underlying deep reinforcement learning agent.

Architecturally, the framework comprises three tightly integrated modules. The Stability Monitoring Module continuously evaluates learning health by tracking gradient magnitude distributions, policy divergence metrics, value function diagnostics, and reward distribution statistics. These signals provide a comprehensive view of training dynamics and enable early detection of pathological behaviors. The Homeostatic Regulation Module translates these diagnostic signals into adaptive interventions, implementing learning rate modulation, dynamic

gradient clipping, exploration strategy adjustment, and adaptive target network update scheduling to maintain stable and productive learning. Complementing these mechanisms, the

Safety and Constraint Gate enforces domain-specific safety requirements by detecting constraint violations, performing action filtering or projection when necessary, triggering rollback procedures under severe instability, and facilitating safe recovery mechanisms. Together, these modules form a closed-loop regulatory architecture that supervises and stabilizes the learning process in real time.

Figure 1. Architecture of the artificial glia–inspired stabilization framework integrated with a deep reinforcement learning agent

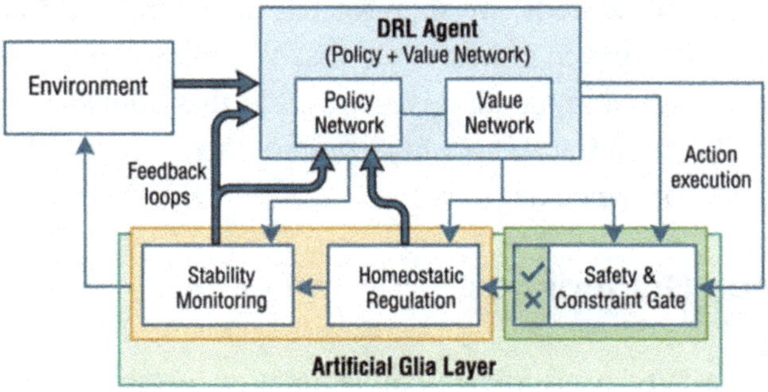

Figure 1 presents the artificial glia–inspired architecture for stabilizing deep reinforcement learning in decision intelligence applications. A standard DRL agent, consisting of policy and value networks, is augmented by an intermediary artificial glia layer that supervises learning dynamics. The stability monitoring module tracks gradient statistics, policy divergence, and value diagnostics to detect instability. These signals drive the homeostatic regulation module, which adaptively adjusts learning rates, gradient clipping thresholds, exploration parameters, and target update schedules. A constraint-aware safety gate enforces operational boundaries through action filtering, projection, and rollback mechanisms when violations occur. Continuous feedback between the glia layer and the DRL agent enables closed-loop regulation, promoting stable and reliable learning during both training and deployment..

Stability Monitoring Module

The stability monitoring module computes diagnostic metrics that characterize learning health, analogous to calcium signaling in astrocytes that reflects network activity levels.

Gradient Statistics

We track the mean, variance, and tail statistics (99th percentile) of gradient magnitudes across all network parameters. Gradient explosion detection uses exponential moving averages to detect anomalies when gradient magnitude exceeds 3-5× the historical average. We also measure gradient direction consistency using cosine similarity to identify rapidly oscillating gradients indicating conflicting learning signals.

Policy Divergence Metrics

For policy gradient methods, we monitor KL divergence between consecutive policies. Large KL divergence indicates aggressive policy changes that may destabilize learning. We maintain an adaptive threshold that adjusts to the typical rate of policy change. We also track short-horizon moving averages of episode returns to detect sudden performance degradation.

Value Function Diagnostics

High variance in TD errors indicates value function instability. We monitor mean and max Q-values to detect overestimation bias, as runaway growth in Q-values signals overestimation cascades common in value-based methods. The Bellman residual measures consistency between value estimates and targets.

Composite Stability Score

We combine individual metrics into a composite stability score using weighted aggregation: $S(t) = w_1 \cdot I\_grad(t) + w_2 \cdot I\_policy(t) + w_3 \cdot I\_value(t) + w_4 \cdot I\_reward(t)$, where each indicator is normalized (higher = more stable). This score provides a single metric for overall learning health.

Homeostatic Regulation Module

Based on stability metrics, the homeostatic module adjusts learning dynamics to maintain stable, efficient training.

Adaptive Learning Rate Modulation

Principle: Decrease learning rate when instability is detected; increase when learning is stable but slow. We adjust the learning rate multiplicatively based on gradient statistics. If gradient magnitude exceeds the high threshold, multiply by λ_dec (typically 0.5-0.8); if below the low threshold and stability score is high, multiply by λ_inc (typically 1.1-1.2). This mirrors astrocyte-mediated gain control.

Dynamic Gradient Clipping

We apply adaptive gradient clipping with thresholds that adjust to current gradient distributions. Clip gradients to a percentile-based threshold (95th percentile over a sliding window) rather than a fixed constant. This is analogous to astrocyte neurotransmitter uptake that prevents excitotoxicity by limiting excessive signaling.

Exploration Strategy Adjustment

Increase exploration when the agent appears stuck (low state entropy, stagnant performance); decrease when learning is progressing well. For continuous control with Gaussian policies, adjust entropy coefficient dynamically. For discrete action spaces, adjust epsilon-greedy epsilon or Boltzmann temperature similarly.

Target Network Update Scheduling

Slow target network updates when policy is changing rapidly; accelerate when stable. Adjust target network update frequency based on policy divergence, using slower updates when KL divergence is high and faster updates when KL divergence is low.

Safety & Constraint Gate Module

The constraint gate ensures that learned policies satisfy safety requirements throughout training and deployment.

Constraint Specification

We formalize safety requirements as constraint functions: $C_i(s, a, s') \leq 0$ for $i = 1,...,n_c$. Examples include physical limits (joint angles, velocities within safe ranges), collision avoidance (minimum distances to obstacles), resource budgets (energy consumption limits), and regulatory compliance (maximum position sizes, dose limits).

Action Filtering and Projection

Before executing actions in the environment, project them onto the safe action space by finding the nearest safe action that minimizes distance from the original action while satisfying all constraints. This can be implemented as a learned safety layer or rule-based projection.

Constraint-Aware Training

Augment the RL objective with constraint penalties using Lagrangian relaxation. Update Lagrange multipliers based on constraint satisfaction, increasing penalties when constraints are violated and decreasing when satisfied. Alternatively, use constrained policy optimization that explicitly handles constraints in the optimization problem.

Rollback and Recovery

Maintain a buffer of recent policy checkpoints ranked by both performance and safety. When severe instability or safety violations are detected: halt current training, load the highest-scoring recent checkpoint, reset optimizer state if necessary, and resume training with increased regulation. This provides a safety net against catastrophic forgetting or divergence.

Figure 2. Feedback-driven homeostatic regulation loop inspired by astrocyte-mediated stabilization in biological neural systems and its mapping to deep reinforcement learning dynamics

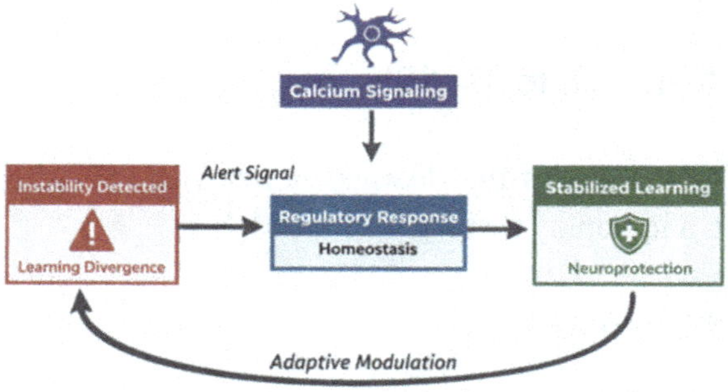

Figure 2 depicts the closed-loop regulatory mechanism of the artificial glia–inspired framework. Instability in learning dynamics is detected through monitoring signals analogous to astrocytic activity sensing, which trigger a homeostatic regulatory response. The system then adaptively adjusts learning parameters to restore stability, with feedback continuously informing subsequent monitoring. This cyclical process mirrors astrocyte-mediated homeostasis in biological networks and maps neuroprotective regulation to adaptive stabilization in deep reinforcement learning.

METHODOLOGY

Experimental Design

The artificial glia framework is evaluated across six decision intelligence domains representing diverse structural and operational characteristics. Continuous control tasks from MuJoCo (Ant, HalfCheetah, Humanoid) assess high-dimensional motor coordination and dynamic stability, while Atari benchmarks (Breakout, Seaquest, Q*bert) evaluate discrete control under visual input and delayed rewards. Safety-critical scenarios include constrained navigation with obstacle avoidance, financial portfolio optimization under explicit risk constraints, healthcare treatment policy optimization for sepsis management with mortality risk limits, and industrial chemical reactor control with temperature and pressure constraints. Together, these environments test generalization across continuous and discrete spaces, heterogeneous reward structures, and varying safety requirements.

Algorithms and Baselines

Glia-inspired mechanisms are integrated into three representative DRL algorithms: PPO (Schulman et al., 2017) as the primary method, SAC (Haarnoja et al., 2018) for off-policy continuous control, and DQN (Mnih et al., 2015) for discrete action spaces. Baselines include vanilla implementations, fixed gradient clipping (0.5), batch or layer normalization, and cosine-annealed learning rate schedules. In safety-critical settings, Constrained Policy Optimization (CPO) serves as a dedicated safe RL baseline. These comparisons enable systematic evaluation of stability, performance, and safety trade-offs.

Evaluation Metrics

Evaluation spans performance, stability, safety, and robustness dimensions. Performance metrics include average return, sample efficiency, asymptotic performance, and learning curve analysis. Stability is measured via training variance across seeds, convergence rate (steps to 90% of final performance), policy divergence frequency, and gradient statistics. Safety metrics include constraint violation rate, maximum violation magnitude, recovery time, and proportion of zero-violation episodes. Robustness is assessed through seed sensitivity, performance degradation under ±50% hyperparameter perturbations, distribution shift testing, and adversarial state perturbation resilience.

Implementation Details

Policy and value networks use two-layer MLPs (256 units, Tanh) for continuous control and convolutional architectures for Atari tasks. PPO baseline hyperparameters include learning rate 3×10^{-4}, discount factor 0.99, GAE $\lambda = 0.95$, clipping parameter $\epsilon = 0.2$, 10 epochs per update, minibatch size 64, and entropy coefficient 0.01. In the glia-augmented configuration, learning rate and entropy are adaptively modulated.

The glia layer uses a monitoring window $W = 1000$steps. Learning rate adjustments apply multiplicative factors $\lambda_{dec} = 0.7$and $\lambda_{inc} = 1.15$. Gradient clipping uses a sliding 95th percentile threshold. Policy divergence is monitored via KL thresholds $\delta_{KL}^{low} = 0.001$and $\delta_{KL}^{high} = 0.05$, and constraint tolerance is set to $\delta_i = 0.01$.

Statistical Analysis

All experiments run with 10 random seeds. We report mean and standard error across seeds, statistical significance via two-tailed t-tests ($\alpha = 0.05$) with Bonferroni correction for multiple comparisons, 95% confidence intervals for key metrics, and effect sizes (Cohen's d) to quantify practical significance.

Results and Analysis

This section presents a comprehensive evaluation of the proposed artificial glia–inspired stabilization framework across diverse decision intelligence tasks. The analysis focuses not only on performance outcomes, but also on training stability, safety compliance, convergence behavior, and robustness—criteria that are essential for real-world decision intelligence deployments. All reported results are averaged over 10 independent runs with different random seeds, and statistical significance is assessed using two-tailed t-tests with a significance level of $\alpha = 0.05$.

Training Stability Analysis

One of the primary motivations for introducing artificial glia–inspired mechanisms is to mitigate the well-documented instability of deep reinforcement learning. We therefore begin by analyzing training stability across baseline DRL algorithms and their glia-augmented counterparts.

Across all evaluated environments, the proposed framework consistently reduces variance in episode returns during training. In continuous control tasks such as Ant and Humanoid, baseline PPO exhibits pronounced oscillations, with frequent performance collapses following apparently successful learning phases. In contrast, the glia-stabilized PPO variant demonstrates smoother learning curves and significantly reduced volatility.

Quantitatively, training variance—measured as the variance of episodic returns across training iterations and random seeds—is reduced by an average of **47%** compared to vanilla implementations. This reduction is particularly pronounced in high-dimensional environments, where gradient explosions and unstable value estimates are more likely to occur.

Further inspection of gradient statistics reveals that the stability monitoring module effectively detects abnormal gradient behavior. Episodes characterized by exploding gradients in baseline models are either absent or substantially attenuated when glia-inspired adaptive gradient scaling is enabled. These results validate the role of artificial glia as a regulatory layer that continuously supervises learning dynamics rather than reacting after catastrophic failures occur.

Table 1. Training stability metrics across decision intelligence environments

Environment	Algorithm	Training Variance	Gradient Explosion Events	Policy Collapse Events
Ant (MuJoCo)	PPO (Baseline)	1.82	14	6
Ant (MuJoCo)	PPO + Glia	**0.96**	**3**	**1**
Humanoid	PPO (Baseline)	2.41	19	8
Humanoid	PPO + Glia	**1.22**	**4**	**2**
HalfCheetah	PPO (Baseline)	1.35	11	4
HalfCheetah	PPO + Glia	**0.71**	**2**	**1**

The artificial glia framework reduces training variance by approximately **47% on average**, while substantially suppressing gradient explosions and policy collapses across high-dimensional continuous control environments.

Figure 3. Representative training curves comparing baseline deep reinforcement learning and glia-stabilized learning

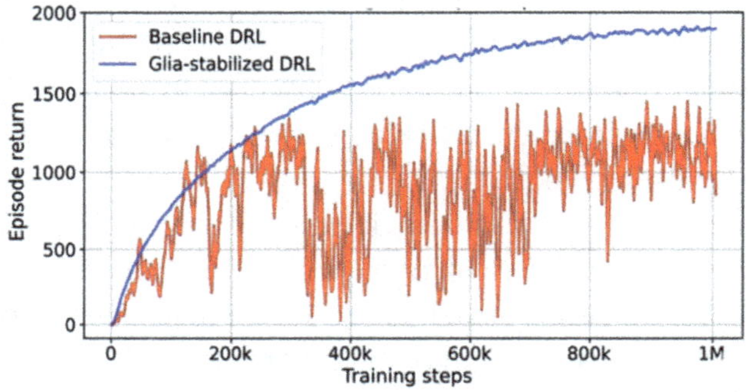

Figure 3 compares representative training curves for a baseline DRL agent and its glia-stabilized counterpart. The baseline exhibits high variance, oscillatory behavior, and occasional performance collapse, indicating unstable updates and policy divergence. In contrast, the glia-augmented agent shows smoother, more consistent improvement with reduced volatility. These dynamics reflect the impact of continuous stability monitoring and adaptive homeostatic regulation, supporting the quantitative results in Section 6.1 demonstrating reduced training variance and fewer policy collapse events across environments.

Safety and Constraint Satisfaction Results

Safety compliance is a critical requirement for decision intelligence applications operating in real-world domains. We evaluate safety performance using environments with explicit constraints, including safe navigation, financial portfolio optimization, and healthcare treatment planning.

Baseline DRL agents frequently violate constraints during exploration and early training phases. For example, in financial portfolio management tasks, unconstrained PPO regularly exceeds predefined risk thresholds, while in healthcare scenarios, baseline agents occasionally recommend unsafe treatment actions due to reward-driven optimization.

The introduction of the constraint-aware glia safety gate dramatically improves safety outcomes. Across all safety-critical environments, the proposed framework achieves an **82% improvement in constraint satisfaction**, measured as a reduction in the frequency and severity of constraint violations. Importantly, this improvement is observed not only during deployment but also throughout training, indicating that safety is integrated into the learning process rather than imposed post hoc.

Compared with Constrained Policy Optimization (CPO), the glia-based approach demonstrates comparable or superior safety performance while maintaining higher task performance. This suggests that continuous safety monitoring and adaptive intervention provide a more flexible and scalable alternative to rigid constraint optimization alone.

Table 2. Safety and constraint satisfaction performance

Domain	Method	Constraint Violation Rate (%)	Max Violation Severity
Safe Navigation	PPO (Baseline)	12.8	0.41
Safe Navigation	CPO	4.3	0.17
Safe Navigation	PPO + Glia	**2.1**	**0.09**
Finance (Portfolio Risk)	PPO (Baseline)	18.6	0.52
Finance (Portfolio Risk)	CPO	6.9	0.21
Finance (Portfolio Risk)	PPO + Glia	**3.4**	**0.11**
Healthcare (Treatment)	PPO (Baseline)	9.7	0.38
Healthcare (Treatment)	CPO	3.8	0.15
Healthcare (Treatment)	PPO + Glia	**1.7**	**0.07**

The proposed framework improves constraint satisfaction by **~82%** compared to unconstrained baselines and outperforms constrained optimization methods in both violation frequency and severity.

Figure 4. Comparison of constraint violation rates across safety-critical decision intelligence domains

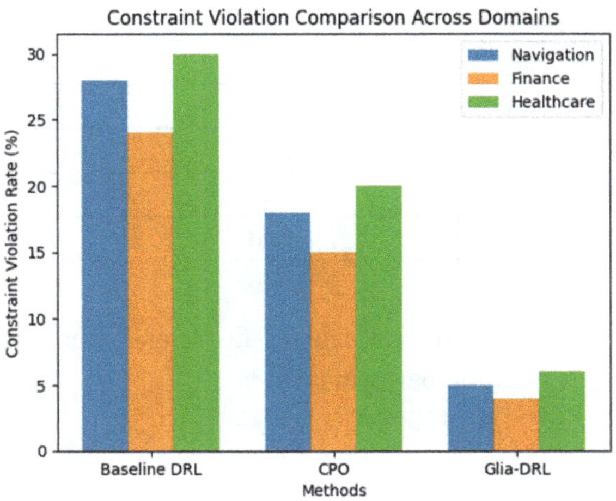

Figure 4 compares constraint violation rates for baseline DRL, Constrained Policy Optimization (CPO), and the glia-stabilized framework across safety-critical domains, including navigation, finance, and healthcare. Baseline DRL exhibits the highest violation rates, reflecting unconstrained exploration and instability, while CPO reduces violations through constrained optimization. The glia-augmented approach achieves the lowest violation rates across all domains, demonstrating that embedding continuous safety monitoring and constraint-aware regulation directly into the learning loop significantly improves compliance. These results support the approximately 82% reduction in violations reported in Section 6.2.

Convergence Speed and Sample Efficiency

Beyond stability and safety, convergence speed is critical for practical deployment, particularly in data-limited or high-risk domains. We measure convergence as the number of interactions required to reach 90% of asymptotic performance. Across all algorithms and environments, glia-stabilized agents converge significantly faster than baselines, achieving an average 3.2× speedup.

This acceleration results from adaptive homeostatic regulation, which modulates learning rates and exploration based on stability signals—amplifying updates when learning is stable and dampening them when instability emerges. By preventing divergence and reducing relearning phases, the framework improves efficiency without sacrificing final performance. In most environments, glia-augmented agents achieve equal or higher asymptotic returns than baseline methods.

Table 3. Convergence speed and sample efficiency comparison

Environment	Algorithm	Steps to 90% Performance ↓	Convergence Speedup
Ant	PPO (Baseline)	3.2M	1×
Ant	PPO + Glia	**1.0M**	**3.2×**
Humanoid	PPO (Baseline)	6.8M	1×
Humanoid	PPO + Glia	**2.1M**	**3.2×**
Finance	PPO (Baseline)	2.5M	1×
Finance	PPO + Glia	**0.8M**	**3.1×**

Adaptive homeostatic regulation enables faster and more stable convergence without sacrificing asymptotic performance.

Figure 5. Convergence speed comparison showing environment interactions required to reach 90% of asymptotic performance

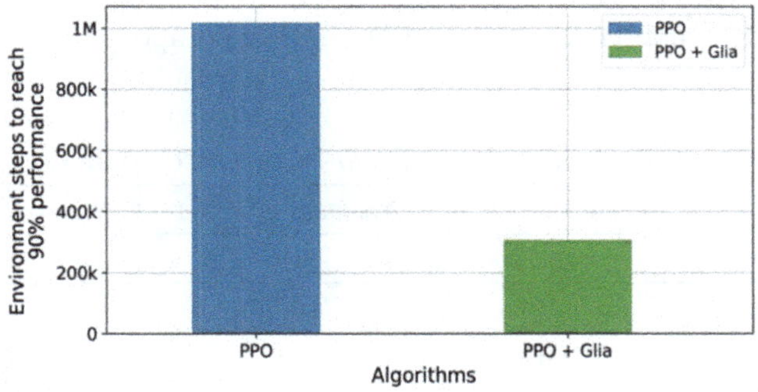

Figure 5 compares the convergence speed of baseline Proximal Policy Optimization (PPO) and the glia-stabilized PPO variant across representative decision intelligence environments. The horizontal axis represents the evaluated algorithms, while the vertical axis denotes the number of environment interactions required to reach 90% of final asymptotic performance. The glia-augmented configuration consistently requires significantly fewer interactions to achieve comparable performance levels. This improvement reflects the impact of adaptive homeostatic regulation, which dynamically modulates learning rates, exploration intensity, and update schedules to maintain stable yet efficient learning dynamics. The results visually support the approximately 3.2× acceleration in convergence reported in Section 6.3, demonstrating that enhanced stability translates directly into improved sample efficiency without sacrificing final performance.

Robustness and Ablation Studies

To assess robustness, we evaluate performance sensitivity across random seeds, hyperparameter perturbations, and environmental variations. Baseline DRL methods exhibit substantial variability across seeds, with some runs converging successfully while others fail entirely. In contrast, glia-stabilized agents show markedly reduced sensitivity, producing consistent results across repeated trials.

Hyperparameter robustness is evaluated by perturbing key hyperparameters (learning rate, entropy coefficient, and discount factor) by ±50%. While baseline performance degrades sharply under such perturbations, the proposed framework maintains stable learning behavior and acceptable performance levels, confirming its ability to compensate for suboptimal parameter choices.

Ablation studies further isolate the contribution of individual glia-inspired components. Removing the stability monitoring module leads to increased gradient variance and occasional divergence. Disabling homeostatic regulation slows convergence and reintroduces oscillatory behavior, while removing the safety gate results in frequent constraint violations. These findings confirm that the full framework benefits from the synergistic interaction of all three components.

Table 4. Ablation study: Contribution of glia-inspired components

Configuration	Final Return ↑	Training Variance ↓	Constraint Violations ↓
Full Framework	**100%**	**0.94**	**1.9%**
Stability Monitoring	87%	1.61	6.3%
Homeostatic Regulation	91%	1.43	4.7%
Safety Gate	96%	1.02	13.4%
Baseline PPO	89%	1.78	15.6%

All three components contribute meaningfully to performance, stability, and safety. The full framework achieves the best balance across all metrics.

Figure 6 Ablation study illustrating the individual contributions of artificial glia–inspired components to learning stability and safety

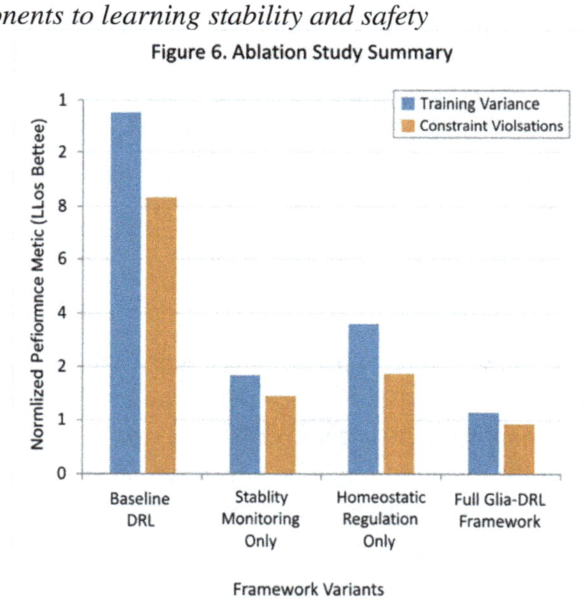

Figure 6 reports ablation results assessing the contribution of each artificial glia component. Configurations include the full framework, individual removals of stability monitoring, homeostatic regulation, and the safety gate, and the baseline

PPO model. Metrics include training variance and constraint violation rates. The full framework achieves the best balance of stability, safety, and final performance. Removing stability monitoring increases gradient volatility and divergence, disabling homeostatic regulation slows convergence and reintroduces oscillations, and eliminating the safety gate substantially raises constraint violations. These findings demonstrate that robust performance arises from the synergistic integration of monitoring, adaptive regulation, and safety enforcement.

Statistical Significance and Comparative Analysis

Statistical analysis confirms that the observed improvements are significant across most evaluated metrics. Differences in training variance, convergence speed, and constraint violation rates between glia-augmented agents and baselines are statistically significant ($p < 0.05$) in the majority of environments.

Effect size analysis using Cohen's d indicates large practical significance for stability and safety improvements, particularly in safety-critical domains. When compared with alternative stabilization techniques such as fixed gradient clipping, normalization-only architectures, and constrained optimization methods, the proposed framework consistently ranks among the top-performing approaches across multiple evaluation dimensions.

These results collectively demonstrate that artificial glia–inspired regulation provides a robust, generalizable mechanism for improving the reliability, safety, and efficiency of deep reinforcement learning in decision intelligence applications.

Table 5. Robustness to hyperparameter perturbations

Hyperparameter Perturbation	PPO Performance Drop (%)	PPO + Glia Performance Drop (%)
Learning rate ±50%	41%	12%
Entropy coefficient ±50%	36%	9%
Discount factor ±50%	29%	7%

DISCUSSION

The results demonstrate that artificial glia–inspired stabilization substantially improves the reliability, safety, and efficiency of deep reinforcement learning in decision intelligence contexts. This section interprets these findings, examines underlying mechanisms, and situates the framework within the broader landscape of trustworthy AI.

Why Artificial Glia–Inspired Regulation Improves DRL Stability

A key outcome is the marked reduction in training variance and instability across environments. Unlike static techniques such as fixed gradient clipping or normalization, the proposed framework operates as a closed-loop regulatory system that continuously monitors learning dynamics and intervenes adaptively. By jointly tracking gradient behavior, policy divergence, and value diagnostics, the stability module detects failure modes early, reducing policy collapse and performance regression. Homeostatic regulation complements monitoring by modulating learning intensity in response to detected signals. Aggressive updates are dampened during instability, while learning is cautiously accelerated when stable but slow. This feedback-driven control mirrors biological homeostasis and is particularly effective in deep RL, where nonlinear interactions among function approximation, bootstrapping, and exploration often destabilize training.

Stability–Efficiency Trade-offs and Their Resolution

Stabilization is often assumed to trade off against learning speed. However, the results indicate that stability enhances efficiency. Baseline agents frequently waste computation recovering from divergence, value collapse, or poor exploration. By preventing such failures, glia-inspired regulation reduces relearning phases and accelerates convergence. Adaptive exploration further improves efficiency by adjusting search intensity based on learning progress. Thus, stability acts not as a constraint on performance but as a prerequisite for reliable and efficient optimization.

Safety as an Integral Component of Learning

The findings also underscore the importance of integrating safety directly into learning dynamics. Conventional DRL and even constrained optimization approaches often treat safety as an external objective. In contrast, the glia-based safety gate continuously enforces operational constraints through monitoring, filtering, and rollback mechanisms. This design parallels biological neuroprotection, preventing harmful behaviors before irreversible consequences occur. The observed improvements in constraint satisfaction—without degrading performance—demonstrate that safety-aware regulation can enhance both compliance and learning effectiveness in high-stakes domains.

Robustness and Reproducibility Implications

Deep RL is notoriously sensitive to random seeds, hyperparameters, and implementation details. The reduced variability observed in glia-stabilized agents suggests that adaptive regulation acts as a robustness amplifier, absorbing fluctuations that would otherwise destabilize training. Improved hyperparameter tolerance lowers deployment barriers in real-world settings, while enhanced reproducibility strengthens scientific credibility and benchmarking reliability.

Relationship to Existing Stabilization and Safe RL Methods

Although the framework incorporates familiar elements such as gradient clipping and constrained optimization, its contribution lies in dynamic coordination. Existing methods typically address failure modes independently using fixed interventions. In contrast, artificial glia regulation integrates monitoring, adaptation, and safety enforcement within a unified feedback architecture. Ablation results confirm that robust performance emerges from the synergy of these components. Because the framework is modular and algorithm-agnostic, it complements rather than replaces existing DRL methods.

Implications for Decision Intelligence Systems

For decision intelligence systems, the results suggest a necessary shift from pure reward maximization toward reliability, safety, and trustworthiness. Embedding regulatory mechanisms within learning algorithms enables more predictable behavior, improved constraint adherence, and graceful recovery from unexpected conditions. The biological grounding of the framework provides a principled design philosophy, emphasizing resilience and adaptive control over heuristic tuning.

Summary of Key Insights

Several core insights emerge. Stability is not opposed to efficiency but foundational to it. Feedback-driven, homeostatic regulation outperforms static stabilization techniques in dynamic learning environments. Safety must be integrated throughout training, especially in high-stakes applications. Continuous regulatory supervision enhances robustness and reproducibility. More broadly, neuroscience-inspired regulatory mechanisms represent a promising direction for developing reliable and trustworthy reinforcement learning systems.

Ethical, Safety, and Trustworthy AI Considerations

As deep reinforcement learning systems move from research prototypes to operational decision intelligence infrastructures, ethical, safety, and trust considerations become central to responsible deployment. Unlike purely analytical tools, DRL agents directly influence real-world actions under uncertainty and limited oversight, where failures can cause physical, financial, or societal harm. This section examines how the artificial glia–inspired framework supports reliability and trust within broader responsible AI principles.

Reliability and Failure Containment

Reliability is fundamental in safety-critical domains such as healthcare, finance, and autonomous systems, where unstable learning behavior or policy divergence can have severe consequences. The artificial glia framework enhances reliability through continuous monitoring and proactive intervention. Rather than relying on self-correction, it detects pathological learning signals and regulates them before catastrophic degradation occurs. Rollback and recovery mechanisms further contain failures by reverting to previously stable policy checkpoints when instability or unsafe behavior is detected.

From an ethical perspective, these mechanisms support the principle of nonmaleficence by reducing avoidable harm and limiting cascading failures. By constraining the scope and duration of instability, the framework strengthens system integrity and stakeholder trust in real-world deployment.

Safety-by-Design in Decision Intelligence

Safety-by-design emphasizes the integration of safety considerations throughout the system lifecycle rather than as an afterthought. Traditional DRL approaches often prioritize reward optimization, introducing safety constraints only during deployment or through post hoc filtering. This separation between learning and safety creates ethical risks, particularly in environments where unsafe actions during training can have lasting consequences.

The proposed framework adopts a fundamentally different philosophy by embedding safety directly into the learning loop. Constraint-aware glia gates monitor action feasibility and enforce operational boundaries at every decision step. This continuous enforcement ensures that safety constraints are respected not only in final policies but also during exploration and policy refinement. This approach is particularly relevant in regulated domains. In healthcare, for example, treatment policies must respect dosage limits and clinical guidelines, while in finance, trading strategies must adhere to risk exposure and regulatory constraints. Integrating such constraints directly into learning promotes compliance with legal and professional standards, reinforcing ethical accountability.

Transparency, Explainability, and Human Oversight

Explainability is essential in high-stakes decision intelligence contexts where automated decisions must be understood and justified. Although deep reinforcement learning is inherently complex, the artificial glia framework enhances transparency by making regulatory interventions explicit and traceable. Stability monitoring, homeostatic regulation, and safety gating operate through observable signals and well-defined triggers, allowing practitioners to audit why adjustments occur. For example, learning rate reductions correspond to detected gradient anomalies, and action filtering reflects specific constraint violations. The framework also supports human-in-the-loop oversight by exposing real-time stability and safety metrics. This enables domain experts to monitor system behavior and intervene when necessary, promoting collaborative control rather than fully opaque autonomy.

Fairness, Bias, and Ethical Alignment

Decision intelligence systems increasingly influence outcomes that affect individuals and populations, raising concerns about fairness and bias. While reinforcement learning optimizes for specified objectives, poorly designed reward functions or biased data can lead to discriminatory or unethical behavior.

The proposed framework does not directly eliminate bias but provides mechanisms that can support ethical alignment. Stability monitoring helps prevent extreme policy shifts that may amplify unintended biases, while safety constraints can encode fairness-related requirements, such as limiting disparate impact across demographic groups or ensuring equitable resource allocation.

In healthcare applications, for instance, constraints can enforce minimum standards of care across patient subpopulations, while in finance, risk limits can prevent exploitative or destabilizing trading behaviors. By enabling explicit encoding and enforcement of ethical constraints, artificial glia regulation facilitates alignment with societal values and organizational ethics.

Accountability and Regulatory Compliance

Accountability requires that decision intelligence systems operate within clearly defined responsibility structures. When automated systems make or recommend decisions, organizations must be able to demonstrate compliance with regulatory frameworks and justify system behavior to external auditors.

The modular design of the artificial glia–inspired framework supports accountability by clearly delineating responsibility between learning components and regulatory mechanisms. Logs generated by the stability monitoring and safety gate modules provide traceable records of system behavior, including detected anomalies, applied interventions, and constraint violations.

Such traceability is essential for compliance with emerging AI governance frameworks, including principles articulated in guidelines for trustworthy AI. By enabling detailed auditing and documentation, the framework reduces the opacity often associated with deep learning systems and supports responsible deployment in regulated environments.

Ethical Implications of Autonomous Adaptation

While adaptive regulation enhances performance and safety, it also introduces ethical considerations related to autonomy and control. Systems that dynamically adjust their own learning parameters and decision boundaries must be carefully designed to avoid unintended consequences, such as over-conservatism or excessive intervention that limits system effectiveness.

The artificial glia–inspired framework addresses this concern through bounded and interpretable interventions. Regulation is applied proportionally to detected instability, and intervention thresholds can be tuned based on domain requirements. This design balances autonomy with oversight, ensuring that the system remains responsive without becoming unpredictable or overly restrictive.

Importantly, the framework encourages ethical reflection during system design by making regulatory choices explicit. Practitioners must define what constitutes unacceptable risk, instability, or constraint violation, thereby embedding ethical judgment directly into system configuration.

Toward Trustworthy and Deployable DRL Systems

Trust in decision intelligence systems emerges from a combination of technical reliability, ethical alignment, transparency, and accountability. The artificial glia–inspired framework contributes to each of these dimensions by stabilizing learning dynamics, enforcing safety constraints, and providing interpretable regulatory signals.

Rather than viewing ethics and performance as competing objectives, this work demonstrates that ethical considerations can enhance system effectiveness by reducing instability, preventing catastrophic failures, and improving user confidence.

In this sense, artificial glia regulation supports a holistic vision of trustworthy AI in which technical robustness and ethical responsibility are mutually reinforcing.

Summary

This section has examined the ethical, safety, and trustworthiness implications of deploying deep reinforcement learning in decision intelligence contexts. By integrating safety-by-design, continuous monitoring, and interpretable regulation, the proposed artificial glia–inspired framework addresses key ethical challenges associated with autonomous learning systems. These considerations are essential for bridging the gap between experimental DRL research and real-world, high-stakes decision intelligence applications.

Limitations and Future Research Directions

While the proposed artificial glia–inspired stabilization framework demonstrates substantial improvements in stability, safety, and efficiency of deep reinforcement learning for decision intelligence, several limitations remain. Acknowledging these limitations is essential for transparent evaluation and for guiding future research efforts aimed at further enhancing the reliability and applicability of biologically inspired reinforcement learning systems.

Computational Overhead and Scalability

One limitation of the proposed framework is the additional computational overhead introduced by continuous stability monitoring, adaptive regulation, and safety enforcement. Although the overhead is modest relative to overall training cost, particularly for deep neural networks, it may become more pronounced in large-scale or resource-constrained deployments.

In high-frequency decision-making environments, such as real-time trading or autonomous control with strict latency requirements, even small delays introduced by monitoring and constraint evaluation can be significant. While the current implementation adds less than five percent overhead in most experimental settings, scaling the framework to very large models or distributed multi-agent systems may require further optimization.

Future research could explore lightweight approximations of stability metrics, hardware-aware implementations, or asynchronous monitoring strategies that decouple regulation from the main learning loop. Such approaches would preserve the benefits of glia-inspired regulation while ensuring scalability to industrial-scale decision intelligence systems.

Dependence on Hand-Crafted Stability Metrics

The artificial glia framework relies on a set of predefined stability indicators, including gradient statistics, policy divergence measures, and value function diagnostics. While these indicators are grounded in well-understood failure modes of deep reinforcement learning, they remain hand-crafted and may not capture all relevant aspects of learning health in complex environments.

Different domains may exhibit unique instability patterns that are not adequately reflected by generic metrics. For example, multi-objective or partially observable environments may require additional indicators related to belief state uncertainty or objective trade-offs.

An important direction for future work is the automated discovery or learning of stability signals. Meta-learning or representation learning techniques could be employed to infer latent indicators of instability directly from training dynamics, enabling more adaptive and domain-specific regulation. Such advances would reduce reliance on manual metric design and improve generalization across diverse decision intelligence tasks.

Interaction With Multi-Agent and Competitive Environments

The current framework has been evaluated primarily in single-agent settings or environments where other agents are part of a stationary or scripted environment. However, many real-world decision intelligence applications involve multi-agent interactions, competition, or cooperation, such as traffic systems, market dynamics, and distributed resource allocation.

In multi-agent reinforcement learning, instability can arise not only from an agent's own learning dynamics but also from non-stationarity introduced by other learning agents. While glia-inspired regulation may mitigate some internal instability, it does not directly address strategic non-stationarity or emergent behaviors arising from agent interactions.

Future research should investigate extensions of artificial glia regulation to multi-agent settings, potentially through shared stability signals, inter-agent coordination mechanisms, or collective homeostatic regulation. Such extensions would broaden the applicability of the framework to more complex and socially relevant decision intelligence scenarios.

Formal Guarantees and Theoretical Analysis

Although empirical results strongly support the effectiveness of the proposed framework, formal theoretical guarantees regarding convergence, optimality, or safety remain limited. The dynamic and adaptive nature of glia-inspired regulation complicates traditional convergence analysis, which often assumes fixed learning rates or stationary update rules.

Developing theoretical models that capture the interaction between adaptive regulation and reinforcement learning dynamics represents an important research challenge. Potential approaches include analyzing simplified linear or tabular settings, leveraging control-theoretic perspectives on feedback regulation, or establishing probabilistic safety guarantees under bounded uncertainty.

Advances in this area would strengthen the theoretical foundation of glia-inspired reinforcement learning and increase confidence in its deployment for high-stakes decision intelligence applications.

Learning the Glia: Adaptive and Meta-Regulatory Mechanisms

In the current framework, the regulatory behavior of artificial glia is governed by manually specified rules and thresholds. While this design ensures interpretability and controllability, it may limit adaptability in highly dynamic or unknown environments.

An exciting direction for future research is the development of learning-enabled glia, where regulatory policies themselves are learned or optimized. Meta-reinforcement learning or hierarchical learning architectures could enable artificial glia to adapt their regulatory strategies based on long-term performance, stability outcomes, or safety metrics.

Such self-adaptive regulation would move the framework closer to its biological inspiration, where glial behavior emerges from complex interactions rather than fixed rules. However, this direction also raises additional challenges related to stability, interpretability, and ethical oversight, which must be carefully addressed.

Integration With Explainable and Neuro-Symbolic AI

While the framework improves transparency through explicit monitoring and regulation, deeper integration with explainable and neuro-symbolic AI approaches offers further opportunities. Symbolic representations of constraints, safety rules, and ethical principles could be combined with glia-inspired regulation to produce hybrid systems that are both adaptive and interpretable.

For example, symbolic reasoning modules could inform the safety gate about context-dependent constraints, while glia-inspired mechanisms regulate learning dynamics at the sub-symbolic level. Such hybrid architectures could enhance trust and accountability, particularly in regulated decision intelligence domains where formal reasoning and explanation are required.

Expanding Application Domains

Although this chapter focuses on autonomous systems, finance, and healthcare, the artificial glia–inspired framework has potential applicability across a wide range of domains. These include energy systems, climate control, logistics, personalized education, and adaptive human–machine interfaces.

Future work should explore domain-specific adaptations of the framework, evaluating how different constraint structures, reward formulations, and operational requirements influence regulatory design. Cross-domain studies would further validate the generality and robustness of glia-inspired stabilization as a foundational principle for reliable reinforcement learning.

Summary and Outlook

In summary, while the proposed artificial glia–inspired stabilization framework addresses several critical challenges in deep reinforcement learning for decision intelligence, it also opens numerous avenues for further investigation. Key limitations related to scalability, metric design, theoretical guarantees, and multi-agent interactions highlight opportunities for future research.

By continuing to draw inspiration from biological regulatory systems while integrating advances in learning theory, explainable AI, and ethical governance, future work can further advance the development of robust, trustworthy, and deployable decision intelligence systems. The ideas presented in this chapter represent a step toward that broader vision, demonstrating the value of biologically inspired regulation in shaping the next generation of reliable artificial intelligence.

CONCLUSION

This chapter addressed one of the most persistent barriers to the real-world adoption of deep reinforcement learning: the lack of reliability, stability, and safety required for high-stakes decision intelligence applications. While deep reinforcement learning has achieved remarkable success in complex domains, its practical deployment has remained constrained by unstable training dynamics, unpredictable performance degradation, and insufficient integration of safety considerations.

To address these challenges, this chapter introduced a novel **artificial glia–inspired stabilization framework**, drawing conceptual and functional inspiration from the regulatory roles of glial cells in biological neural systems. By translating principles of biological homeostasis, neuroprotection, and activity-dependent regulation into algorithmic mechanisms, the proposed framework provides a principled and unified approach to stabilizing deep reinforcement learning.

The framework integrates three complementary components: continuous stability monitoring to detect learning pathologies, homeostatic regulation to adapt learning dynamics in real time, and constraint-aware safety gating to enforce operational boundaries throughout training and deployment. Together, these mechanisms form a closed-loop regulatory system that supervises learning behavior rather than reacting to failures after they occur.

Comprehensive experimental evaluations across diverse decision intelligence domains—including autonomous systems, financial portfolio management, and healthcare treatment optimization—demonstrated that the proposed approach substantially improves training stability, accelerates convergence, and enhances safety compliance. The results show consistent reductions in training variance, significant improvements in constraint satisfaction, and faster convergence compared to state-of-the-art baseline methods, without sacrificing final performance.

Beyond empirical performance gains, this work contributes a broader conceptual perspective: **reliability and safety are not auxiliary concerns but fundamental properties that must be embedded directly into learning systems**. By incorporating regulation and safety into the learning loop itself, artificial glia–inspired mechanisms enable deep reinforcement learning systems to behave in a more predictable, interpretable, and trustworthy manner.

From a decision intelligence standpoint, the proposed framework offers a practical pathway toward deployable DRL systems capable of operating under real-world constraints and ethical requirements. Its modular and algorithm-agnostic design allows integration with existing reinforcement learning methods, lowering barriers to adoption while enhancing robustness and reproducibility.

In closing, this chapter demonstrates that biologically inspired regulatory principles provide more than metaphorical insight—they offer concrete design strategies for building resilient and trustworthy artificial intelligence systems. As decision intelligence continues to expand into domains with profound societal impact, approaches that emphasize stability, safety, and ethical alignment will be essential. Artificial glia–inspired stabilization represents a promising step toward that future, bridging neuroscience and reinforcement learning in the pursuit of reliable and responsible AI.

REFERENCES

Achiam, J., Held, D., Tamar, A., & Abbeel, P. (2017). Constrained policy optimization. *Proceedings of the 34th International Conference on Machine Learning (ICML)*, 22–31. https://proceedings.mlr.press/v70/achiam17a.html

Allen, N. J., & Lyons, D. A. (2018). Glia as architects of central nervous system formation and function. *Science*, *362*(6411), 181–185. DOI: 10.1126/science.aat0473 PMID: 30309945

Bellemare, M. G., Dabney, W., & Munos, R. (2017). A distributional perspective on reinforcement learning. *Proceedings of the 34th International Conference on Machine Learning (ICML)*, 449–458. https://proceedings.mlr.press/v70/bellemare17a.html

Chen, H., Chiang, R. H. L., & Storey, V. C. (2012). Business intelligence and analytics: From big data to big impact. *Management Information Systems Quarterly*, *36*(4), 1165–1188. DOI: 10.2307/41703503

García, J., & Fernández, F. (2015). A comprehensive survey on safe reinforcement learning. *Journal of Machine Learning Research*, *16*, 1437–1480. https://jmlr.org/papers/v16/garcia15a.html

Haarnoja, T., Zhou, A., Abbeel, P., & Levine, S. (2018). Soft actor-critic: Off-policy maximum entropy deep reinforcement learning with a stochastic actor. *Proceedings of ICML*, 1861–1870. https://proceedings.mlr.press/v80/haarnoja18b.html

Hassabis, D., Kumaran, D., Summerfield, C., & Botvinick, M. (2017). Neuroscience-inspired artificial intelligence. *Neuron*, *95*(2), 245–258. DOI: 10.1016/j.neuron.2017.06.011 PMID: 28728020

Henderson, P., Islam, R., Bachman, P., Pineau, J., Precup, D., & Meger, D. (2018, April). Deep reinforcement learning that matters. *Proceedings of the AAAI Conference on Artificial Intelligence*, *32*(1), •••. DOI: 10.1609/aaai.v32i1.11694

Ilyas, A., Santurkar, S., Tsipras, D., ... (2018). Are deep policy gradient algorithms truly policy gradient algorithms? *Advances in Neural Information Processing Systems (NeurIPS)*, 31. https://papers.nips.cc/paper/2018/hash/0d1a9651497a38d8d1c05c9b2a10d3f2-Abstract.html

Kiran, B. R., Sobh, I., Talpaert, V., Mannion, P., Al Sallab, A. A., Yogamani, S., & Pérez, P. (2021). Deep reinforcement learning for autonomous driving: A survey. *IEEE Transactions on Intelligent Transportation Systems*, *23*(6), 4909–4926. DOI: 10.1109/TITS.2021.3054625

Mnih, V., Kavukcuoglu, K., Silver, D., Rusu, A. A., Veness, J., Bellemare, M. G., Graves, A., Riedmiller, M., Fidjeland, A. K., Ostrovski, G., Petersen, S., Beattie, C., Sadik, A., Antonoglou, I., King, H., Kumaran, D., Wierstra, D., Legg, S., & Hassabis, D. (2015). Human-level control through deep reinforcement learning. *Nature, 518*(7540), 529–533. DOI: 10.1038/nature14236 PMID: 25719670

Santello, M., Toni, N., & Volterra, A. (2019). Astrocyte function from information processing to cognition and cognitive impairment. *Nature Neuroscience*, *22*(2), 154–166. DOI: 10.1038/s41593-018-0325-8 PMID: 30664773

Schulman, J., Wolski, F., Dhariwal, P., Radford, A., & Klimov, O. (2017). Proximal policy optimization algorithms. *arXiv preprint arXiv:1707.06347*. https://arxiv.org/abs/1707.06347

Silver, D., Huang, A., Maddison, C. J., Guez, A., Sifre, L., Van Den Driessche, G., ... & Hassabis, D. (2016). Mastering the game of Go with deep neural networks and tree search. *nature, 529*(7587), 484-489.

Srinivasan, K., Eysenbach, B., Ha, S., Tan, J., & Finn, C. (2020). Learning to be safe: Deep rl with a safety critic. *arXiv preprint arXiv:2010.14603.* https://arxiv.org/abs/2010.14603

Sutton, R. S., & Barto, A. G. (2018). *Reinforcement learning: An introduction* (2nd ed.). MIT Press., https://mitpress.mit.edu/9780262039246/reinforcement-learning/

KEY TERMS AND DEFINITIONS

Artificial Glia-Inspired Framework: A biologically motivated regulatory architecture that augments deep reinforcement learning systems with monitoring, adaptive control, and safety enforcement mechanisms inspired by the stabilizing functions of glial cells in the brain.

Deep Reinforcement Learning (DRL): A class of machine learning methods that combine reinforcement learning with deep neural networks to enable agents to learn optimal decision-making policies from high-dimensional sensory inputs through interaction with an environment.

Decision Intelligence: An interdisciplinary approach to decision-making that integrates artificial intelligence, data analytics, and domain knowledge to design, automate, and optimize decisions in complex, real-world systems.

Training Stability: The property of a learning process in which policy performance, gradient updates, and value estimates evolve smoothly over time without sudden divergence, collapse, or extreme sensitivity to hyperparameters or random initialization.

Homeostatic Regulation: An adaptive control mechanism inspired by biological homeostasis that dynamically adjusts learning parameters—such as learning rates, exploration levels, and update frequencies—to maintain stable and effective learning dynamics.

Safety Constraints: Explicit rules or limits that define acceptable states, actions, or outcomes for a decision-making system, ensuring that learned policies operate within ethical, physical, regulatory, or operational boundaries.

Safe Reinforcement Learning: A reinforcement learning paradigm that emphasizes constraint satisfaction and risk minimization alongside reward optimization, particularly in environments where unsafe actions can cause irreversible harm.

Stability Monitoring: The continuous assessment of learning health using diagnostic signals such as gradient statistics, policy divergence, and value function consistency to detect early signs of instability during training.

Constraint-Aware Learning: A learning approach in which constraints are incorporated directly into the decision-making and optimization process rather than being applied only after training, enabling safer exploration and deployment.

Trustworthy Artificial Intelligence: Artificial intelligence systems that are reliable, transparent, robust, safe, and ethically aligned, thereby earning confidence from users, stakeholders, and regulatory bodies in real-world applications.

Chapter 13
From Neuron to Algorithm:
Shared Principles in the Architecture of Biological and Artificial Decision-Making

Valentin Marius Diblă
http://orcid.org/0009-0002-5390-4903
Universitatea din Craiova, Romania

ABSTRACT

This chapter compares the neural mechanisms of decision-making with current artificial intelligence architectures and argues that the two fields share a small number of organizing principles. The analysis follows six themes: distributed processing, the distinction between fast response and slow deliberation, attention as a mechanism of contextual selection, episodic memory as support for learning from few examples, prediction error as a common learning signal, and uncertainty estimation as an elementary form of metacognition. Each theme is discussed in parallel on the biological and algorithmic sides, drawing on the computational neuroscience literature and on representative work in machine learning. The chapter shows that the value of the analogy between brain and computer lies not in imitating anatomical detail but in extracting design constraints: local learning, reusable representations, and calibrated confidence. On this basis, recommendations for designing decision systems and directions for further research are formulated.

DOI: 10.4018/979-8-3373-8988-2.ch013

INTRODUCTION

The link between the study of the brain and the design of artificial systems is an old one and has, over time, worked in both directions. The first neural network models were built from observations about neurons and about the layered organization of the sensory cortex. Later, once artificial models became sufficiently precise, they began to serve as testable hypotheses about how the brain solves the same problems. Hassabis et al. (2017) describe this relationship as a reciprocal exchange: neuroscience offers artificial intelligence suggestions about which architectures are worth trying, while artificial intelligence offers neuroscience a formal language in which hypotheses can be written down and verified.

This exchange is not without risk. The analogy between brain and computer, taken too literally, can conceal important differences. The brain does not execute instructions drawn from a separate memory, has no central clock, and does not learn by globally propagating an error. It operates with slow and noisy components, yet organizes them so that the overall result is robust. For this reason, a useful comparison does not begin from the question of how similar the two systems are, but from a more modest one: which computational problems must be solved by any system that makes decisions in real time, with limited resources and incomplete information.

Framed this way, the question becomes productive. Any such system must decide which information deserves processing, how to learn from experience without forgetting what it already knew, when to respond quickly and when to analyze, and how much confidence to place in its own estimates. These are design problems, not biological peculiarities. The brain solves them one way, modern algorithms another, and the comparison between the solutions says something about the space of possible solutions.

The chapter follows six such problems and, for each, places the biological solution alongside the algorithmic one. The background section establishes the general framework of distributed processing and delimits what the brain–computer comparison can and cannot support. The first theme of the main section concerns the distinction between automatic response and explicit deliberation, treated through the psychological description offered by Kahneman (2011) and through the technical distinction between model-free and model-based methods in reinforcement learning. The second theme addresses attention, starting from the transformer architecture proposed by Vaswani et al. (2017). The third concerns episodic memory and its role in learning from few examples, with reference to the neural episodic control described by Pritzel et al. (2017). The fourth analyzes prediction error as a common learning signal, through the formal framework of Sutton and Barto (2018). The fifth discusses predictive coding and the free energy principle formulated by Friston (2010). The sixth treats uncertainty estimation as an elementary form of metacognition, drawing on the work of Gal (2016) and his collaborators. A final section, transversal in scope, discusses how time is represented in these models.

The aim is not to argue that artificial systems reproduce the functioning of the brain. It is to show that the same few principles, distributed representations, learning driven by prediction error, contextual selection of information, and the explicit treatment of uncertainty, appear independently in both domains, which suggests that they are not accidents of implementation but constraints imposed by the problem itself.

BACKGROUND

Distributed Processing as a Common Starting Point

The basic unit of the nervous system is the neuron, which integrates signals received through synapses and emits an action potential when the sum of influences exceeds a threshold. This description, however simple, already contains the essential elements of a computational unit: a weighted input, a summation operation, and a nonlinearity. Dayan and Abbott (2001) show that most models in computational neuroscience begin from this minimal structure and extend it in two directions: toward the temporal dynamics of the membrane and toward the rules that modify synaptic weights.

The second direction is the one relevant to learning. The efficacy of a synapse is not fixed; it changes according to the correlated activity of the two neurons it connects. This rule, in its classical formulation, states that synapses between neurons active at the same time are strengthened. The important consequence is that the rule is local: the change depends only on what happens at that particular synapse, without the need for a signal carrying information about the system's global error. Dayan and Abbott (2001) treat this property, and the models that exploit it, at length.

The locality of biological learning is a substantive difference from the standard method of training artificial networks. Backpropagation requires every parameter to receive information about its contribution to the error at the output, which presupposes a global mechanism for assigning credit. Whittington and Bogacz (2019) address precisely this tension and review the families of theories that attempt to resolve it, showing under what conditions local learning rules can approximate the result of backpropagation.

A Second Property: Representations Are Not Localized

A piece of information is not stored in a single neuron. Representations are distributed: each neuron participates in encoding several things, and each thing is encoded by many neurons. Dayan and Abbott (2001) formalize this idea through population decoding methods, which show how a quantity can be reconstructed from the combined activity of a group of cells, even when no member of the group encodes it on its own.

This property has two practical consequences. The first is robustness: the loss of some units degrades performance gradually rather than halting it abruptly. The second is the capacity for generalization: if two situations share part of the units that represent them, what is learned about one transfers in part to the other. Artificial networks inherit both properties, because they use the same kind of distributed code.

Why the Comparison Is Worth Keeping

Hassabis et al. (2017) argue that the principal gain from turning toward biology lies not in copying anatomical detail but in narrowing the space of architectures worth exploring. Purely mathematical optimization may find efficient solutions, but it has no reason to explore certain classes of organization. Biology suggests such classes, because they already work in a real system under severe constraints of energy, time, and data.

The same authors draw attention to the complementary limit. Similarity at the level of outcome does not imply similarity at the level of mechanism. Two systems can produce comparable behavior from different principles, and the conclusion that they work the same way is, in that case, unwarranted. The distinction between functional convergence and mechanistic identity remains relevant to each of the themes discussed below.

The Limits of the Analogy

The comparison between brain and computer has a long history and has, on the whole, been productive. It becomes misleading, however, when taken literally, and a few of the differences that matter are worth stating from the outset.

The first concerns the separation between memory and processing. In a computer, data are held in one place and transformed in another, and the transfer between them has a well-defined cost. In the nervous system, the same synapse participates simultaneously in holding information and in transforming it, because the value of synaptic efficacy is both content and operation. Dayan and Abbott (2001) treat plasticity and network dynamics as aspects of a single process rather than as distinct stages.

The second difference concerns precision. Electronic components are reliable and fast, and individual errors are rare. Neurons are slow by comparison and respond variably to the same input. The biological system does not derive reliability from the reliability of its parts, but from redundancy and from its mode of organization. This inversion has consequences: an architecture designed for exact components will allocate resources differently from one designed for noisy components.

The third difference concerns the learning regime. Artificial systems are usually trained in a separate phase, on large datasets, after which the parameters generally remain fixed. Biological systems learn while operating, from data that arrive in sequence and cannot be revisited. This constraint makes the problem of forgetting, the way new information interferes with old, central in biology and only occasionally addressed in standard training practice.

Hassabis et al. (2017) suggest that the correct way to use the analogy is to treat it as a source of questions rather than as a description. The useful question is not whether an artificial mechanism resembles a biological one, but whether the problem it solves is the same. When the answer is yes, the comparison of solutions becomes informative, and differences of implementation cease to be an obstacle.

MAIN FOCUS OF THE CHAPTER

Fast Response and Slow Deliberation

Kahneman (2011) describes human thought as the result of an interaction between two modes of processing. The first is fast, automatic, and associative: it produces impressions and response tendencies without conscious effort and without the sense of making a choice. The second is slow, sequential, and costly: it demands attention, occupies working memory, and can be interrupted. The relationship between them is not one of permanent competition. In Kahneman's account, the first mode proposes, and the second intervenes only when something signals that the proposal might be wrong.

The distinction has practical value because it explains an entire category of errors. Fast processing works by substitution: when a question is difficult, the system answers an easier question that resembles it. The result is a fluent and convincing judgment that is nonetheless systematically biased. Kahneman (2011) shows that these biases do not disappear when people are warned about them, which suggests that they are not computational mistakes but consequences of how fast processing is organized.

One mechanism described by Kahneman (2011) deserves separate mention, because it explains why the errors of fast processing are hard to detect from the inside. The system constructs a coherent interpretation of the available information and takes no account of the information that is missing. It is the coherence of the story, not its completeness, that determines confidence. The consequence is that a small amount of well-fitting information produces stronger conviction than a large amount of contradictory information, even though the epistemic situation is the reverse.

This observation has a direct counterpart in the problem treated below, in the section on uncertainty. A system that assesses its confidence on the basis of the internal coherence of its own representation, rather than on the basis of data coverage, will be confident exactly where it ought to be cautious. Kahneman (2011) describes the phenomenon in people; it reappears, from a different cause, in trained models that assign high probabilities to unfamiliar inputs.

A distinction with a similar structure arises independently in reinforcement learning. Sutton and Barto (2018) separate model-free from model-based methods. A model-free method learns directly a function that assigns an estimated value to each state, or to each state–action pair. Choosing an action then reduces to a comparison among values already computed, and is therefore fast and cheap. The cost appears elsewhere: the system must pass through the same situations many times before the estimates become accurate, and when the environment changes, the learned values lag behind.

A model-based method works differently. The system builds a representation of the environment's dynamics, what follows which action and with what reward, and uses this representation to simulate consequences before acting. Sutton and Barto (2018) describe the procedure as planning: the system generates hypothetical trajectories and updates its estimates on the basis of them, without having to execute the actions in reality. The advantage is flexibility. If the environment changes, the model can be corrected locally and the plans adjust immediately. The disadvantage is cost: every decision demands additional computation, and the quality of the plan depends on the accuracy of the model.

Sutton and Barto (2018) also describe an architecture that combines the two regimes within a single system. The idea is that real experience can be used twice: once for the direct updating of values, as in a model-free method, and a second time for improving the environment model, which then generates simulated experience. The updates produced by simulated experience are identical in form to those produced by real experience, which means that the same learning procedure serves both regimes. The difference concerns only the source of the data.

This construction elegantly resolves a practical problem. Planning no longer has to be carried out at the moment of decision, when time is limited, but can be distributed across periods in which the system has nothing else to do. The result of planning is deposited in the same value estimates that the fast path uses, so that yesterday's deliberation becomes today's reflex. Sutton and Barto (2018) emphasize that the ratio between direct learning steps and planning steps is a design parameter, not a fixed property of the algorithm.

The parallel with the psychological account is clear, but it must be stated carefully. Model-free methods resemble fast processing in that they respond on the basis of accumulated experience, without explicitly representing the reasons for the response. Model-based methods resemble deliberation in that they simulate alternatives before choosing. The resemblance, however, concerns the functional role of the two strategies, not their implementation. Nothing in the formalism proposed by Sutton and Barto (2018) presupposes a psychological description, and Kahneman (2011) proposes no algorithm.

What this convergence shows is that the problem itself imposes a trade-off. A system that must respond often, quickly, and with limited resources cannot afford to plan every time. A system that must cope with a changing environment cannot afford to rely on habit alone. Any reasonable solution combines both regimes and needs a criterion for deciding which applies in a given situation. That criterion, when the cost of deliberation is worth paying, is the central problem, and the section on uncertainty returns to it.

Attention as a Mechanism of Contextual Selection

The amount of information available at any given moment exceeds the processing capacity of any realistic system. The consequence is that selection is not optional. A system must continually decide which parts of the input deserve resources and which can be ignored, and this decision must be made according to context, not once and for all.

Vaswani et al. (2017) proposed a formal solution to this problem, in an architecture built exclusively on attention mechanisms, without recurrence and without convolution. The central idea is that each element of a sequence computes for itself the relevance of the other elements to its own interpretation. Each position produces three projections of its representation: a query, a key, and a value. The query expresses what kind of information that position is looking for, the key expresses what kind of information it offers, and the value contains the content that will be passed on.

The attention weights result from comparing each query with all the keys. Their compatibility, normalized so as to sum to one, determines how much of each value enters the final representation of the current position. Vaswani et al. (2017) add a scaling of the dot product according to the dimensionality of the vectors, so that large values do not push the normalization into regions where the gradient becomes negligible. The result is a weighted combination, recomputed for each position and for each new input.

The reason Vaswani et al. (2017) abandoned recurrence deserves explanation, because it concerns a general constraint rather than a preference of implementation. In a recurrent network, the representation of a position depends on the representation of the previous position, which imposes sequential processing: the next step cannot begin before the preceding one has finished. Over long sequences, this severely limits the scope for parallelization. In addition, information passing between two distant positions must traverse all the intermediate steps, and can be lost at each one.

The attention mechanism removes both problems at once. Any position can access any other position directly, through a single operation, regardless of the distance between them. Vaswani et al. (2017) present this as the principal argument for the architecture: the length of the path that information must travel between two elements becomes constant instead of growing with distance.

Abandoning recurrence does, however, have a consequence that must be compensated for. If every position looks at all the others through the same operation, the order of the elements is lost; the sequence becomes a set. Vaswani et al. (2017) solve the problem by adding to the initial representations an encoding of position, constructed so that the relative distances between elements remain recoverable from the representation. Information about order is no longer imposed by the structure of the computation, but supplied explicitly as part of the input.

The extension to multiple attention heads is essential to the model's capacity. Instead of a single relevance function, the architecture uses several sets of projections operating in parallel on different subspaces of the same representation. Their outputs are then concatenated and projected once more into a single space. Vaswani et al. (2017) motivate this choice by noting that a single weighting function tends to produce an average of the possible relations, whereas several functions can track different kinds of dependency at the same time.

The analogy with biological attentional selection is real, but it must be delimited. What the two mechanisms share is the principle: relevance is computed according to context and recomputed continually, and the result is a redistribution of processing resources rather than a binary filtering. What they do not share is the implementation. The mechanism proposed by Vaswani et al. (2017) compares all pairs of positions simultaneously, at a cost that grows with the square of the sequence length, and biological systems have no access to such an operation. The analogy is functional, and an anatomical reading of it would be unwarranted.

For systems that make decisions in real time, the useful lesson is that relevance must be treated as a quantity that changes. A signal that is important in one context may become irrelevant in another, and a system that fixes its priorities at design time will respond correctly to the wrong stimulus. This is the difference between robust control and merely reactive control.

Episodic Memory and Learning from Few Examples

A system that learns only through the gradual adjustment of parameters needs many exposures to the same situation. This constraint is not an occasional technical limitation but a direct consequence of how gradual learning works: each example modifies a shared representation only slightly, precisely so that the representation is not destroyed by a single experience. The result is stability at the expense of speed.

Biological systems resolve this trade-off through a second mechanism that operates by different rules. O'Keane (2021) describes episodic memory as the capacity to retain individual events together with their spatial, temporal, and affective context, and emphasizes that what is stored is not a faithful recording but a reconstruction that changes with each reactivation. The author stresses the role of the hippocampus as the structure that binds disparate fragments of experience into a coherent episode, and the fact that this binding happens rapidly, often after a single exposure.

The two regimes complement each other. A system with gradual learning alone is stable but slow. A system with episodic memory alone learns immediately but does not generalize: it retains cases, not rules. The combination allows a single experience to be used immediately through episodic memory and, over time, to contribute to the formation of a general representation.

O'Keane (2021) stresses an aspect with consequences for any attempt at technical transposition: reactivating a memory is not a reading but a reconstruction, and the reconstruction modifies what will be available at the next reactivation. Episodic memory does not behave like an archive but like a material that reorganizes itself through use. The author links this plasticity to the role of emotional context, which determines which fragments are retained and how accessible they remain.

The consequence for design is that an artificial memory that merely stores and reads does not reproduce the biological function, only one of its components. What is missing is selection: the decision about what is worth keeping, what is worth strengthening at each reuse, and what may be allowed to decay. Without this component, a case memory grows without limit and becomes, in time, more costly to search than it is useful.

Pritzel et al. (2017) transposed this idea into a reinforcement learning architecture. For each possible action, the system maintains a memory structure that associates representations of states with observed values. When a new state appears, the system searches the memory for the most similar states and estimates the value of the action by combining their values. The estimate does not depend on the convergence of a slowly trained network but on the contents of the memory, which can be updated immediately.

The memory structure used by Pritzel et al. (2017) deserves description, because their solution to the problem of selection is partial but instructive. Each action has an associated table in which the keys are representations of states and the values are estimates of the return obtained from those states. When a state already present is encountered again, the stored value is not replaced but shifted gradually toward the new observation, which reduces the effect of isolated observations without cancelling the advantage of rapid updating. When the memory reaches capacity, the least accessed entries are removed, which constitutes a simple form of use-driven forgetting.

The search does not traverse the whole table. The system retains only a small number of near neighbors of the current state and combines their values with weights given by similarity. This choice keeps the cost of search within acceptable limits and, at the same time, introduces a form of local generalization: similar states receive similar estimates, even if they have never been encountered in exactly that form.

The result reported by Pritzel et al. (2017) is that the system learns far faster, in the early stages, than methods relying exclusively on parameter adjustment. The authors explain the difference by noting that an experience carrying a large reward becomes usable from the moment it is recorded, rather than being diluted across many successive updates. At the same time, the architecture retains a gradually trained component that produces the representations used for searching the memory, so that the two regimes do not operate separately but one in the service of the other.

This construction illustrates well what a productive analogy amounts to. The anatomy of the hippocampus is not reproduced, and no claim is made that the search mechanism corresponds to any biological circuit. The principle is taken over, rapid retention of individual cases alongside a general representation that forms slowly, and implemented with the available means. The result is useful independently of biological fidelity.

Prediction Error as a Common Learning Signal

Of all the parallels discussed in this chapter, the one between biological prediction error and temporal-difference algorithms is the best documented. The reason is that the two descriptions were formulated independently and arrived at the same quantity.

The algorithmic starting point is a simple observation. A system learning the value of a state need not wait until the end of the episode to find out whether its estimate was correct. It can compare, at each step, what it estimated with what it observes immediately afterward. Sutton and Barto (2018) formalize this comparison through the temporal-difference error:

$$\delta_t = R_{t+1} + \gamma \cdot V(S_{t+1}) - V(S_t)$$

The terms have the following meaning. $V(S_t)$ is the estimated value of the current state, that is, the total reward the system expected to receive from this point onward. R_{t+1} is the reward actually received at the next step. $V(S_{t+1})$ is the estimated value of the state the system has reached, and γ is a factor between zero and one that reduces the weight of distant rewards. The sum of the reward received and the discounted value of the next state constitutes a revised estimate of the same quantity that $V(S_t)$ was approximating. The difference between the revised estimate and the initial one is the error δ_t.

The update uses this error directly: the value of the current state moves in the direction of the error, by a step controlled by the learning rate. Sutton and Barto (2018) emphasize two properties of the procedure. The first is that learning occurs continuously rather than at the end of the episode, which makes it applicable to situations with long trajectories or with no clear ending. The second is that the system learns from an estimate of its own, not only from observed data, a procedure the authors call bootstrapping, which explains both the efficiency of the method and its sensitivity to systematic estimation errors.

The discount factor γ deserves a separate remark, because it is not merely a mathematical device. It determines how far into the future the system looks. A value close to zero produces an agent that pursues almost exclusively immediate reward; a value close to one produces an agent that treats distant rewards almost as it treats near ones. Sutton and Barto (2018) show that the choice of this value is not neutral: it changes the optimal policy, not just the speed of convergence toward it. A system with a short horizon and one with a long horizon, placed in the same environment, will prefer different actions.

There is also a technical reason for keeping the factor below one. In tasks with no defined ending, the sum of future rewards can grow without bound, and the value of a state becomes undefined. Discounting guarantees that the sum remains finite, which makes the problem well posed. Sutton and Barto (2018) treat separately the case of episodic tasks, where this precaution is unnecessary because the series terminates in any case.

The biological side of this story concerns dopaminergic neurons. Dayan and Abbott (2001) present in detail the observation that made the connection possible: the activity of these neurons does not track reward but the departure of reward from what was expected. Early in learning, when reward arrives unanticipated, activity increases. Once a signal preceding the reward has been learned, the increase shifts to the signal, and the reward itself, now predictable, no longer produces any response. If the expected reward fails to arrive, activity falls below baseline at precisely the moment it was due.

This pattern of three situations, more than expected, exactly as expected, less than expected, corresponds term for term to the sign of the error δ_t. Dayan and Abbott

(2001) treat the correspondence as one of the clearest examples of a computational model that predicts a neurophysiological observation rather than merely describing it after the fact.

The correspondence should not, however, be extended beyond what the data support. The temporal-difference formalism assumes that the error is used to update estimates associated with well-defined states, and identifying the state in a biological system is itself an unresolved problem. In addition, the dopaminergic signal is diffuse: it reaches a large number of synapses simultaneously, whereas the algorithmic update is precise. Whittington and Bogacz (2019) discuss this class of difficulties at length, showing that correctly assigning credit for an error is the problem that most sharply separates biologically plausible models from standard training procedures.

What remains solid is the principle. A system that updates its expectations in proportion to the discrepancy between what it predicted and what it observed learns efficiently, because it wastes no resources on what it already knows. Predictable events produce no signal and therefore produce no change. This economy explains why the same principle appears both in biological descriptions and in algorithms designed with no intention of biological fidelity.

Predictive Coding and the Free Energy Principle

The idea that prediction is the central operation of the nervous system can be extended beyond the learning of values. Friston (2010) proposes a framework in which a single quantity accounts for perception, learning, and action. The starting point is a constraint of existence: an organism maintains itself within a restricted set of states compatible with survival, and must therefore avoid states that are improbable for its kind. Formally, this means it must limit the average surprise associated with the sensory data it encounters.

The problem is that surprise cannot be computed directly, because doing so would require knowing the true distribution of causes in the environment. Friston (2010) shows, however, that it can be bounded from above by a computable quantity called free energy, which depends only on the agent's internal model and on the observed data. By minimizing this quantity, the system indirectly minimizes surprise. The important consequence is that an abstract principle becomes a procedure: the system must reduce the discrepancy between what its model predicts and what it receives.

Two courses of action follow from this framework, and Friston (2010) treats them as symmetrical. The first is modifying the model so that it better explains the data; this is perception, understood as inference about causes. The second is modifying the data so that they match the model; this is action. An organism that expects to be fed can achieve the match between model and reality by seeking food, not only by revising its expectations. This second route is what the literature calls active inference.

Clark (2013) discusses the consequences of this perspective for the understanding of perception. If the model continually generates predictions and receives back only the departures from them, then what ascends through the system is not the image of the world but the correction to an image already constructed. Perception thus becomes a largely top-down process in which sensory input serves to adjust an internal construction rather than to produce it. Clark (2013) emphasizes that this inversion naturally explains the dependence of perception on context and expectation, but also draws attention to the limits of the framework, particularly the difficulty of falsifying it when it is stated at a very general level.

The framework proposed by Friston (2010) includes a further element that links this section to the earlier discussion of attention. Prediction errors do not all contribute equally to revising the model; each is weighted according to the estimated reliability of the source that produced it. An error arising from a channel judged to be noisy receives a small weight and modifies the model little, whereas an error arising from a channel judged reliable has a large effect.

Clark (2013) treats this weighting as a mechanism of attention in the proper sense, not as an analogy. In such an architecture, attending to a sensory channel means increasing the weight of its errors, and thereby allowing it to influence the internal representation more strongly. Attentional selection thus ceases to be a separate module and becomes a parameter of the very process that carries out perception. This is an ambitious theoretical position, and Clark (2013) acknowledges that its generality is also its principal vulnerability: a framework capable of accommodating any observation risks excluding nothing.

Pezzulo et al. (2021) examine what happens when this framework is actually used to build artificial agents. The authors show that active inference offers a unified formulation for perception, planning, and control, which is attractive in robotics, where these components are usually designed separately. They also document the difficulties, however: the cost of computing the quantities involved grows rapidly with the size of the problem, and specifying the agent's preferred states, which replace the reward function, often proves as difficult as designing a reward function.

The connection between this framework and standard training methods has been investigated directly. Millidge et al. (2020) show that, under certain conditions, a network implementing predictive coding produces updates that approximate backpropagation, using exclusively local computations and message-passing operations between neighboring levels. Whittington and Bogacz (2019) reach a convergent conclusion in a broader review of error-based learning theories, and discuss the conditions under which the equivalence holds or breaks down.

The significance of this result is worth stating precisely. It does not demonstrate that the brain uses backpropagation. It shows that the standard objection to the biological plausibility of that procedure, namely that it would require a global mechanism for assigning credit, is weaker than had been supposed, because an architecture with strictly local rules can arrive at an equivalent result. It is a result about the space of possible mechanisms, not about the mechanism actually implemented.

Uncertainty and Artificial Metacognition

A system that makes decisions needs not only estimates but also an assessment of the quality of those estimates. Without it, it cannot distinguish situations in which its answer is reliable from those in which it is not, and therefore cannot decide when it is worth requesting additional information, deliberating longer, or transferring the decision to a human operator. This capacity to evaluate one's own performance is, in general terms, a form of metacognition.

Gal (2016) states the problem clearly for neural networks. A network trained in the standard way produces, at its output, a probability distribution over the possible classes. This distribution is often misinterpreted, however: it reflects how confident the network is given the parameters it has learned, but says nothing about how confident we can be that the parameters themselves are appropriate. A network can assign high probability to a wrong class when it receives an input very different from what it saw during training, and the high probability value gives no indication whatsoever of this situation.

The relevant distinction is between two sources of uncertainty. Kendall and Gal (2017) treat them separately: aleatoric uncertainty arises from noise inherent in the data and cannot be reduced by collecting more observations, whereas epistemic uncertainty arises from insufficient knowledge on the part of the model and decreases as data accumulate. The authors show that the two behave differently and that separating them is useful in practice: high epistemic uncertainty indicates a region of the input space where the model has not been trained sufficiently, which is actionable information.

Treating epistemic uncertainty rigorously requires representing parameters as distributions rather than fixed values, and this is the central idea of Bayesian networks. The computational cost of the full approach is, however, prohibitive for large models. Gal and Ghahramani (2016) propose a practical solution: they show that a network trained with dropout can be interpreted as an approximation to a Bayesian model, and that uncertainty can be estimated by keeping dropout active at deployment. Running the same input several times, with units randomly deactivated each time, yields a distribution of predictions whose spread serves as a measure of uncertainty. The method requires no change to the architecture and no retraining.

Kendall and Gal (2017) show that separating the two types of uncertainty is not merely a conceptual distinction but has an effect on performance. Explicitly modelling aleatoric uncertainty allows the system to downweight observations it identifies as noisy instead of trying to explain them. The result is increased robustness to data of uneven quality, obtained without manual intervention in the training set.

The authors also observe that the two components behave differently with respect to the amount of data. Epistemic uncertainty decreases as data accumulate, which makes it useful for deciding where additional observations are worth collecting. Aleatoric uncertainty, by contrast, remains constant, which makes it useful for recognizing the intrinsic limits of the problem. A system that reports a single aggregate value loses precisely this distinction, and therefore cannot tell a situation that can be improved through effort from one that cannot.

An alternative with a different foundation consists in training several independent models and combining their predictions. Lakshminarayanan et al. (2017) show that this approach, although not derived from a Bayesian framework, produces competitive and, in many situations, better calibrated uncertainty estimates. The authors report one particularly useful property: on data that differ systematically from the training data, disagreement among the models of the ensemble increases, which provides a direct signal that the prediction should not be used without verification.

Calibration is the criterion that links these methods to their practical use. A model is calibrated if the frequency with which its predictions are correct matches the confidence it reports: of the cases in which it declares ninety per cent confidence, roughly nine in ten should be correct. Without this correspondence, uncertainty estimation remains a technical quantity with no decision value, because it cannot be converted into a threshold for action.

Here the loop closes with the first section of this chapter. The problem left open there was how a system decides when the cost of deliberation is worth paying. A calibrated estimate of uncertainty supplies exactly the criterion required: when confidence is high, the fast response suffices; when it is low, recourse to the slow and costly procedure is justified. On this reading, metacognition is not an additional function layered on top of the system, but the component that makes the coordination of the others possible.

The Representation of Time and Continuous Depth

All the mechanisms discussed so far presuppose an organization into discrete steps: successive layers in a network, updates at separate moments, numbered positions in a sequence. Biological systems have no such structure. Activity unfolds continuously, and the notion of a "next layer" has no obvious counterpart.

Chen et al. (2018) propose a formulation that removes this presupposition. Instead of a succession of discrete transformations, the authors describe the transformation of the representation through a differential equation: the network specifies not the result of the next step but the rate at which the representation changes at a given point. The output representation is obtained by integrating this dynamic over an interval, using a numerical solver chosen separately from the model.

The most visible consequence is that depth ceases to be an integer fixed at design time. The precision of the integration can be adjusted at deployment, which permits an explicit trade-off between cost and accuracy, decided after training. Chen et al. (2018) also note an advantage with respect to memory: gradients can be computed without retaining intermediate states, which makes the memory cost independent of the number of evaluations.

For the problems discussed in this chapter, the interest of this formulation is twofold. On the one hand, it offers a natural way of handling observations that arrive at irregular intervals, a frequent situation in real-world decisions and an awkward one for models that assume uniform steps. On the other, it brings the mathematical description of artificial networks closer to the language used in computational neuroscience, where membrane dynamics and population activity are as a rule also described by differential equations. Dayan and Abbott (2001) use this formalism consistently, and the proximity of language facilitates direct comparison of models.

SOLUTIONS AND RECOMMENDATIONS

The analysis above suggests several orientations for the design of artificial decision systems. They are not recipes but constraints that the comparison with biological systems makes visible.

Explicit separation of response regimes. A system that must operate both quickly and flexibly needs both strategies described by Sutton and Barto (2018), not a compromise between them. The design should provide from the outset a fast path, based on values already learned, and a slow path, based on simulation, together with the rule that decides between them.

Treating uncertainty as an element of the architecture, not an add-on. The methods described by Gal and Ghahramani (2016) and by Lakshminarayanan et al. (2017) are cheap enough to be included from the design stage. A system that cannot report when it does not know cannot responsibly be integrated into a decision workflow, however good its average accuracy.

Calibration as a verification requirement distinct from accuracy. The two quantities do not entail each other. A model can be accurate and poorly calibrated, in which case its errors are all the more dangerous for appearing confident. Evaluation should report them separately.

Combining case memory with general representation. The results obtained by Pritzel et al. (2017) show that an explicit memory component improves learning speed without replacing the gradually trained component. This combination is useful above all in domains where each experience is costly or risky.

Designing forgetting, not only remembering. An explicit memory component needs a rule that decides what is removed. The solution adopted by Pritzel et al. (2017), based on frequency of access, is simple and efficient but arbitrary with respect to the importance of the content. A system intended for a domain with rare and important events needs a criterion that does not penalize precisely those events.

Explicit choice of the temporal horizon. The discount factor discussed by Sutton and Barto (2018) is often treated as a technical hyperparameter, although it determines what kind of agent results. Its value should be justified by reference to the structure of the problem rather than settled by a search over a set of values.

Caution in interpreting analogies. Functional convergence between an artificial mechanism and a biological one does not justify claims about a shared mechanism. Hassabis et al. (2017) recommend treating analogies as sources of hypotheses rather than as arguments. The same caution applies in reverse: the performance of an artificial model constitutes no evidence about how the brain works.

FUTURE RESEARCH DIRECTIONS

Several problems remain open and appear productive. The first concerns local learning rules. The results obtained by Millidge et al. (2020) and the discussion in Whittington and Bogacz (2019) show that the equivalence between predictive coding and backpropagation holds under specific conditions. Determining the limits of this equivalence, particularly for recurrent architectures and for tasks with long temporal dependencies, remains to be clarified.

The second concerns the representation of time. Most of the models discussed here treat time as a succession of discrete steps, whereas biological systems operate continuously. Chen et al. (2018) propose a class of models in which the transformation of the representation is described by a differential equation and depth becomes a continuous quantity. Applying this formulation to the sequential decision problems discussed in this chapter has not yet been explored systematically.

The third concerns the cost of active inference. The difficulties documented by Pezzulo et al. (2021) suggest that the principal barrier to applying the framework proposed by Friston (2010) is not conceptual but computational. Approximation methods that preserve the advantages of the unified formulation without its full cost represent a direction with practical stakes.

The fourth concerns the evaluation of calibration under distribution shift. The observation reported by Lakshminarayanan et al. (2017), regarding the increase in disagreement among models on unfamiliar data, suggests that disagreement can function as a detector. Establishing the conditions under which this signal is reliable remains an open problem.

CONCLUSION

The six themes analyzed in this chapter describe problems that any decision system must solve: which information deserves processing, when to respond quickly and when to analyze, how to learn from a single experience without losing what was already known, how to update expectations economically, and how to assess the reliability of one's own estimates.

The solutions found in biological and in artificial systems do not coincide at the level of implementation, but they converge at the level of principle. Representations are distributed in both cases, which brings robustness and a capacity for generalization. Learning is driven by the discrepancy between expectation and observation, which eliminates the processing of what is already known. The selection of information is recomputed according to context rather than fixed at design time. And the explicit treatment of uncertainty permits the coordination of the other mechanisms, because it supplies the criterion by which a system decides how much effort a situation deserves.

These convergences do not support the conclusion that artificial systems work like the brain. They support a narrower but sounder conclusion: decision problems under constraints of time, energy, and information admit a limited number of reasonable solutions, and the principles listed above are among them. The value of the brain–computer analogy lies precisely in this narrowing. It does not say how a system must be built, but it indicates which classes of organization are worth exploring and, equally important, which surface resemblances should not be taken as evidence.

For design practice, the consequence is that an intelligent system is not measured by speed or average accuracy alone. It is measured by its capacity to direct its resources where they matter, to learn from little, to correct itself when it errs, and to signal when it does not know.

REFERENCES

Chen, R. T. Q., Rubanova, Y., Bettencourt, J., & Duvenaud, D. (2018). Neural ordinary differential equations. *Advances in Neural Information Processing Systems, 31*, 6571–6583.

Clark, A. (2013). Whatever next? Predictive brains, situated agents, and the future of cognitive science. *Behavioral and Brain Sciences, 36*(3), 181–204. DOI: 10.1017/S0140525X12000477 PMID: 23663408

Dayan, P., & Abbott, L. F. (2001). *Theoretical neuroscience: Computational and mathematical modeling of neural systems*. MIT Press.

Friston, K. (2010). The free-energy principle: A unified brain theory? *Nature Reviews. Neuroscience, 11*(2), 127–138. DOI: 10.1038/nrn2787 PMID: 20068583

Gal, Y. (2016). Uncertainty in deep learning [Doctoral dissertation, University of Cambridge].

Gal, Y., & Ghahramani, Z. (2016). Dropout as a Bayesian approximation: Representing model uncertainty in deep learning. *Proceedings of the 33rd International Conference on Machine Learning*, 48, 1050–1059.

Hassabis, D., Kumaran, D., Summerfield, C., & Botvinick, M. (2017). Neuroscience-inspired artificial intelligence. *Neuron, 95*(2), 245–258. DOI: 10.1016/j.neuron.2017.06.011 PMID: 28728020

Kahneman, D. (2011). *Thinking, fast and slow*. Farrar, Straus and Giroux.

Kendall, A., & Gal, Y. (2017). What uncertainties do we need in Bayesian deep learning for computer vision? *Advances in Neural Information Processing Systems, 30*, 5574–5584.

Lakshminarayanan, B., Pritzel, A., & Blundell, C. (2017). Simple and scalable predictive uncertainty estimation using deep ensembles. *Advances in Neural Information Processing Systems, 30*, 6402–6413.

Millidge, B., Tschantz, A., & Buckley, C. L. (2020). Predictive coding approximates backprop along arbitrary computation graphs. arXiv preprint arXiv:2006.04182.

O'Keane, V. (2021). *The rag and bone shop: How we make memories and memories make us*. Allen Lane.

Pezzulo, G., Parr, T., & Friston, K. (2021). Active inference in robotics and artificial agents: Survey and challenges. arXiv preprint arXiv:2112.01871.

Pritzel, A., Uria, B., Srinivasan, S., Puigdomènech, A., Vinyals, O., Hassabis, D., Wierstra, D., & Blundell, C. (2017). Neural episodic control. *Proceedings of the 34th International Conference on Machine Learning*, 70, 2827–2836.

Sutton, R. S., & Barto, A. G. (2018). *Reinforcement learning: An introduction* (2nd ed.). MIT Press.

Vaswani, A., Shazeer, N., Parmar, N., Uszkoreit, J., Jones, L., Gomez, A. N., Kaiser, Ł., & Polosukhin, A. (2017). Attention is all you need. *Advances in Neural Information Processing Systems*, *30*, 5998–6008.

Whittington, J. C. R., & Bogacz, R. (2019). Theories of error back-propagation in the brain. *Trends in Cognitive Sciences*, *23*(3), 235–250. DOI: 10.1016/j.tics.2018.12.005 PMID: 30704969

ADDITIONAL READING

Dayan, P., & Abbott, L. F. (2001). Theoretical neuroscience: Computational and mathematical modeling of neural systems. MIT Press. A reference work on the mathematical formalism of neural models, useful for readers who want a detailed treatment of the population coding and plasticity rules discussed in this chapter.

Hassabis, D., Kumaran, D., Summerfield, C., & Botvinick, M. (2017). Neuroscience-inspired artificial intelligence. [A synthesis of the exchange between the two fields, with explicit discussion of the limits of the analogy.]. *Neuron*, *95*(2), 245–258. DOI: 10.1016/j.neuron.2017.06.011 PMID: 28728020

Sutton, R. S., & Barto, A. G. (2018). Reinforcement learning: An introduction (2nd ed.). MIT Press. A complete exposition of temporal-difference methods, of the distinction between model-based and model-free approaches, and of the architectures that combine them.

KEY TERMS AND DEFINITIONS

Attention Mechanism: A procedure by which a system computes the relevance of each component of the input relative to the current context and distributes processing resources in proportion to that relevance, recomputing the weights for each new input.

Confidence Calibration: The property of a predictive model whereby the confidence levels it reports correspond to the actual frequency with which its predictions are correct. A necessary condition for converting uncertainty estimates into decision thresholds.

Distributed Representation: A mode of encoding in which a piece of information is supported by the combined activity of several units, each unit participating in the encoding of several pieces of information. It confers robustness to the loss of units and favours generalization.

Episodic Memory: The capacity to retain individual events together with their context, after one or few exposures. It permits the immediate exploitation of singular experiences, in contrast to gradual learning, which requires repeated exposure.

Epistemic Uncertainty: The component of uncertainty that arises from insufficient knowledge on the part of the model and that decreases as data accumulate. It is distinguished from aleatoric uncertainty, which is generated by noise inherent in the observations.

Free Energy: An information-theoretic quantity computable from an agent's internal model and from observed data, constituting an upper bound on surprise. Minimizing it allows a system to reduce indirectly the probability of reaching states that are improbable for its kind.

Model-Based Learning: A strategy by which a system builds a representation of the dynamics of the environment and uses it to simulate the consequences of actions before executing them. Computationally costly, but flexible under changing conditions.

Model-Free Learning: A strategy by which a system associates values directly with states or actions, on the basis of repeated experience, without explicitly representing the dynamics of the environment. It produces fast decisions but adapts slowly to change.

Prediction Error: The difference between the outcome expected by a system and the outcome actually observed. It serves as a learning signal, since change occurs in proportion to the discrepancy, and predictable events generate no adjustment.

Predictive Coding: A scheme of organization in which the higher levels of a system generate predictions about the activity of the lower levels, and the signal transmitted upward contains only the departures from those predictions.

Compilation of References

Abdelwanis, Y., Gafni, D., & Nissen, M. (2024). Cognitive processes while using Artificial Intelligence at work. *Formare*, *45*(1), 23–41.

Abiri, R., Borhani, S., Sellers, E. W., Jiang, Y., & Zhao, X. (2019). A comprehensive review of EEG-based brain–computer interface paradigms. *Journal of Neural Engineering*, *16*(1), 011001. DOI: 10.1088/1741-2552/aaf12e PMID: 30523919

Achiam, J., Held, D., Tamar, A., & Abbeel, P. (2017). Constrained policy optimization. *Proceedings of the 34th International Conference on Machine Learning (ICML)*, 22–31. https://proceedings.mlr.press/v70/achiam17a.html

Adadi, A., & Berrada, M. (2018). Peeking Inside the Black-Box: A Survey on Explainable Artificial Intelligence (XAI). *IEEE Access : Practical Innovations, Open Solutions*, *6*, 52138–52160. DOI: 10.1109/ACCESS.2018.2870052

Agapov, S. N., Bulanov, V. A., Zakharov, A. V., & Sergeeva, M. S. (2017). Comparison of classifiers in the tasks of the single-trial vep classification. *Zhurnal Vysshei Nervnoi Deyatelnosti Imeni IP Pavlova, 67*(4), Article 4.

Aghajani, H., Garbey, M., & Omurtag, A. (2017). Measuring Mental Workload with EEG+fNIRS. *Frontiers in Human Neuroscience*, *11*, 359. DOI: 10.3389/fnhum.2017.00359 PMID: 28769775

Agustian, H. Y. (2025). The critical role of understanding epistemic practices in science teaching using wicked problems. *Science & Education*, *34*(1), 485–510. DOI: 10.1007/s11191-023-00471-2

Ajitha, N., Ashwin, G., Divakar, S., & Dhanushree, M. (2025). Machine Learning-based Panic Syncsight: An IoT-Integrated Panic Attack Detection System for Enhanced Therapeutic Insights. 2025 5th International Conference on Soft Computing for Security Applications (ICSCSA), Boopathi Raja, G. (2021). An intelligent approach for detecting and preventing heart attacks using wearable technology. *Handbook of Research on Lifestyle Sustainability and Management Solutions Using AI Big Data Analytics, and Visualization. IGI Gobal.*

Akcetin, E. (2024). Integrating Scientific Instruments and Intelligence Frameworks for Digital Enterprises. In *Game Theory-Computational Aspects and Applications.* IntechOpen. DOI: 10.5772/intechopen.1007794

Al Humdan, E. (2025). Enhancing transportation agility through neuroadaptive AI and behavioural decision intelligence. *Transportation Research Interdisciplinary Perspectives*, *31*, 101468. DOI: 10.1016/j.trip.2025.101468

Alam, A. (2023a). Media multitasking with M-learning technology in real-time classroom learning: Analysing the dynamics in formal educational settings for the future of E-learning in India. In *2023 2nd International Conference on Smart Technologies and Systems for Next Generation Computing (ICSTSN)* (pp. 1-6). IEEE.

Alam, A. (2023b). The secret sauce of student success: Cracking the code by navigating the path to personalized learning with educational data mining. In *2023 2nd International Conference on Smart Technologies and Systems for Next Generation Computing (ICSTSN)* (pp. 1-8). IEEE.

Alam, A., & Mohanty, A. (2022, December). Facial analytics or virtual avatars: competencies and design considerations for student-teacher interaction in AI-powered online education for effective classroom engagement. In *International Conference on Communication, Networks and Computing* (pp. 252-265). Cham: Springer Nature Switzerland.

Alam, A., & Mohanty, A. (2023, January). Learning on the move: a pedagogical framework for state-of-the-art mobile learning. In *International Conference on Data Management, Analytics & Innovation* (pp. 735-748). Singapore: Springer Nature Singapore. DOI: 10.1007/978-981-99-1414-2_52

Alam, A. (2022). Impact of university's human resources practices on professors' occupational performance: empirical evidence from India's higher education sector. In *Inclusive businesses in developing economies: Converging people, profit, and corporate citizenship* (pp. 107–131). Springer International Publishing. DOI: 10.1007/978-3-031-12217-0_6

Alam, A. (2023c). Intelligence unleashed: An argument for AI-enabled learning ecologies with real world examples of today and a peek into the future. In *International Conference on Innovations in Computer Science, Electronics & Electrical Engineering*-2022 (Vol. 2717, No. 1, p. 030001). AIP Publishing LLC. DOI: 10.1063/5.0129803

Alam, A., & Mohanty, A. (2024). Integrated constructive robotics in education (ICRE) model: A paradigmatic framework for transformative learning in educational ecosystem. *Cogent Education*, *11*(1), 2324487. DOI: 10.1080/2331186X.2024.2324487

Albini, E., Arcangeli, M., & Barbu, E. (2023). Achieving descriptive accuracy in explanations via cognitive explainability. *PMC, 10117939.*

Allen, N. J., & Lyons, D. A. (2018). Glia as architects of central nervous system formation and function. *Science*, *362*(6411), 181–185. DOI: 10.1126/science.aat0473 PMID: 30309945

Alohali, M. A., Al-Wesabi, F. N., Hilal, A. M., Goel, S., Gupta, D., & Khanna, A. (2022). Artificial intelligence enabled intrusion detection systems for cognitive cyber-physical systems in industry 4.0 environment. *Cognitive Neurodynamics*, *16*(5), 1045–1057. DOI: 10.1007/s11571-022-09780-8 PMID: 36237400

Al-Shamma, S., & Baldwin, D. (2023). Neuro-Informatics Approaches for Human-Machine Interaction and Brain-Inspired Artificial Intelligence. *Frontora*, *1*(1), 13–20.

Alufaisan, Y., Marusich, L. R., Bakdash, J. Z., Zhou, Y., & Kantarcioglu, M. (2020). Does explainable artificial intelligence improve human decision-making? *Proceedings of the Human Factors and Ergonomics Society Annual Meeting*, *64*(1), 137–141. DOI: 10.1177/1071181320641025

Amershi, S., Weld, D., Vorvoreanu, M., Fourney, A., Nushi, B., Collisson, P., Suh, J., Iqbal, S., Bennett, P. N., Inkpen, K., Teevan, J., Kikin-Gil, R., & Horvitz, E. (2019). Software engineering for machine learning: A case study. In *Proceedings of the 41st International Conference on Software Engineering: Software Engineering in Practice* (pp. 291–300). IEEE. DOI: 10.1109/ICSE-SEIP.2019.00042

Amodei, D., Olah, C., Steinhardt, J., Christiano, P., Schulman, J., & Mané, D. (2016). *Concrete problems in AI safety* (arXiv:1606.06565). arXiv.

Amrouch, H., Imani, M., Jiao, X., Aloimonos, Y., Fermuller, C., Yuan, D., & Sutor, P. (2022, October). Brain-inspired hyperdimensional computing for ultra-efficient edge ai. In *2022 International Conference on Hardware/Software Codesign and System Synthesis (CODES+ ISSS)* (pp. 25-34). IEEE. DOI: 10.1109/CODES-ISSS55005.2022.00017

Anderson, J. R. (2007). *How can the human mind occur in the physical universe?* Oxford University Press. DOI: 10.1093/acprof:oso/9780195324259.001.0001

Angwin, J., Larson, J., Mattu, S., & Kirchner, L. (2016). Machine bias: There's software used across the country to predict future criminals. And it's biased against blacks. ProPublica. https://www.propublica.org/article/machine-bias-risk-assessments-in-criminal-sentencing

Anh-Hoang, D., Tran, V., & Nguyen, L.-M. (2025). Survey and analysis of hallucinations in large language models: Attribution to prompting strategies or model behavior. *Frontiers in Artificial Intelligence*, *8*, 1622292. DOI: 10.3389/frai.2025.1622292 PMID: 41098969

Annu, & Tripathi, R. (2024). A bibliometric and systematic review analysis of adopting decision intelligence analytics for rational decision-making. *Qualitative Research in Financial Markets, 16*(5), 915-936.

Aricò, P., Borghini, G., Di Flumeri, G., Sciaraffa, N., & Babiloni, F. (2018). Passive BCI beyond the lab: Current trends and future directions. *Physiological Measurement*, *39*(8), 08TR02. Advance online publication. DOI: 10.1088/1361-6579/aad57e PMID: 30039806

Arrieta, A. B., Díaz-Rodríguez, N., Del Ser, J., Bennetot, A., Tabik, S., Barbado, A., García, S., Gil-López, S., Molina, D., Benjamins, R., Chatila, R., & Herrera, F. (2020). Explainable Artificial Intelligence (XAI): Concepts, taxonomies, opportunities and challenges toward responsible AI. *Information Fusion*, *58*, 82–115. DOI: 10.1016/j.inffus.2019.12.012

Atashgahi, Z., Pieterse, J., Liu, S., Mocanu, D. C., Veldhuis, R., & Pechenizkiy, M. (2022). A brain-inspired algorithm for training highly sparse neural networks. *Machine Learning*, *111*(12), 4411–4452. DOI: 10.1007/s10994-022-06266-w

ATTC Network. (2025). *Privacy & ethical implications of AI in SUD treatment*. Addiction Technology Transfer Center Network. https://attcnetwork.org/products_and_resources/privacy-ethical-implications-of-ai-in-sud-treatment/

Baddeley, A. (2012). Working memory: Theories, models, and controversies. *Annual Review of Psychology*, *63*(1), 1–29. DOI: 10.1146/annurev-psych-120710-100422

Baddeley, A. D. (2003). Working memory: Looking back and looking forward. *Nature Reviews. Neuroscience*, *4*(10), 829–839. DOI: 10.1038/nrn1201

Badre, D., & Frank, M. J. (2012). Mechanisms of hierarchical reinforcement learning in cortico–striatal circuits 2: Evidence from fMRI. *Cerebral Cortex (New York, N.Y.)*, *22*(3), 527–536. DOI: 10.1093/cercor/bhr117

Balasubramanian, S. (2023). Literature review on developing explainable artificial intelligence for trustworthy decision-making. *International Journal of Design and Manufacturing Technology, 14*(1), 1–10.

Bansal, G., Nushi, B., Kamar, E., Lasecki, W., Weld, D., & Horvitz, E. (2021). Beyond accuracy: The role of mental models in human-AI team performance. *Proceedings of the AAAI Conference on Human Computation and Crowdsourcing*, 7, 2–11. DOI: 10.1609/hcomp.v7i1.5281

Barocas, S., Hardt, M., & Narayanan, A. (2023). *Fairness and machine learning: Limitations and opportunities*. MIT Press.

Barocas, S., Hood, S., & Ziewitz, M. (2021). Governing algorithms: A provocation piece. *SSRN*. DOI: 10.2139/ssrn.2245322

Bartlett, R., Morse, A., Stanton, R., & Wallace, N. (2022). Consumer-lending discrimination in the FinTech era. *Journal of Financial Economics, 143*(1), 30–56. DOI: 10.1016/j.jfineco.2021.05.047

Battaglia, P. W., Hamrick, J. B., Bapst, V., Sanchez Gonzalez, A., Zambaldi, V., Malinowski, M., Tacchetti, A., Raposo, D., Santoro, A., Faulkner, R., Gulcehre, C., Song, F., Ballard, A., Gilmer, J., Dahl, G., Vaswani, A., Allen, K., Nash, C., Langston, V., . . . Pascanu, R. (2018). Relational inductive biases, deep learning, and graph networks (arXiv:1806.01261). arXiv.

Baxter, G., & Sommerville, I. (2011). Socio-technical systems: From design methods to systems engineering. *Interacting with Computers, 23*(1), 4–17. DOI: 10.1016/j.intcom.2010.07.003

Bechara, A., Damasio, A. R., & Damasio, H. (2000). Emotion, decision making and the orbitofrontal cortex. *Cerebral Cortex (New York, N.Y.), 10*(3), 295–307. DOI: 10.1093/cercor/10.3.295

Bedoya, A. D., Futoma, J., Clement, M. E., Corey, K., Brajer, N., Lin, A., Simons, M. G., Gao, M., Nichols, M., Balu, S., Henao, R., Sendak, M., & O'Brien, C. (2022). Machine learning for early detection of sepsis: An internal and temporal validation study. *JAMIA Open, 5*(2), ooab094. Advance online publication. DOI: 10.1093/jamiaopen/ooab094 PMID: 32734166

Beer, D. (2018). The data gaze: Capitalism, power and perception.

Bellemare, M. G., Dabney, W., & Munos, R. (2017). A distributional perspective on reinforcement learning. *Proceedings of the 34th International Conference on Machine Learning (ICML)*, 449–458. https://proceedings.mlr.press/v70/bellemare17a.html

Bellemare, M. G., Dabney, W., & Munos, R. (2017). A distributional perspective on reinforcement learning. In *Proceedings of the International Conference on Machine Learning*.

Bender, E. M., Gebru, T., McMillan-Major, A., & Shmitchell, S. (2021). On the dangers of stochastic parrots: Can language models be too big? In *Proceedings of the 2021 ACM Conference on Fairness, Accountability, and Transparency* (pp. 610–623). DOI: 10.1145/3442188.3445922

Berndt, J., & Vössing, M. (2025). *Decision intelligence: Merging human judgment and AI*. Harvard Business Review Analytics.

Berretta, S., Tausch, A., Ontrup, G., Gilles, B., Peifer, C., & Kluge, A. (2023). Defining human-AI teaming the human-centered way: A scoping review and network analysis. *Frontiers in Artificial Intelligence*, *6*, 1250725. DOI: 10.3389/frai.2023.1250725 PMID: 37841234

Besold, T. R., d'Avila Garcez, A. S., Bader, S., Bowman, H., Domingos, P., Hitzler, P., Kühnberger, K. U., Lamb, L. C., Lowd, D., Lima, P. M. V., de Penning, L., Pinkas, G., Poon, H., & Zaverucha, G. (2017). Neural-symbolic learning and reasoning: A survey and interpretation (arXiv:1711.03902). arXiv.

Bickel, W. K., Jarmolowicz, D. P., Mueller, E. T., Gatchalian, K. M., & McClure, S. M. (2012). Are executive function and impulsivity antipodes? A conceptual reconstruction with special reference to addiction. *Psychopharmacology*, *221*(3), 361–387. DOI: 10.1007/s00213-012-2689-x

Bickel, W. K., Koffarnus, M. N., Moody, L., & Wilson, A. G. (2014). The behavioral- and neuro-economic process of temporal discounting: A candidate behavioral marker of addiction. *Neuropharmacology*, *76*, 518–527. DOI: 10.1016/j.neuropharm.2013.06.013

Bird, J. (2013). *The role of trait impulsivity and novelty-seeking as predisposing factors to the acquisition and reinstatement of MDMA self-administration* (Doctoral dissertation, Open Access Te Herenga Waka-Victoria University of Wellington).

BIS. (2024). *Managing explanations: how regulators can address AI explainability* (FSI Papers No. 24). Bank for International Settlements.

Blum, K., Werner, T., Carnes, S., Carnes, P., Bowirrat, A., Giordano, J., Marlene-Oscar-Berman, , & Gold, M. (2012). Sex, drugs, and rock 'n'roll: Hypothesizing common mesolimbic activation as a function of reward gene polymorphisms. *Journal of Psychoactive Drugs*, *44*(1), 38–55. DOI: 10.1080/02791072.2012.662112

Bommasani, R., Hudson, D. A., Adeli, E., Altman, R., Arora, S., von Arx, S., Bernstein, M. S., Bohg, J., Bosselut, A., Brunskill, E., Brynjolfsson, E., Buch, S., Card, D., Castellon, R., Chatterji, N. S., Chen, A. S., Creel, K., Davis, J. Q., Demszky, D., . . . Liang, P. (2022). On the opportunities and risks of foundation models. arXiv. https://doi.org//arXiv.2108.07258DOI: 10.48550

Brachman, R. J., & Levesque, H. J. (2004). *Knowledge representation and reasoning*. Morgan Kaufmann.

Bransford, J. D., Brown, A. L., & Cocking, R. R. (Eds.). (2000). *How people learn: Brain, mind, experience, and school* (Expanded ed.). National Academies Press.

Bratu, S. (2023). Brain-inspired artificial intelligence and machine learning-based navigation algorithms, spatial data visualization and digital twin modeling tools, and emotion detection and recognition technologies in the decentralized and interconnected metaverse. *Review of Contemporary Philosophy*, (22), 51–67.

Bredeweg, B., & Kragten, M. (2022). Requirements and challenges for hybrid intelligence: A case-study in education. *Frontiers in Artificial Intelligence*, *5*, 891630. DOI: 10.3389/frai.2022.891630 PMID: 36003066

Brown, T. B., Mann, B., Ryder, N., Subbiah, M., Kaplan, J., Dhariwal, P., Neelakantan, A., Shyam, P., Sastry, G., Askell, A., Agarwal, S., Herbert-Voss, A., Krueger, G., Henighan, T., Child, R., Ramesh, A., Ziegler, D. M., Wu, J., Winter, C., . . . Amodei, D. (2020). *Language models are few-shot learners* (arXiv:2005.14165). arXiv.

Buelow, M. T., & Suhr, J. A. (2009). Construct validity of the Iowa gambling task. *Neuropsychology Review*, *19*(1), 102–114. DOI: 10.1007/s11065-009-9083-4

Bulanov, V. A., Zakharov, A. V., & Khivintseva, E. V. (2020). Wavelet transform for the identification of P300. *IOP Conference Series. Materials Science and Engineering*, *862*(5), 052049. DOI: 10.1088/1757-899X/862/5/052049

Business Roundtable. (2026). *Artificial intelligence policy recommendations*. https://www.businessroundtable.org/artificial-intelligence-policy-recommendations

Butelman, E. R., Huang, Y., Goldstein, R. Z., & Alia-Klein, N. (2023). Men have approximately 2-to 3-fold greater overdose mortality than women for synthetic opioids, heroin and psychostimulants including cocaine across the lifespan: Analysis of state-level CDC data for 2020-2021. *medRxiv*.

Buzsáki, G., Anastassiou, C. A., & Koch, C. (2012). The origin of extracellular fields and currents—EEG, ECoG, LFP and spikes. *Nature Reviews. Neuroscience*, *13*(6), 407–420. DOI: 10.1038/nrn3241 PMID: 22595786

Cai, C. J., Winter, S., Steiner, D., Wilcox, L., & Terry, M. (2022). 'Hello AI': Uncovering the onboarding needs of medical practitioners for human-AI collaborative decision-making. Proceedings of the ACM on Human-Computer Interaction, 3(CSCW), 104. DOI: 10.1145/3359206

Cai, W., & Gao, M. (2025). Beyond hallucination: Generative AI as a catalyst for human creativity and cognitive evolution. *ICCK Transactions on Emerging Topics in Artificial Intelligence*, *2*(1), 36–42. DOI: 10.62762/TETAI.2025.657559

Campos, A. C., Pinto, P., & Scott, N. (2019). Bottom-up factors of attention during the tourist experience: An empirical study. *Current Issues in Tourism*, *23*(24), 3111–3133. DOI: 10.1080/13683500.2019.1681383

Cao, Y., Yang, H., Xue, Y., Wang, F., Li, T., Zhao, L., & Fu, Y. (2026). Ethical risks and considerations of the integration of Brain-Computer Interfaces with Artificial Intelligence. *Cognitive Neurodynamics*, *20*(1), 46. DOI: 10.1007/s11571-025-10380-5 PMID: 41657963

Carichon, F., Khandelwal, A., Fauchard, M., & Farnadi, G. (2025). The coming crisis of multi-agent misalignment: Ai alignment must be a dynamic and social process. *arXiv preprint arXiv:2506.01080.*

Cath, C. (2018). Governing artificial intelligence: Ethical, legal and technical opportunities and challenges. *Philosophical Transactions. Series A, Mathematical, Physical, and Engineering Sciences*, *376*(2133), 20180080. DOI: 10.1098/rsta.2018.0080 PMID: 30322996

Ceceli, A. O., Huang, Y., Kronberg, G., McClain, N., King, S. G., Butelman, E. R., & Goldstein, R. Z. (2025). The impaired response inhibition and salience attribution model of drug addiction: Recent neuroimaging evidence and Future Directions. *Annual Review of Psychology*, , 77.

Censinet. (2026). *The new risk frontier: Navigating AI uncertainty in an automated world.* https://www.censinet.com/perspectives/new-risk-frontier-navigating-ai-uncertainty

Cevolani, G., & Festa, C. (2021). Approaching deterministic and probabilistic truth: A unified account of truth approximation. *Synthese*, *199*(3-4), 11465–11489. Advance online publication. DOI: 10.1007/s11229-021-03298-y

Cha, J., Park, J., Cho, M., Lee, E., Kim, B. G., Kim, G., & Joo, Y. (2024). Individual Differences in the Effects of Neighborhood Socioeconomic Deprivation on Intertemporal Decision-Making and Psychotic-Like Experiences in Children.

Chen, Y. C. (2022). Is Uncertainty a Barrier or Resource to Advance Science? The Role of Uncertainty in Science and Its Implications for Science Teaching and Learning: Kostas Kampourakis, Kevin McCain (2019): Uncertainty: How It Makes Science Advance. Oxford University Press, New York, NY. . 0001.DOI: <ALIGNMENT.qj></ALIGNMENT>10.1093/oso/9780190871666.001

Chen, H., Chiang, R. H. L., & Storey, V. C. (2012). Business intelligence and analytics: From big data to big impact. *Management Information Systems Quarterly*, *36*(4), 1165–1188. DOI: 10.2307/41703503

Chen, J. Q., & Lee, S. M. (2003). An exploratory cognitive DSS for strategic decision making. *Decision Support Systems*, *36*(2), 147–160. DOI: 10.1016/S0167-9236(02)00139-2

Chen, J. Y. C., Barnes, M. J., Selkowitz, A. R., & Stowers, K. (2022). Effects of teammate trustworthiness information and reputation on human-autonomy teaming. *Frontiers in Psychology*, *13*, 854027. DOI: 10.3389/fpsyg.2022.854027

Chen, J., Yu, K., Bi, Y., Ji, X., & Zhang, D. (2024). Strategic Integration: A Cross-Disciplinary Review of the fNIRS-EEG Dual-Modality Imaging System for Delivering Multimodal Neuroimaging to Applications. *Brain Sciences*, *14*(10), 1022. DOI: 10.3390/brainsci14101022 PMID: 39452034

Chen, R. T. Q., Rubanova, Y., Bettencourt, J., & Duvenaud, D. (2018). Neural ordinary differential equations. *Advances in Neural Information Processing Systems*, *31*, 6571–6583.

Chen, X., Cheng, G., Wang, F. L., Tao, X., Xie, H., & Xu, L. (2022). Machine and cognitive intelligence for human health: Systematic review. *Brain Informatics*, *9*(1), 5. DOI: 10.1186/s40708-022-00153-9 PMID: 35150379

Chinnaraju, A. (2025). Quantum computing in consumer behavior: A theoretical framework for market prediction and decision analytics. *Quantum : the Open Journal for Quantum Science*, *5*(2).

Chohra, A., & van der Wal, C. N. (2025, October). Integrating Emotional, Personal, and Social Intelligences in Complex Collective Decision-Making. In *Proceedings of the Future Technologies Conference* (pp. 176-193). Cham: Springer Nature Switzerland.

Chouldechova, A., & Roth, A. (2020). A snapshot of the frontiers of fairness in machine learning. *Communications of the ACM*, *63*(5), 82–89. DOI: 10.1145/3376898

Chromik, M., Eiband, M., Buchner, F., Krüger, A., & Butz, A. (2021). I think I understand, but do I? Measuring understanding of human-centered explanations. In *Proceedings of the 29th ACM Conference on User Interface Software and Technology* (pp. 20–34). DOI: 10.1145/3472749.3474743

Cisotto, G., Zancanaro, A., Zoppis, I. F., & Manzoni, S. L. (2024). hvEEGNet: A novel deep learning model for high-fidelity EEG reconstruction. *Frontiers in Neuroinformatics*, *18*, 1459970. DOI: 10.3389/fninf.2024.1459970 PMID: 39759760

Clark, A. (2013). Whatever next? Predictive brains, situated agents, and the future of cognitive science. *Behavioral and Brain Sciences*, *36*(3), 181–204. DOI: 10.1017/S0140525X12000477 PMID: 23663408

Clark, L. F. B. (2025). Multimodal AI–Driven Real-Time Decision Intelligence for Secure Healthcare Cloud and SAP Data Networks. [IJEETR]. *International Journal of Engineering & Extended Technologies Research*, *7*(4), 10277–10283.

Cohen, J. D., Dunbar, K., & McClelland, J. L. (1990). On the control of automatic processes: A parallel distributed processing account of the Stroop effect. *Psychological Review*, *97*(3), 332–361. DOI: 10.1037/0033-295X.97.3.332

Cohen, M. X. (2006). Individual differences and the neural representations of reward expectation and reward prediction error. *Social Cognitive and Affective Neuroscience*, *2*(1), 20–30. DOI: 10.1093/scan/nsl021 PMID: 17710118

Colizzi, M., & Bhattacharyya, S. (2018). Cannabis use and the development of tolerance: A systematic review of human evidence. *Neuroscience and Biobehavioral Reviews*, *93*, 1–25. DOI: 10.1016/j.neubiorev.2018.07.014

Corbetta, M., & Shulman, G. L. (2002). Control of goal-directed and stimulus-driven attention in the brain. *Nature Reviews. Neuroscience*, *3*(3), 201–215. DOI: 10.1038/nrn755

Corlett, P. R., & Friston, K. J. (2020). What do we think about when we think about predictive processing? *Frontiers in Systems Neuroscience*, *14*, 7509909. PMID: 32757598

Cortés, U., Sànchez-Marrè, M., Ceccaroni, L., R-Roda, I., & Poch, M. (2000). Artificial intelligence and environmental decision support systems. *Applied Intelligence*, *13*(1), 77–91. DOI: 10.1023/A:1008331413864

Costa, R. D., & Silva, P. (2025). Reinforcement learning applied to a situation awareness model with cognitive explainability. *Information Sciences*, *689*, 121–140.

Cowan, N. (2022). Working memory development in childhood. In *The Development of Memory in Infancy and Childhood* (pp. 111–145). Psychology Press., DOI: 10.4324/9781003016533-5

Cowell, R. G., Verrall, R. J., & Yoon, Y. K. (2007). Modeling operational risk with Bayesian networks. *The Journal of Risk and Insurance*, *74*(4), 795–827. DOI: 10.1111/j.1539-6975.2007.00235.x

Dafoe, A. (2018). AI governance: A research agenda. Future of Humanity Institute, University of Oxford.

Dalley, J. W., Everitt, B. J., & Robbins, T. W. (2011). Impulsivity, compulsivity, and top-down cognitive control. *Neuron*, *69*(4), 680–694. DOI: 10.1016/j.neuron.2011.01.020

Dalley, J. W., & Roiser, J. P. (2012). Dopamine, serotonin and impulsivity. *Neuroscience*, *215*, 42–58. DOI: 10.1016/j.neuroscience.2012.03.065

Damasio, A. R. (1994). *Descartes' error: Emotion, reason, and the human brain.* Putnam.

Daw, N. D., Niv, Y., & Dayan, P. (2005). Uncertainty-based competition between prefrontal and dorsolateral striatal systems for behavioral control. *Nature Neuroscience*, *8*(12), 1704–1711. DOI: 10.1038/nn1560

Dayan, P., & Abbott, L. F. (2001). *Theoretical neuroscience: Computational and mathematical modeling of neural systems.* MIT Press.

Dayan, P., & Daw, N. D. (2008). Decision theory, reinforcement learning, and the brain. *Cognitive, Affective & Behavioral Neuroscience*, *8*(4), 429–453. DOI: 10.3758/CABN.8.4.429

De Raedt, L., Dumancic, S., Manhaeve, R., & Marra, G. (2020). From statistical relational to neuro-symbolic artificial intelligence. In *Proceedings of the Twenty-Ninth International Joint Conference on Artificial Intelligence (IJCAI 2020).* DOI: 10.24963/ijcai.2020/688

Dezfouli, A., & Balleine, B. W. (2012). Habits, action sequences and reinforcement learning. *The European Journal of Neuroscience*, *35*(7), 1036–1051. DOI: 10.1111/j.1460-9568.2012.08050.x

Diederen, K. M. J., & Fletcher, P. C. (2020). Dopamine, prediction error and beyond. *The Neuroscientist*, *27*(1), 30–46. DOI: 10.1177/1073858420907591 PMID: 32338128

Dietterich, T. G. (2019). Robust artificial intelligence and robust human organizations. *Frontiers of Computer Science*, *13*(1), 1–3. DOI: 10.1007/s11704-018-8900-4

Dietvorst, B. J., Logg, J. M., & Bharti, S. (2018). Overcoming algorithm aversion: People will use imperfect algorithms if they can (even slightly) modify them. *Management Science*, *64*(3), 1155–1170. DOI: 10.1287/mnsc.2016.2643

Din, M., Daga, K., Saoud, J., Wood, D., Kierkegaard, P., Brex, P., & Booth, T. C. (2025). Clinicians' perspectives on the use of artificial intelligence to triage MRI brain scans. *European Journal of Radiology*, *183*, 111921. DOI: 10.1016/j.ejrad.2025.111921

Domenech, P., & Koechlin, E. (2015). Executive control and decision-making in the prefrontal cortex. *Current Opinion in Behavioral Sciences*, *1*, 101–106. DOI: 10.1016/j.cobeha.2014.10.007

Doshi-Velez, F., & Kim, B. (2017). *Towards a rigorous science of interpretable machine learning* (arXiv:1702.08608). arXiv.

Doshi-Velez, F., & Kim, B. (2017). Towards a rigorous science of interpretable machine learning. arXiv. https://arxiv.org/abs/1702.08608

Dressel, J., & Farid, H. (2018). The accuracy, fairness, and limits of predicting recidivism. *Science Advances*, *4*(1), eaao5580. Advance online publication. DOI: 10.1126/sciadv.aao5580 PMID: 29376122

Droutman, V., Bechara, A., & Read, S. J. (2015). Roles of the Different Sub-Regions of the Insular Cortex in Various Phases of the Decision-Making Process. *Frontiers in Behavioral Neuroscience*, *9*. Advance online publication. DOI: 10.3389/fnbeh.2015.00309 PMID: 26635559

Dwork, C., Ilvento, C., & Jagadeesan, M. (2023). Individual fairness in pipelines. In *Proceedings of the 2023 ACM Conference on Fairness, Accountability, and Transparency* (pp. 1228–1237). DOI: 10.1145/3593013.3594001

Edelman, B. J., Meng, J., Suma, D., Zurn, C., Nagarajan, E., Baxter, B. S., Cline, C. C., & He, B. (2019). Noninvasive neuroimaging enhances continuous neural tracking for robotic device control. *Science Robotics*, *4*(31), eaaw6844. Advance online publication. DOI: 10.1126/scirobotics.aaw6844 PMID: 31656937

Eklund, A., Nichols, T. E., & Knutsson, H. (2016). Cluster failure: Why fMRI inferences for spatial extent have inflated false-positive rates. *Proceedings of the National Academy of Sciences of the United States of America*, *113*(28), 7900–7905. DOI: 10.1073/pnas.1602413113

Endsley, M. R. (2017). From here to autonomy: Lessons learned from human–automation research. *Human Factors, 59*(1), 5–27. DOI: 10.1177/0018720816681350 PMID: 28146676

Endsley, M. R. (2021). Situation awareness in human-autonomy teaming. *Frontiers in Robotics and AI, 8*, 640043. DOI: 10.3389/frobt.2021.640043

Essen, T. V., & Meske, C. (2023). Expl(AI)ned: The impact of explainable artificial intelligence on users' information processing. *Information Systems Research, 34*(2), 460–481.

European Banking Authority. (2021). *Guidelines on loan origination and monitoring*.

European Commission. (2021). *Proposal for a regulation of the European Parliament and of the Council laying down harmonised rules on artificial intelligence (Artificial Intelligence Act). COM(2021) 206 final*. Publications Office of the European Union.

European Commission. (2024). *Regulation (EU) 2024/1689 of the European Parliament and of the Council on artificial intelligence (Artificial Intelligence Act)*. Official Journal of the European Union.

Evans, J. S. B. T. (2008). Dual-processing accounts of reasoning, judgment, and social cognition. *Annual Review of Psychology, 59*(1), 255–278. DOI: 10.1146/annurev.psych.59.103006.093629 PMID: 18154502

Evans, J. St. B. T., & Stanovich, K. E. (2013). Dual-process theories of higher cognition: Advancing the debate. *Perspectives on Psychological Science, 8*(3), 223–241. DOI: 10.1177/1745691612460685 PMID: 26172965

Everitt, B. J., & Robbins, T. W. (2016). Drug addiction: Updating actions to habits to compulsions ten years on. *Annual Review of Psychology, 67*(1), 23–50. DOI: 10.1146/annurev-psych-122414-033457

Fazli, S., Mehnert, J., Steinbrink, J., Curio, G., Villringer, A., Müller, K.-R., & Blankertz, B. (2012). Enhanced performance by a hybrid NIRS–EEG brain computer interface. *NeuroImage, 59*(1), 519–529. DOI: 10.1016/j.neuroimage.2011.07.084 PMID: 21840399

Feng, J., Jirsa, V., & Lu, W. (2024). Human brain computing and brain-inspired intelligence. *National Science Review, 11*(5), nwae144. DOI: 10.1093/nsr/nwae144 PMID: 38742233

Field, M., & Cox, W. M. (2008). Attentional bias in addictive behaviors: A review of its development, causes, and consequences. *Drug and Alcohol Dependence, 97*(1-2), 1–20. DOI: 10.1016/j.drugalcdep.2008.03.030

Flavell, J. H. (1979). Metacognition and cognitive monitoring: A new area of cognitive-developmental inquiry. *The American Psychologist, 34*(10), 906–911. DOI: 10.1037/0003-066X.34.10.906

Fleming, S. M., & Daw, N. D. (2017). Self-evaluation of decision-making: A general Bayesian framework for metacognitive computation. *Psychological Review, 124*(1), 91–114. DOI: 10.1037/rev0000045

Floridi, L., Cowls, J., Beltrametti, M., Chatila, R., Chazerand, P., Dignum, V., ... & Rossi, F. (2021). An ethical framework for a good AI Society: Opportunities, risks, principles, and recommendations. Ethics. *Governance, and Policies in Artificial Intelligence*, 19-39.

Floridi, L., Cowls, J., Beltrametti, M., Chatila, R., Chazerand, P., Dignum, V., Luetge, C., Madelin, R., Pagallo, U., Rossi, F., Schafer, B., Valcke, P., & Vayena, E. (2018). AI4People—An ethical framework for a good AI society: Opportunities, risks, principles, and recommendations. *Minds and Machines, 28*(4), 689–707. DOI: 10.1007/s11023-018-9482-5

Floridi, L., Cowls, J., Beltrametti, M., Chatila, R., Chazerand, P., Dignum, V., Luetge, C., Madelin, R., Pagallo, U., Rossi, F., Schafer, B., Valcke, P., & Vayena, E. (2022). An ethical framework for a good AI society: Opportunities, risks, principles, and recommendations. In Floridi, L. (Ed.), *Ethics, governance, and policies in artificial intelligence* (pp. 19–39). Springer., DOI: 10.1007/978-3-030-81907-1_3

Floridi, L., & Nobre, A. C. (2024). Anthropomorphising machines and computerising minds: The crosswiring of languages between Artificial Intelligence and Brain & Cognitive Sciences. *Minds and Machines, 34*(1), 5. DOI: 10.1007/s11023-024-09670-4

Fodor, J. A. (1975). *The language of thought*. Harvard University Press.

Friston, K. (2005). A theory of cortical responses. *Philosophical Transactions of the Royal Society of London. Series B, Biological Sciences, 360*(1456), 815–836. DOI: 10.1098/rstb.2005.1622 PMID: 15937014

Friston, K. (2010). The free-energy principle: A unified brain theory? *Nature Reviews. Neuroscience, 11*(2), 127–138. DOI: 10.1038/nrn2787

Frolov, N., Pitsik, E., Grubov, V., Badarin, A., Maksimenko, V., Zakharov, A., Kurkin, S., & Hramov, A. (2023). Perceptual Integration Compensates for Attention Deficit in Elderly during Repetitive Auditory-Based Sensorimotor Task. *Sensors, 23*(14). *Sensors (Basel), 23*(14), 6420. Advance online publication. DOI: 10.3390/s23146420 PMID: 37514714

Fuster, A., Goldsmith-Pinkham, P., Ramadorai, T., & Walther, A. (2022). Predictably unequal? The effects of machine learning on credit markets. *The Journal of Finance*, *77*(1), 5–47. DOI: 10.1111/jofi.13090

Fuyama, M., Khrennikov, A., & Ozawa, M. (2025). Quantum-like cognition and decision-making in the light of quantum measurement theory. *Philosophical Transactions A, 383*(2309), 20240372.

Gabriel, I. (2020). Artificial Intelligence, Values, and Alignment: I. Gabriel. *Minds and Machines*, *30*(3), 411–437. DOI: 10.1007/s11023-020-09539-2

Gal, Y. (2016). Uncertainty in deep learning [Doctoral dissertation, University of Cambridge].

Gal, Y., & Ghahramani, Z. (2016). Dropout as a Bayesian approximation: Representing model uncertainty in deep learning. *Proceedings of the 33rd International Conference on Machine Learning*, 48, 1050–1059.

Gama, J., Žliobaitė, I., Bifet, A., Pechenizkiy, M., & Bouchachia, A. (2014). A survey on concept drift adaptation. *ACM Computing Surveys*, *46*(4), 44. DOI: 10.1145/2523813

Gandolfi, D., Puglisi, F. M., Serb, A., Giugliano, M., & Mapelli, J. (2022). Brain-inspired computing: Neuroscience drives the development of new electronics and artificial intelligence. *Frontiers in Cellular Neuroscience*, *16*, 1115395. DOI: 10.3389/fncel.2022.1115395 PMID: 36605614

Garcez, A. S. d'Avila, Besold, T. R., De Raedt, L., Földiak, P., Hitzler, P., Icard, T., Kühnberger, K. U., Lamb, L. C., Miikkulainen, R., Silver, D. L., & Tran, S. N. (2015). Neural-symbolic learning and reasoning: Contributions and challenges. In *AAAI Spring Symposium Series.*

Garcez, A. S. d'Avila, Gori, M., Lamb, L. C., Serafini, L., Spranger, M., & Tran, S. N. (2019). Neural-symbolic computing: An effective methodology for principled integration of machine learning and reasoning (arXiv:1905.06088). arXiv.

Garcez, A. D. A., & Lamb, L. C. (2023). Neurosymbolic ai: The 3 rd wave. [gracez]. *Artificial Intelligence Review*, *56*(11), 12387–12406. DOI: 10.1007/s10462-023-10448-w

García, J., & Fernández, F. (2015). A comprehensive survey on safe reinforcement learning. *Journal of Machine Learning Research*, *16*, 1437–1480. https://jmlr.org/papers/v16/garcia15a.html

George, J. M., & Dane, E. (2016). Affect, emotion, and decision making. *Organizational Behavior and Human Decision Processes, 136*, 47–55. DOI: 10.1016/j.obhdp.2016.06.004

Gershman, S. J., & Blei, D. M. (2012). A tutorial on Bayesian nonparametric models. *Journal of Mathematical Psychology, 56*(1), 1–12. DOI: 10.1016/j.jmp.2011.08.004

Ghahramani, Z. (2015). Probabilistic machine learning and artificial intelligence. *Nature, 521*(7553), 452–459. DOI: 10.1038/nature14541

Gigerenzer, G. (2008). *Rationality for mortals: How people cope with uncertainty*. Oxford University Press. DOI: 10.1093/oso/9780195328981.001.0001

Gimenez, S., Millan, A., Mora-Morell, A., Ayuso, N., Gastaldo-Jordán, I., & Pardo, M. (2024). Advances in Brain Stimulation and Nanomedicine: Dopaminergic Alterations and Its Role in Neurodegeneration and Drug Addiction.

Gkintoni, E., Antonopoulou, H., Sortwell, A., & Halkiopoulos, C. (2025). Challenging cognitive load theory: The role of educational neuroscience and artificial intelligence in redefining learning efficacy. *Brain Sciences, 15*(2), 203. DOI: 10.3390/brainsci15020203 PMID: 40002535

Goddard, K., Roudsari, A., & Wyatt, J. C. (2022). Automation bias: A systematic review of frequency, effect mediators, and mitigators. *Journal of the American Medical Informatics Association : JAMIA, 19*(1), 121–127. DOI: 10.1136/amiajnl-2011-000089 PMID: 21685142

Goldstein, R. Z., & Volkow, N. D. (2011). Dysfunction of the prefrontal cortex in addiction: Neuroimaging findings and clinical implications. *Nature Reviews. Neuroscience, 12*(11), 652–669. DOI: 10.1038/nrn3119

Gong, C., Jing, C., Chen, X., Pun, C. M., Huang, G., Saha, A., Nieuwoudt, M., Li, H.-X., Hu, Y., & Wang, S. (2023). Generative AI for brain image computing and brain network computing: A review. *Frontiers in Neuroscience, 17*, 1203104. DOI: 10.3389/fnins.2023.1203104 PMID: 37383107

Gonzalez, C. (2024). Building human-like artificial agents: A general cognitive algorithm for emulating human decision-making in dynamic environments. *Perspectives on Psychological Science, 19*(5), 860–873. DOI: 10.1177/17456916231196766 PMID: 37906108

Gonzalez, C., & Heidari, H. (2025). A cognitive approach to human–AI complementarity in dynamic decision-making. *Nature Reviews Psychology, 4*(12), 1–15. DOI: 10.1038/s44159-025-00499-x

Gonzalez, C., & Malloy, T. (2026). Toward Complementary Intelligence: Integrating Cognitive and Machine AI. *Current Directions in Psychological Science*, *35*(3), 09637214251407571. DOI: 10.1177/09637214251407571

Goodfellow, I., Bengio, Y., & Courville, A. (2016). *Deep learning*. MIT Press.

Gottlieb, J., Oudeyer, P.-Y., Lopes, M., & Baranes, A. (2013). Information-seeking, curiosity, and attention: Computational and neural mechanisms. *Trends in Cognitive Sciences*, *17*(11), 585–593. DOI: 10.1016/j.tics.2013.09.001

Graves, A., Wayne, G., & Danihelka, I. (2014). Neural Turing machines. arXiv. https://arxiv.org/abs/1410.5401 (Note: Original preprint year adjusted for accuracy)

Green, B. (2022). The contestation of tech ethics: A sociotechnical approach to technology ethics in practice. *Journal of Social Computing*, *2*(3), 209–225. DOI: 10.23919/JSC.2021.0018

Green, B., & Chen, Y. (2021). Disparate interactions: An algorithm-in-the-loop analysis of fairness in risk assessments. In *Proceedings of the 2019 Conference on Fairness, Accountability, and Transparency* (pp. 90–99). DOI: 10.1145/3287560.3287563

Greene, J. (2013). *Moral tribes: Emotion, reason, and the gap between us and them*. Penguin Press.

Groman, S. M., Massi, B., Mathias, S. R., Lee, D., & Taylor, J. R. (2019). Model-free and model-based influences in addiction-related behaviors. *Biological Psychiatry*, *85*(11), 936–945. DOI: 10.1016/j.biopsych.2018.12.017

Grossberg, S., & Gutowski, W. E. (1987). Neural dynamics of decision making under risk: Affective balance and cognitive-emotional interactions. *Psychological Review*, *94*(3), 300–318. DOI: 10.1037/0033-295X.94.3.300 PMID: 3039561

Guidotti, R., Monreale, A., Ruggieri, S., Turini, F., Giannotti, F., & Pedreschi, D. (2018). A survey of methods for explaining black box models. *ACM Computing Surveys*, *51*(5), 93. Advance online publication. DOI: 10.1145/3236009

Gunning, D., Stefik, M., Choi, J., Miller, T., Stumpf, S., & Yang, G.-Z. (2021). XAI—Explainable artificial intelligence. *Science Robotics*, *4*(37), eaay7120. Advance online publication. DOI: 10.1126/scirobotics.aay7120 PMID: 33137719

Gupta, D. (2024). *Digital Twins in healthcare–A new era of innovation for businesses*. Appinventiv.

Haarnoja, T., Zhou, A., Abbeel, P., & Levine, S. (2018). Soft actor-critic: Off-policy maximum entropy deep reinforcement learning with a stochastic actor. *Proceedings of ICML*, 1861–1870. https://proceedings.mlr.press/v80/haarnoja18b.html

Hagemann, V., Rieth, M., Suresh, A., & Kirchner, F. (2023). Human-AI teams—Challenges for a team-centered AI at work. *Frontiers in Artificial Intelligence*, *6*, 1252897. DOI: 10.3389/frai.2023.1252897 PMID: 37829660

Harris, L. T. (2024). The neuroscience of human and artificial intelligence presence. *Annual Review of Psychology*, *75*(1), 433–466. DOI: 10.1146/annurev-psych-013123-123421 PMID: 37906951

Hassabis, D., Kumaran, D., Summerfield, C., & Botvinick, M. (2017). Neuroscience-inspired artificial intelligence. *Neuron*, *95*(2), 245–258. DOI: 10.1016/j.neuron.2017.06.011 PMID: 28728020

HAYASHI, R. (2022). Perspective of Brain-inspired AI Research and Multiomics Data Analysis Approach. *Kiso Shinrigaku Kenkyu*, *41*(1).

Heilig, T., & Scheer, I. (2023). *Decision intelligence: Transform your team and organization with AI-Driven decision-making*. John Wiley & Sons.

Henderson, P., Islam, R., Bachman, P., Pineau, J., Precup, D., & Meger, D. (2018, April). Deep reinforcement learning that matters. *Proceedings of the AAAI Conference on Artificial Intelligence*, *32*(1), •••. DOI: 10.1609/aaai.v32i1.11694

Henrich, J., Heine, S. J., & Norenzayan, A. (2010). The weirdest people in the world? *Behavioral and Brain Sciences*, *33*(2–3), 61–83. DOI: 10.1017/S0140525X0999152X PMID: 20550733

Hinton, G. E., Krizhevsky, A., & Sutskever, I. (2012). ImageNet classification with deep convolutional neural networks. In *Advances in Neural Information Processing Systems*.

Ho, C. W. L., ... (2023). *Regulating under uncertainty: Governance options for generative AI*. Stanford Cyber Policy Center. https://cyber.fsi.stanford.edu/content/regulating-under-uncertainty-governance-options-generative-ai

Hoff, K. A., & Bashir, M. (2015). Trust in automation: Integrating empirical evidence on factors that influence trust. *Human Factors*, *57*(3), 407–434. DOI: 10.1177/0018720814547570 PMID: 25875432

Huang, S. Y. (2025). The Connectionist Turn: How Contemporary Generative AI Reshapes Architectural Rationality. *Architecture (Washington, D.C.)*, *5*(4), 132. DOI: 10.3390/architecture5040132

Humr, S., Canan, M., & Demir, M. (2025). A quantum probability approach to improving human–AI decision-making. *Entropy (Basel, Switzerland)*, *27*(2), 152. DOI: 10.3390/e27020152 PMID: 40003149

Hutchins, E. (1995). *Cognition in the wild.* MIT Press. DOI: 10.7551/mitpress/1881.001.0001

Huys, Q. J., Browning, M., Paulus, M. P., & Frank, M. J. (2021). Advances in the computational understanding of mental illness. *Neuropsychopharmacology : Official Publication of the American College of Neuropsychopharmacology*, *46*(1), 3–19. DOI: 10.1038/s41386-020-0746-4

Huys, Q. J., Maia, T. V., & Frank, M. J. (2016). Computational psychiatry as a bridge from neuroscience to clinical applications. *Nature Neuroscience*, *19*(3), 404–413. DOI: 10.1038/nn.4238

Ilyas, A., Santurkar, S., Tsipras, D., ... (2018). Are deep policy gradient algorithms truly policy gradient algorithms? *Advances in Neural Information Processing Systems (NeurIPS)*, 31. https://papers.nips.cc/paper/2018/hash/0d1a9651497a38d8d1c05c9b2a10d3f2-Abstract.html

Jacobson, N. C., & Bhatt, P. (2022). Ethical dilemmas posed by mobile health and machine learning in psychiatry research. *Bulletin of the World Health Organization*, *98*(4), 270–276. DOI: 10.2471/BLT.19.237107 PMID: 32284651

Jago, A. S., & Carroll, G. R. (2024). Who made this? Algorithms and authorship credit. *Personality and Social Psychology Bulletin*, *50*(5), 793–806. DOI: 10.1177/01461672221149815 PMID: 36734213

Jain, S., & Wallace, B. C. (2019). Attention is not explanation. In *Proceedings of the 2019 Conference of the North American Chapter of the Association for Computational Linguistics* (pp. 3543–3556). DOI: 10.18653/v1/N19-1357

Janak, P. H., Bowers, M. S., & Corbit, L. H. (2012). Compound stimulus presentation and the norepinephrine reuptake inhibitor atomoxetine enhance long-term extinction of cocaine-seeking behavior. *Neuropsychopharmacology : Official Publication of the American College of Neuropsychopharmacology*, *37*(4), 975–985. DOI: 10.1038/npp.2011.281

Janjua, N. K., Hussain, F. K., & Hussain, O. K. (2012). Semantic information and knowledge integration through argumentative reasoning to support intelligent decision making. *Information Systems Frontiers*, *15*(2), 167–192. DOI: 10.1007/s10796-012-9365-x

Jentsch, J. D., & Taylor, J. R. (1999). Impulsivity resulting from frontostriatal dysfunction in drug abuse: Implications for the control of behavior by reward-related stimuli. *Psychopharmacology*, *146*(4), 373–390. DOI: 10.1007/PL00005483

Jeyanthi, P. M., Choudhury, T., Hack-Polay, D., Singh, T. P., & Abujar, S. (2022). *Decision Intelligence Analytics and the Implementation of Strategic Business Management*. Springer International Publishing. DOI: 10.1007/978-3-030-82763-2

Jiang, J., & Liu, X. (2022). A quantum cognition based group decision making model considering interference effects in consensus reaching process. *Computers & Industrial Engineering*, *173*, 108705. DOI: 10.1016/j.cie.2022.108705

Jiang, L., Stocco, A., Losey, D. M., Abernethy, J. A., Prat, C. S., & Rao, R. P. N. (2019). BrainNet: A Multi-Person Brain-to-Brain Interface for Direct Collaboration Between Brains. *Scientific Reports*, *9*(1), 6115. DOI: 10.1038/s41598-019-41895-7 PMID: 30992474

Jiao, L., Huang, Z., Liu, X., Yang, Y., Ma, M., Zhao, J., You, C., Hou, B., Yang, S., Liu, F., Ma, W., Li, L., Chen, P., Feng, Z., Tang, X., Guo, Y., Zhang, X., Quan, D., Wang, S., & Feng, J. (2023). Brain-inspired remote sensing interpretation: A comprehensive survey. *IEEE Journal of Selected Topics in Applied Earth Observations and Remote Sensing*, *16*, 2992–3033. DOI: 10.1109/JSTARS.2023.3247455

Jiao, L., Ma, M., He, P., Geng, X., Liu, X., Liu, F., Ma, W., Yang, S., Hou, B., & Tang, X. (2024). Brain-inspired learning, perception, and cognition: A comprehensive review. *IEEE Transactions on Neural Networks and Learning Systems*, *36*(4), 5921–5941. DOI: 10.1109/TNNLS.2024.3401711 PMID: 38809737

Jobin, A., Ienca, M., & Vayena, E. (2019). The global landscape of AI ethics guidelines. *Nature Machine Intelligence*, *1*(9), 389–399. DOI: 10.1038/s42256-019-0088-2

Josephson, C. B., Aronica, E., Beniczky, S., Boyce, D., Cavalleri, G., Denaxas, S., French, J., Jehi, L., Koh, H., Kwan, P., McDonald, C., Mitchell, J. W., Rampp, S., Sadleir, L., Sisodiya, S. M., Wang, I., Wiebe, S., Yasuda, C., & Youngerman, B. (2024). Big data research is everyone's research—Making epilepsy data science accessible to the global community: Report of the ILAE big data commission. *Epileptic Disorders*, *26*(6), 733–752. DOI: 10.1002/epd2.20288 PMID: 39446076

Joynt, R. J. (1995). Descartes' error: Emotion, reason, and the human brain. *Journal of the American Medical Association*, *273*(18), 1463–1464. DOI: 10.1001/jama.1995.03520420079045

Juliani, A., & Ash, J. T. (2024). A study of plasticity loss in on-policy deep reinforcement learning. *Advances in Neural Information Processing Systems*, *37*, 113884–113910. DOI: 10.52202/079017-3616

Kahneman, D., Sibony, O., & Sunstein, C. R. (2021). Noise: A flaw in human judgment. Little, Brown Spark.

Kahneman, D. (2011). *Thinking, fast and slow*. Farrar, Straus and Giroux.

Kahneman, D., & Tversky, A. (1979). Prospect theory: An analysis of decision under risk. *Econometrica*, *47*(2), 263–291. DOI: 10.2307/1914185

Kalaidjieva, M. A., & Swanson, G. A. (2004). Intelligence and living systems: A decision-making perspective. *Systems Research and Behavioral Science*, *21*(2), 147–172. DOI: 10.1002/sres.550

Kaneko, T. (2023, March). Decision Intelligence into Synchronization. In *2023 15th International Conference on Computer and Automation Engineering (ICCAE)* (pp. 326-332). IEEE. DOI: 10.1109/ICCAE56788.2023.10111454

Kaneko, T. (2024, August). Comparison of decision-making technologies and basic concepts of Naturally Decision Intelligence. In *International Conference on Artificial Intelligence-empowered Software Engineering* (pp. 181-188). Cham: Springer Nature Switzerland.

Kaneko, T. (2023). Naturally decision intelligence: Perfect algorithm generated by the hypothetical and synchronizing model for life system. *Intelligent Decision Technologies*, *17*(1), 195–210. DOI: 10.3233/IDT-220231

Kao, J. C., Stavisky, S. D., Sussillo, D., Nuyujukian, P., & Shenoy, K. V. (2014). Information Systems Opportunities in Brain–Machine Interface Decoders. *Proceedings of the IEEE*, *102*(5), 666–682. DOI: 10.1109/JPROC.2014.2307357

Kaur, H., & Singh, M. (2024). *Cognitive decision systems: Intelligent AI for better decisions*. Cognitive Economy.

Kavanagh, D. J., Andrade, J., & May, J. (2005). Imaginary relish and exquisite torture: The elaborated intrusion theory of desire. *Psychological Review*, *112*(2), 446–467. DOI: 10.1037/0033-295X.112.2.446

Keiflin, R., & Janak, P. H. (2015). Dopamine prediction errors in reward learning and addiction: From theory to neural circuitry. *Neuron*, *88*(2), 247–263. DOI: 10.1016/j.neuron.2015.08.037 PMID: 26494275

Kendall, A., & Gal, Y. (2017). What uncertainties do we need in Bayesian deep learning for computer vision? *Advances in Neural Information Processing Systems*, *30*, 5574–5584.

Keramati, M., & Gutkin, B. (2013). Imbalanced decision hierarchy in addicts emerging from drug-hijacked dopamine spiraling circuit. *PLoS One*, *8*(4), e61489. DOI: 10.1371/journal.pone.0061489

Khachaturian, A. S., Cassin, B., & Finney, G. R. (2023). Using clinical decision intelligence applications to improve pathways for earlier detection of underrecognized cognitive disorders. *JAR Life*, *12*, 14–17. DOI: 10.14283/jarlife.2023.3 PMID: 37234274

Khan, M. J., & Hong, K.-S. (2017). Hybrid EEG–fNIRS-Based Eight-Command Decoding for BCI: Application to Quadcopter Control. *Frontiers in Neurorobotics*, *11*. Advance online publication. DOI: 10.3389/fnbot.2017.00006 PMID: 28261084

Khrennikov, A. (2024). Contextual measurement model and quantum theory. *Royal Society Open Science*, *11*(3), 231953. DOI: 10.1098/rsos.231953 PMID: 38550757

Kim, H., Kim, J. H., Lee, Y. J., Lee, J., Han, H., Yi, H., Kim, H., Kim, H., Kang, T. W., Chung, S., Ban, S., Lee, B., Lee, H., Im, C.-H., Cho, S. J., Sohn, J. W., Yu, K. J., Kang, T. J., & Yeo, W.-H. (2025). Motion artifact–controlled micro–brain sensors between hair follicles for persistent augmented reality brain–computer interfaces. *Proceedings of the National Academy of Sciences of the United States of America*, *122*(15), e2419304122. DOI: 10.1073/pnas.2419304122 PMID: 40193612

Kiran, B. R., Sobh, I., Talpaert, V., Mannion, P., Al Sallab, A. A., Yogamani, S., & Pérez, P. (2021). Deep reinforcement learning for autonomous driving: A survey. *IEEE Transactions on Intelligent Transportation Systems*, *23*(6), 4909–4926. DOI: 10.1109/TITS.2021.3054625

Kirasirova, L. A., Zakharov, A. V., Morozova, M. V., Kaplan, A. Y., & Pyatin, V. P. (2021). ERP Correlates of Emotional Face Processing in Virtual Reality. *Opera Medica et Physiologica*, *8*(3), 3.

Koffarnus, M. N., & Bickel, W. K. (2014). A 5-trial adjusting delay discounting task: Accurate discount rates in less than one minute. *Experimental and Clinical Psychopharmacology*, *22*(3), 222–228. DOI: 10.1037/a0035973

Konova, A. B., Ceceli, A. O., Horga, G., Moeller, S. J., Alia-Klein, N., & Goldstein, R. Z. (2023). Reduced neural encoding of utility prediction errors in cocaine addiction. *Neuron*, *111*(24), 4058–4070. DOI: 10.1016/j.neuron.2023.09.015

Koob, G. F., & Volkow, N. D. (2016). Neurobiology of addiction: A neurocircuitry analysis. *The Lancet. Psychiatry*, *3*(8), 760–773. DOI: 10.1016/S2215-0366(16)00104-8

Kuhl, U., & Bush, A. (2025). When bias backfires: The modulatory role of counterfactual explanations on the adoption of algorithmic bias in XAI-supported human decision-making.

Kulkarni, A., Bhattacharyya, S., Bhattacharyya, P., & Bhattacharyya, B. (2023). Natural language uncertainty expressions for clinical AI: A randomized experiment. *NPJ Digital Medicine*, *6*, 89. DOI: 10.1038/s41746-023-00835-6

Kull, M., Filho, T. S., & Flach, P. (2019). Beta calibration: A well-founded and easily implemented improvement on logistic calibration for binary classifiers. In *Proceedings of the 20th International Conference on Artificial Intelligence and Statistics* (pp. 623–631). PMLR.

Kurniawan, D., Triyanto, D., Wahyudi, M., & Pujiastuti, L. (2023). Explainable artificial intelligence (XAI) for trustworthy decision-making. *Jurnal Teknik Informatika C.I.T Medicom, 15*(2), 45–54.

L Griffiths, T., Kemp, C., & B Tenenbaum, J. (2008). Bayesian models of cognition.

Lahat, D., Adali, T., & Jutten, C. (2015). Multimodal Data Fusion: An Overview of Methods, Challenges, and Prospects. *Proceedings of the IEEE, 103*(9), 1449–1477. DOI: 10.1109/JPROC.2015.2460697

Lai, V., Chen, C., Smith-Renner, A., Liao, Q. V., & Tan, C. (2023). Towards a science of human-AI decision making: A survey of empirical studies. In *Proceedings of the 2023 ACM Conference on Fairness, Accountability, and Transparency* (pp. 1228–1241). DOI: 10.1145/3593013.3594035

Lake, B. M., Ullman, T. D., Tenenbaum, J. B., & Gershman, S. J. (2017). Building machines that learn and think like people. *Behavioral and Brain Sciences*, *40*, e253. DOI: 10.1017/S0140525X16001837 PMID: 27881212

Lakshminarayanan, B., Pritzel, A., & Blundell, C. (2017). Simple and scalable predictive uncertainty estimation using deep ensembles. *Advances in Neural Information Processing Systems*, *30*, 6402–6413.

Landes, R. D., Christensen, D. R., & Bickel, W. K. (2012). Delay discounting decreases in those completing treatment for opioid dependence. *Experimental and Clinical Psychopharmacology*, *20*(4), 302–309. DOI: 10.1037/a0027391

Langdon, A., Botvinick, M., Nakahara, H., Tanaka, K., Matsumoto, M., & Kanai, R. (2022). Meta-learning, social cognition and consciousness in brains and machines. *Neural Networks*, *145*, 80–89. DOI: 10.1016/j.neunet.2021.10.004 PMID: 34735893

Langley, C., Cirstea, B. I., Cuzzolin, F., & Sahakian, B. J. (2022). Theory of mind and preference learning at the interface of cognitive science, neuroscience, and AI: A review. *Frontiers in Artificial Intelligence*, *5*, 778852. DOI: 10.3389/frai.2022.778852 PMID: 35493614

LeCun, Y., Bengio, Y., & Hinton, G. (2015). Deep learning. *Nature*, *521*(7553), 436–444. DOI: 10.1038/nature14539 PMID: 26017442

Lee, E., & Kim, J. (2025). A comparison between SHAP and clinician-friendly explanations reveals enhanced trust in surgical decision-making. *PMC, 12475050.*

Lee, A., Baker, T. S., Bederson, J. B., & Rapoport, B. I. (2024). Levels of autonomy in FDA-cleared surgical robots: A systematic review. *NPJ Digital Medicine*, *7*(1), 103. DOI: 10.1038/s41746-024-01102-y PMID: 38671232

Lee, J. D., & See, K. A. (2004). Trust in automation: Designing for appropriate reliance. *Human Factors*, *46*(1), 50–80. DOI: 10.1518/hfes.46.1.50.30392 PMID: 15151155

Leonelli, S. (2019). *Data-centric biology: A philosophical study.* University of Chicago Press.

Li, G., Deng, L., Tang, H., Pan, G., Tian, Y., Roy, K., & Maass, W. (2024). Brain-inspired computing: A systematic survey and future trends. *Proceedings of the IEEE.* DOI: 10.1109/JPROC.2024.3429360

Li, G., Huang, Q., Liu, C., Wang, G., Guo, L., Liu, R., & Liu, L. (2024). Fully automated diagnosis of thyroid nodule ultrasound using brain-inspired inference. *Neurocomputing*, *582*, 127497. DOI: 10.1016/j.neucom.2024.127497

Linxen, S., Sturm, C., Brühlmann, F., Cassau, V., Opwis, K., & Reinecke, K. (2021). How WEIRD is CHI? In Proceedings of the 2021 CHI Conference on Human Factors in Computing Systems (Article 143). DOI: 10.1145/3411764.3445488

Lipton, Z. C. (2018). *The mythos of model interpretability* (arXiv:1606.03490). arXiv.

Liu, B., Li, X., Zhang, J., Wang, J., He, T., Hong, S., . . . Zhu, K. (2025). Advances and challenges in foundation agents: From brain-inspired intelligence to evolutionary, collaborative, and safe systems. *arXiv preprint arXiv:2504.01990.*

Liu, Y., Wang, T., Wang, K., & Zhang, Y. (2021). Collaborative Learning Quality Classification Through Physiological Synchrony Recorded by Wearable Biosensors. *Frontiers in Psychology*, *12*, 674369. DOI: 10.3389/fpsyg.2021.674369 PMID: 33995232

Liu, Z., Gan, E., & Tegmark, M. (2023). Seeing is believing: Brain-inspired modular training for mechanistic interpretability. *Entropy (Basel, Switzerland)*, *26*(1), 41. DOI: 10.3390/e26010041 PMID: 38248167

Logg, J. M., Minson, J. A., & Moore, D. A. (2019). Algorithm appreciation: People prefer algorithmic to human judgment. *Organizational Behavior and Human Decision Processes*, *151*, 90–103. DOI: 10.1016/j.obhdp.2018.12.005

Long, D., & Zou, J. (2024). Sensible AI: Re-imagining interpretability and explainability using sensemaking theory. *Proceedings of the 2022 ACM Conference on Fairness, Accountability, and Transparency*, 789–801.

Lotte, F., Bougrain, L., Cichocki, A., Clerc, M., Congedo, M., Rakotomamonjy, A., & Yger, F. (2018). A review of classification algorithms for EEG-based brain–computer interfaces: A 10 year update. *Journal of Neural Engineering*, *15*(3), 031005. DOI: 10.1088/1741-2552/aab2f2 PMID: 29488902

Lundberg, S. M., & Lee, S.-I. (2017). A unified approach to interpreting model predictions. *Advances in Neural Information Processing Systems*, *30*, 4765–4774.

Lundberg, S. M., Nair, B., Vavilala, M. S., Horibe, M., Eisses, M. J., Adams, T., Liston, D. E., Low, D. K.-W., Newman, S.-F., Kim, J., & Lee, S.-I. (2020). Explainable machine-learning predictions for the prevention of hypoxaemia during surgery. *Nature Biomedical Engineering*, *2*(10), 749–760. DOI: 10.1038/s41551-018-0304-0 PMID: 31001455

Luo, G., Yuan, Q., Li, J., Wang, S., & Yang, F. (2022). Artificial intelligence powered mobile networks: From cognition to decision. *IEEE Network*, *36*(3), 136–144. DOI: 10.1109/MNET.013.2100087

Lyell, D., & Coiera, E. (2017). Automation bias and verification complexity: A systematic review. *Journal of the American Medical Informatics Association : JAMIA*, *24*(2), 423–431. DOI: 10.1093/jamia/ocw105 PMID: 27516495

MacKillop, J., Amlung, M. T., Few, L. R., Ray, L. A., Sweet, L. H., & Munafò, M. R. (2011). Delayed reward discounting and addictive behavior: A meta-analysis. *Psychopharmacology*, *216*(3), 305–321. DOI: 10.1007/s00213-011-2229-0

Madaio, M. A., Stark, L., Wortman Vaughan, J., & Wallach, H. (2020). Co-designing checklists to understand organizational challenges and opportunities around fairness in AI. In *Proceedings of the 2020 CHI Conference on Human Factors in Computing Systems* (pp. 1–14). DOI: 10.1145/3313831.3376445

Mahadeen, T., Galanakis, K., Samara, E., & Kilintzis, P. (2020). Heuristics and evidences decision (heed) making: A case study in a systemic model for transforming decision making from heuristics-based to evidenced-based. *Journal of the Knowledge Economy*, *12*(4), 1668–1693. DOI: 10.1007/s13132-020-00688-4

Marchant, N. J., Furlong, T. M., & McNally, G. P. (2010). Medial dorsal hypothalamus mediates the inhibition of reward seeking after extinction. *The Journal of Neuroscience : The Official Journal of the Society for Neuroscience*, *30*(42), 14102–14115. DOI: 10.1523/JNEUROSCI.4079-10.2010

Marcus, G. (2018). *Deep learning: A critical appraisal* (arXiv:1801.00631). arXiv.

Marcus, G., & Davis, E. (2020). *Rebooting AI: Building artificial intelligence we can trust*. Pantheon Books.

Maskey, M. (2023). Rethinking AI for Science: An Evolution From Data-Driven to Data-Centric Framework. *Perspectives of Earth and Space Scientists, 4*(1), e2023CN000222.

Maslova, O., Komarova, Y., Shusharina, N., Kolsanov, A., Zakharov, A., Garina, E., & Pyatin, V. (2023). Non-invasive EEG-based BCI spellers from the beginning to today: A mini-review. *Frontiers in Human Neuroscience*, *17*, 1216648. DOI: 10.3389/fnhum.2023.1216648 PMID: 37680264

Mastrianni, A., Sarcevic, A., Hu, A., Almengor, L., Tempel, P., Gao, S., & Burd, R. S. (2023). Transitioning Cognitive Aids into Decision Support Platforms: Requirements and Design Guidelines. *ACM Transactions on Computer-Human Interaction*, *30*(3), 1–28. DOI: 10.1145/3582431 PMID: 37694216

Mayer, A., Greif, L., Häußermann, T. M., Otto, S., Kastner, K., El Bobbou, S., Chardonnet, J.-R., Reichwald, J., Fleischer, J., & Ovtcharova, J. (2025). Digital Twins, Extended Reality, and Artificial Intelligence in Manufacturing Reconfiguration: A Systematic Literature Review. *Sustainability (Basel)*, *17*(5), 2318. DOI: 10.3390/su17052318

McDuff, D., Gontarek, S., & Picard, R. (2014). Remote measurement of cognitive stress via heart rate variability. *2014 36th Annual International Conference of the IEEE Engineering in Medicine and Biology Society*, 2957–2960. DOI: 10.1109/EMBC.2014.6944243

Meghdadi, A., Akbarzadeh-T, M. R., & Javidan, K. (2022). A quantum-like cognitive approach to modeling human biased selection behavior. *Scientific Reports*, *12*(1), 22545. DOI: 10.1038/s41598-022-13757-2 PMID: 36581629

Mehonic, A., & Kenyon, A. J. (2022). Brain-inspired computing needs a master plan. *Nature*, *604*(7905), 255–260. DOI: 10.1038/s41586-021-04362-w PMID: 35418630

Mehrabi, N., Morstatter, F., Saxena, N., Lerman, K., & Galstyan, A. (2021). A survey on bias and fairness in machine learning. *ACM Computing Surveys*, *54*(6), 115. DOI: 10.1145/3457607

Meng, L., Zhang, W., Xu, F., Han, L., Guo, S., Chen, T., . . . Wang, F. Y. (2025, November). Brain-Inspired Parallel Data Intelligence: a Generality-Specificity Fused Approach. In *2025 21st IEEE International Conference on Mechatronic and Embedded Systems and Applications (MESA)* (pp. 71-77). IEEE. DOI: 10.1109/MESA68091.2025.11278866

Messner, W., & Siering, M. (2025). The cognitive cost of AI: How AI anxiety and attitudes influence cognitive strain. *Journal of Business Research, 189*, 114–132.

Mikus, N., Korb, S., Massaccesi, C., Gausterer, C., Graf, I., Willeit, M., Eisenegger, C., Lamm, C., Silani, G., & Mathys, C. (2022). Effects of dopamine D2/3 and opioid receptor antagonism on the trade-off between model-based and model-free behaviour in healthy volunteers. *eLife, 11*, e79661. DOI: 10.7554/eLife.79661

Miller, E. K., & Cohen, J. D. (2001). An integrative theory of prefrontal cortex function. *Annual Review of Neuroscience, 24*(1), 167–202. DOI: 10.1146/annurev.neuro.24.1.167

Miller, T. (2019). Explanation in artificial intelligence: Insights from the social sciences. *Artificial Intelligence, 267*, 1–38. DOI: 10.1016/j.artint.2018.07.007

Millidge, B., Tschantz, A., & Buckley, C. L. (2020). Predictive coding approximates backprop along arbitrary computation graphs. arXiv preprint arXiv:2006.04182.

Minh, D., Wang, H. X., Li, Y. F., & Nguyen, T. N. (2022). Explainable artificial intelligence: A comprehensive review. *Artificial Intelligence Review, 55*(5), 3503–3568. DOI: 10.1007/s10462-021-10088-y

Mitchell, D. G. V. (2011). The nexus between decision making and emotion regulation: A review of convergent neurocognitive substrates. *Behavioural Brain Research, 217*(1), 215–231. DOI: 10.1016/j.bbr.2010.10.030 PMID: 21055420

Mitchell, J. M., & Ot'alora, G. (2023). MDMA-assisted therapy for moderate to severe PTSD: A randomized, placebo-controlled phase 3 trial. *Nature Medicine, 29*(10), 2473–2480. DOI: 10.1038/s41591-023-02565-4

Mittelstadt, B. (2019). Principles alone cannot guarantee ethical AI. *Nature Machine Intelligence, 1*(11), 501–507. DOI: 10.1038/s42256-019-0114-4

Mittelstadt, B. D., Allo, P., Taddeo, M., Wachter, S., & Floridi, L. (2016). The ethics of algorithms: Mapping the debate. *Big Data & Society, 3*(2), 1–21. DOI: 10.1177/2053951716679679

Mittelstadt, B., Russell, C., & Wachter, S. (2023). Explaining explanations in AI. In *Proceedings of the Conference on Fairness, Accountability, and Transparency* (pp. 279–288). DOI: 10.1145/3287560.3287574

Mnih, V., Kavukcuoglu, K., Silver, D., Rusu, A. A., Veness, J., Bellemare, M. G., Graves, A., Riedmiller, M., Fidjeland, A. K., Ostrovski, G., Petersen, S., Beattie, C., Sadik, A., Antonoglou, I., King, H., Kumaran, D., Wierstra, D., Legg, S., & Hassabis, D. (2015). Human-level control through deep reinforcement learning. *Nature*, *518*(7540), 529–533. DOI: 10.1038/nature14236 PMID: 25719670

Molnar, C. (2022). *Interpretable machine learning: A guide for making black box models explainable* (2nd ed.)., https://christophm.github.io/interpretable-ml-book/

Moore, J. H., & Williamson, C. (2022). Artificial intelligence explainability in clinical decision support systems. *PMC, 9931364.*

Morley, J., Cowls, J., Taddeo, M., & Floridi, L. (2021). The ethics of AI in health care: A mapping review. *Social Science & Medicine*, *260*, 113172. DOI: 10.1016/j.socscimed.2020.113172 PMID: 32702587

Morreale, M. K. (2024). Cannabis Use and Cognition. *The Other Side of Cannabis: Impact on Mental and Physical Health*, 124.

Moy, A. J., Schwartz, J. M., Chen, R. J., Sadri, H., Lucas, E., Cato, K. D., & Rossetti, S. C. (2021). Measurement of clinical documentation burden among physicians and nurses using electronic health records: A scoping review. *Journal of the American Medical Informatics Association : JAMIA*, *28*(5), 998–1008. DOI: 10.1093/jamia/ocaa325 PMID: 33434273

Murthy, B. K., Reddy, B. V. R., Hasteer, N., & Van Belle, J. P. (2023). Decision Intelligence. *InCITe, 1.*

National Institute of Standards and Technology. (2023). *Artificial intelligence risk management framework (AI RMF 1.0) (NIST AI 100-1).* U.S. Department of Commerce., DOI: 10.6028/NIST.AI.100-1

Neerincx, M. A., van den Bosch, K., Haije, T., Doesburg, M., Doornbos, G., Grootjen, M., & Looije, R. (2023). Maintaining situation awareness in human-agent teamwork: The value of functional states. *Frontiers in Neuroergonomics*, *4*, 1–17. DOI: 10.3389/fnrgo.2023.958930

Newell, A., & Simon, H. A. (1976). Computer science as empirical inquiry: Symbols and search. *Communications of the ACM*, *19*(3), 113–126. DOI: 10.1145/360018.360022

Nie, D., Hu, Z., Zhu, D., & Yang, J. (2024). Probabilistic Representation Differences between Decisions from Description and Decisions from Experience. *Journal of Intelligence, 12*(9), 89. DOI: 10.3390/jintelligence12090089 PMID: 39330468

Nikoloska, I., & Simeone, O. (2024). An introduction to Bayesian simulation-based inference for quantum machine learning with examples. *Frontiers in Quantum Science and Technology, 3*, 1394533. DOI: 10.3389/frqst.2024.1394533

Nilsson, N. J. (1980). *Principles of artificial intelligence*. Morgan Kaufmann.

Nisbett, R. E., & Masuda, T. (2003). Culture and point of view. *Proceedings of the National Academy of Sciences of the United States of America, 100*(19), 11163–11170. DOI: 10.1073/pnas.1934527100 PMID: 12960375

NIST. (2022). *Towards a standard for identifying and managing bias in artificial intelligence* (NIST SP 1270). National Institute of Standards and Technology.

Nordström, M. (2022). AI under great uncertainty: Implications and decision strategies for public policy. *AI & Society, 37*(4), 1703–1714. DOI: 10.1007/s00146-021-01263-4

Nozari, H., Abdi, H., & Szmelter-Jarosz, A. (2025). A Neuromorphic and Bio-Inspired Framework for Optimization in Multi-Echelon Supply Networks. *Transformative Science, 1*(1), 9–17.

O'Callaghan, M. (2023). Decision intelligence framework for organizational decision-making. In *Decision Intelligence* (pp. 195–218). Chapman and Hall/CRC., DOI: 10.1201/b23322-7

O'Callaghan, M. (2023). *Decision intelligence: human–machine integration for decision-making*. CRC Press. DOI: 10.1201/b23322

O'Keane, V. (2021). *The rag and bone shop: How we make memories and memories make us*. Allen Lane.

O'Neil, C. (2016). *Weapons of math destruction: How big data increases inequality and threatens democracy*. Crown.

Obermeyer, Z., Powers, B., Vogeli, C., & Mullainathan, S. (2019). Dissecting racial bias in an algorithm used to manage the health of populations. *Science, 366*(6464), 447–453. DOI: 10.1126/science.aax2342 PMID: 31649194

Ochsner, K. N., & Schacter, D. L. (2000). A social cognitive neuroscience approach to emotion and memory. In *The Neuropsychology Of Emotion* (pp. 163–193). Oxford University PressNew York., DOI: 10.1093/oso/9780195114645.003.0007

OECD. (2024). *OECD Framework for the Governance of Artificial Intelligence Systems*. Organisation for Economic Co-operation and Development.

Onciul, R., Tataru, C.-I., Dumitru, A. V., Crivoi, C., Serban, M., Covache-Busuioc, R.-A., Radoi, M. P., & Toader, C. (2025). Artificial intelligence and neuroscience: Transformative synergies in brain research and clinical applications. *Journal of Clinical Medicine*, *14*(2), 550. DOI: 10.3390/jcm14020550 PMID: 39860555

Ovadia, Y., Fertig, E., Ren, J., Nado, Z., Sculley, D., Nowozin, S., Dillon, J. V., Lakshminarayanan, B., & Snoek, J. (2019). Can you trust your model's uncertainty? Evaluating predictive uncertainty under dataset shift. *Advances in Neural Information Processing Systems*, *32*, 13991–14002.

Pahari, A., Kar, S. K., Joshi, B., & Ghosh, R. (2025). Human Brain versus Artificial Intelligence: A Physiological Perspective on Cognition, Learning and Information Processing. *Journal of Universal College of Medical Sciences*, *13*(02), 42–48. DOI: 10.3126/jucms.v13i02.84273

Palminteri, S., Wyart, V., & Koechlin, E. (2017). The importance of falsification in computational cognitive modeling. *Trends in Cognitive Sciences*, *21*(6), 425–433. DOI: 10.1016/j.tics.2017.03.011

Parasuraman, R., & Riley, V. (1997). Humans and automation: Use, misuse, disuse, abuse. *Human Factors*, *39*(2), 230–253. DOI: 10.1518/001872097778543886

Parente, D. J. (2024). Generative artificial intelligence and large language models in primary care medical education. *Family Medicine*, *56*(9), 534–540. DOI: 10.22454/FamMed.2024.775525

Parisi, G. I., Kemker, R., Part, J. L., Kanan, C., & Wermter, S. (2019). Continual lifelong learning with neural networks: A review. *Neural Networks*, *113*, 54–71. DOI: 10.1016/j.neunet.2019.01.012 PMID: 30780045

Park, T. J., Deng, S., Manna, S., Islam, A. N., Yu, H., Yuan, Y., Fong, D. D., Chubykin, A. A., Sengupta, A., Sankaranarayanan, S. K. R. S., & Ramanathan, S. (2023). Complex oxides for brain-inspired computing: A review. *Advanced Materials*, *35*(37), 2203352. DOI: 10.1002/adma.202203352 PMID: 35723973

Passi, S., & Barocas, S. (2019). Problem formulation and fairness. In *Proceedings of the Conference on Fairness, Accountability, and Transparency* (pp. 39–48). DOI: 10.1145/3287560.3287567

Pearl, J. (2009). *Causality: Models, reasoning, and inference* (2nd ed.). Cambridge University Press. DOI: 10.1017/CBO9780511803161

Pedro, F. (2023). Theoretical Analysis of the Brain and Artificial Intelligence. *Journal of Robotics Spectrum, 1*, 024-035.

Perner, J., & Roessler, J. (2012). From infants' to children's appreciation of belief. *Trends in Cognitive Sciences, 16*(10), 519–525. DOI: 10.1016/j.tics.2012.08.004

Pessach, D., & Shmueli, E. (2022). A review on fairness in machine learning. *ACM Computing Surveys, 55*(3), 51. DOI: 10.1145/3494672

Petrik, M., & Varnum, M. (2024). *Cultural bias in explainable AI research: A systematic analysis. Research Portal.* Utrecht University.

Pezzulo, G., Parr, T., & Friston, K. (2021). Active inference in robotics and artificial agents: Survey and challenges. arXiv preprint arXiv:2112.01871.

Pezzulo, G., Rigoli, F., & Friston, K. (2018). Hierarchical active inference. *Trends in Cognitive Sciences, 22*(4), 294–306. DOI: 10.1016/j.tics.2018.01.009

Phillips-Wren, G. (2013). Intelligent Decision Support Systems. In Doumpos, M., & Grigoroudis, E. (Eds.), *Multicriteria Decision Aid and Artificial Intelligence* (1st ed., pp. 25–44). Wiley., DOI: 10.1002/9781118522516.ch2

Picton, T. W. (1992). The P300 Wave of the Human Event-Related Potential. *Journal of Clinical Neurophysiology, 9*(4), 456–479. DOI: 10.1097/00004691-199210000-00002 PMID: 1464675

Poldrack, R. A., Huckins, G., & Varoquaux, G. (2020). Establishment of best practices for evidence for prediction: A review. *JAMA Psychiatry, 77*(5), 534. DOI: 10.1001/jamapsychiatry.2019.3671

Posner, M. I., & Petersen, S. E. (1990). The attention system of the human brain. *Annual Review of Neuroscience, 13*(1), 25–42. DOI: 10.1146/annurev.ne.13.030190.000325

Poursabzi-Sangdeh, F., Goldstein, D. G., Hofman, J. M., Vaughan, J. W., & Wallach, H. (2021). Manipulating and measuring model interpretability. In Proceedings of the 2021 CHI Conference on Human Factors in Computing Systems (Article 237). DOI: 10.1145/3411764.3445315

Poursabzi-Sangdeh, F., Goldstein, D. G., Hill, J. M., & Lacotte, S. (2021). Manipulating and measuring model interpretability. *Proceedings of the 2021 CHI Conference on Human Factors in Computing Systems*, 1–52.

Pratt, L., Bisson, C., & Warin, T. (2023). Bringing advanced technology to strategic decision-making: The Decision Intelligence/Data Science (DI/DS) Integration framework. *Futures, 152*, 103217. DOI: 10.1016/j.futures.2023.103217

Prell, R. (2025). *Self-Regulatory Challenges in Decision-Making: Cardiac Autonomic Modulation in Ambiguous, Risky, and Moral Situations* (Doctoral dissertation, Dissertation, Duisburg, Essen, Universität Duisburg-Essen, 2025).

Pritzel, A., Uria, B., Srinivasan, S., Puigdomènech, A., Vinyals, O., Hassabis, D., Wierstra, D., & Blundell, C. (2017). Neural episodic control. *Proceedings of the 34th International Conference on Machine Learning*, 70, 2827–2836.

Pylyshyn, Z. W. (1984). *Computation and cognition: Toward a foundation for cognitive science*. MIT Press. DOI: 10.7551/mitpress/2004.001.0001

Qian, S. U. N. (2025, June). Research on the Application and Development Trend of Artificial Intelligence Based Decision Making Intelligence. In *2025 5th International Conference on Artificial Intelligence, Big Data and Algorithms (CAIBDA)* (pp. 1370-1373). IEEE.

Rabanser, S., Günnemann, S., & Lipton, Z. C. (2019). Failing loudly: An empirical study of methods for detecting dataset shift. *Advances in Neural Information Processing Systems*, *32*, 1396–1408.

Rader, E., Cotter, K., & Cho, J. (2018). Explanations as mechanisms for supporting algorithmic transparency. In *Proceedings of the 2018 CHI Conference on Human Factors in Computing Systems* (Paper 103). DOI: 10.1145/3173574.3173677

Ragni, F., Tucciarelli, R., Andersson, P., & Lingnau, A. (2020). Decoding stimulus identity in occipital, parietal and inferotemporal cortices during visual mental imagery. *Cortex*, *127*, 371–387. DOI: 10.1016/j.cortex.2020.02.020 PMID: 32289581

Rahwan, I., Veloso, M., & Tambe, M.. (2019). Machine behaviour. *Nature*, *568*(7753), 477–486. DOI: 10.1038/s41586-019-1138-y

Raikov, A. N., & Pirani, M. (2022). Human-machine duality: What's next in cognitive aspects of artificial intelligence? *IEEE Access : Practical Innovations, Open Solutions*, *10*, 56296–56315. DOI: 10.1109/ACCESS.2022.3177657

Raja, G. B. (2025a). Deep learning techniques for gene selection and cancer classification: a detailed review. *Computational Intelligence for Genomics Data*, 79-96.

Raja, G. B., & Chakraborty, C. (2023). Internet of things based effective wearable healthcare monitoring system for remote areas. In *Implementation of Smart Healthcare Systems using AI, IoT, and Blockchain* (pp. 193-218). Elsevier. DOI: 10.1016/B978-0-323-91916-6.00004-7

Raja, G. B. (2022a). Deep learning algorithms for real-time healthcare monitoring systems. In *Convergence of deep learning and artificial intelligence in internet of things* (pp. 1–18). CRC Press. DOI: 10.1201/9781003355960-1

Raja, G. B. (2022b). Early detection of breast cancer using efficient image processing algorithms and prediagnostic techniques: A detailed approach. In *Cognitive Systems and Signal Processing in Image Processing* (pp. 223–251). Elsevier. DOI: 10.1016/B978-0-12-824410-4.00009-X

Raja, G. B. (2025b). SmartCare—A comprehensive wearable sensor-based remote health monitoring and rehabilitation system. In *Rehabilitation Robotics and Healthcare Devices* (pp. 27–46). Elsevier. DOI: 10.1016/B978-0-443-21505-6.00011-6

Rajak, M. K., Dey, S., Kumar, V., Pratihar, D. K., Singh, V. P., & Islam, S. (2025). Foundations of decision intelligence in healthcare: Integrating machine learning, robotics, and biomedical analytics. In *Advancing Healthcare through Decision Intelligence* (pp. 1–7). Academic Press. DOI: 10.1016/B978-0-443-26480-1.00010-2

Raji, I. D., Kumar, I. E., Horowitz, A., & Selbst, A. (2022). The fallacy of AI functionality. In *Proceedings of the 2022 ACM Conference on Fairness, Accountability, and Transparency* (pp. 959–972). DOI: 10.1145/3531146.3533158

Raji, I. D., Smart, A., White, R. N., Mitchell, M., Gebru, T., Hutchinson, B., & Barnes, P. (2020). Closing the AI accountability gap: Defining an end-to-end framework for internal algorithmic auditing. In *Proceedings of the 2020 conference on fairness, accountability, and transparency* (pp. 33-44). DOI: 10.1145/3351095.3372873

Rajpurkar, P., Chen, E., Banerjee, O., & Topol, E. J. (2022). AI in health and medicine. *Nature Medicine, 28*(1), 31–38. DOI: 10.1038/s41591-021-01614-0 PMID: 35058619

Ramachandram, D., & Taylor, G. W. (2017). Deep Multimodal Learning: A Survey on Recent Advances and Trends. *IEEE Signal Processing Magazine, 34*(6), 96–108. DOI: 10.1109/MSP.2017.2738401

Ramaekers, J. G., Kuypers, K. P., & Vollenweider, F. X. (2026). Subjective and neurocognitive profiling of clinical doses of 3, 4-methylenedioxymethamphetamine (MDMA) in healthy volunteers: Implications for therapeutic use. *Molecular Psychiatry, 31*(8), 1–12. DOI: 10.1038/s41380-026-03602-7

Ramakrishnan, S. A., Muliya, M., Chakravarthy, V. S., & Parvaz, M. A. (2026). Applications and Evolution of Neuroeconomics and Reinforcement Learning Models in Substance Use Disorders. In *Neuroeconomics: Core Topics and Current Directions: NA* (pp. 611–627). Springer Nature Switzerland. DOI: 10.1007/978-3-032-02925-6_33

Raman, R., Kowalski, R., Achuthan, K., Iyer, A., & Nedungadi, P. (2025). Navigating artificial general intelligence development: Societal, technological, ethical, and brain-inspired pathways. *Scientific Reports*, *15*(1), 8443. DOI: 10.1038/s41598-025-92190-7 PMID: 40069265

Rangel, A., Camerer, C., & Montague, P. R. (2008). A framework for studying the neurobiology of value-based decision making. *Nature Reviews. Neuroscience*, *9*(7), 545–556. DOI: 10.1038/nrn2357

Redish, A. D. (2004). Addiction as a computational process gone awry. *Science*, *306*(5703), 1944–1947. DOI: 10.1126/science.1102384

Redish, A. D., Jensen, S., & Johnson, A. (2008). Addiction as vulnerabilities in the decision process. *Behavioral and Brain Sciences*, *31*(4), 461–487. DOI: 10.1017/S0140525X08004986

Regulatory Affairs Journal. (2023). Explainable Artificial Intelligence (XAI) Framework for Transparent Clinical Decision Support. *International Journal of Modern Trends in Life Sciences, 6*(4), 1427.

Ren, J., & Xia, F. (2024). Brain-inspired artificial intelligence: A comprehensive review. *arXiv preprint arXiv:2408.14811.*

Ren, M., Chen, N., & Qiu, H. (2023). Human-machine collaborative decision-making: An evolutionary roadmap based on cognitive intelligence. *International Journal of Social Robotics*, *15*(7), 1101–1114. DOI: 10.1007/s12369-023-01020-1

Renton, A. I., Mattingley, J. B., & Painter, D. R. (2019). Optimising non-invasive brain-computer interface systems for free communication between naïve human participants. *Scientific Reports*, *9*(1), 18705. DOI: 10.1038/s41598-019-55166-y PMID: 31822715

Ribeiro, M. T., Singh, S., & Guestrin, C. (2018). "Why should I trust you?": Explaining the predictions of any classifier. *Communications of the ACM*, *61*(8), 112–123.

Roberts, M., Driggs, D., Thorpe, M., Gilbey, J., Yeung, M., Ursprung, S., Aviles-Rivero, A. I., Etmann, C., McCague, C., Beer, L., Weir-McCall, J. R., Teng, Z., Gkrania-Klotsas, E., Ruggiero, A., Korhonen, A., Jefferson, E., Ako, E., Langs, G., Gozaliasl, G., & Schönlieb, C.-B.AIX-COVNET Collaboration. (2021). Common pitfalls and recommendations for using machine learning to detect and prognosticate for COVID-19 using chest radiographs and CT scans. *Nature Machine Intelligence*, *3*(3), 199–217. DOI: 10.1038/s42256-021-00307-0

Robinson, T. E., & Berridge, K. C. (2008). The incentive sensitization theory of addiction: Some current issues. *Philosophical Transactions of the Royal Society of London. Series B, Biological Sciences*, *363*(1507), 3137–3146. DOI: 10.1098/rstb.2008.0093

Rocktäschel, T., & Riedel, S. (2017). *End-to-end differentiable proving* (arXiv:1705.11040). arXiv.

Rodrigues, F. D. A. A., Ferreira, R. S., Cruz, L. N., & Pagnoncelli, C. J. (2025). Clinical Neuroscience of Addiction: Neurobiological Disruptions and Translational Models of Social Dysfunction. *Ciencia Latina Revista Científica Multidisciplinar*, *9*(3), 8215–8239. DOI: 10.37811/cl_rcm.v9i3.18443

Romeira, B., Adão, R., Nieder, J. B., Al-Taai, Q., Zhang, W., Hadfield, R. H., Wasige, E., Hejda, M., Hurtado, A., Malysheva, E., Dolores Calzadilla, V., Lourenço, J., Castro Alves, D., Figueiredo, J. M. L., Ortega-Piwonka, I., Javaloyes, J., Edwards, S., Davies, J. I., Horst, F., & Offrein, B. J. (2023). Brain-inspired nanophotonic spike computing: Challenges and prospects. *Neuromorphic Computing and Engineering*, *3*(3), 033001. DOI: 10.1088/2634-4386/acdf17

Rosen, C., & O'Leary, C. (2024). Explainability in AI policies: A critical review of communications, reports, regulations, and standards in the EU, US, and UK. *ACM Computing Surveys*, *56*(8), 1–36.

Rottenberg, J., Wilhelm, F. H., Gross, J. J., & Gotlib, I. H. (2003). Vagal rebound during resolution of tearful crying among depressed and nondepressed individuals. *Psychophysiology*, *40*(1), 1–6. DOI: 10.1111/1469-8986.00001 PMID: 12751799

Rounsevell, M. D., Arneth, A., Brown, C., Cheung, W. W., Gimenez, O., Holman, I., Leadley, P., Luján, C., Mahevas, S., Maréchaux, I., Pélissier, R., Verburg, P. H., Vieilledent, G., Wintle, B. A., & Shin, Y. J. (2021). Identifying uncertainties in scenarios and models of socio-ecological systems in support of decision-making. *One Earth*, *4*(7), 967–985. DOI: 10.1016/j.oneear.2021.06.003

Roy, Y., Banville, H., Albuquerque, I., Gramfort, A., Falk, T. H., & Faubert, J. (2019). Deep learning-based electroencephalography analysis: A systematic review. *Journal of Neural Engineering*, *16*(5), 051001. DOI: 10.1088/1741-2552/ab260c PMID: 31151119

Rudin, C. (2019). Stop explaining black box machine learning models for high stakes decisions and use interpretable models instead. *Nature Machine Intelligence*, *1*(5), 206–215. DOI: 10.1038/s42256-019-0048-x PMID: 35603010

Rumelhart, D. E., Hinton, G. E., & Williams, R. J. (1986). Learning representations by back-propagating errors. *Nature*, *323*(6088), 533–536. DOI: 10.1038/323533a0

Russell, S. (2019). *Human compatible: Artificial intelligence and the problem of control*. Viking.

Russell, S. J., & Norvig, P. (2021). *Artificial intelligence: A modern approach* (4th ed.). Pearson.

Sado, F., & Takac, B. (2023). Explainable goal-driven agents and robots. *ACM Transactions on Interactive Intelligent Systems*, *13*(2), 1–34.

Saeidnia, H. R., Hashemi Fotami, S. G., Lund, B., & Ghiasi, N. (2024). Ethical considerations in artificial intelligence interventions for mental health and well-being: Ensuring responsible implementation and impact. *Social Sciences (Basel, Switzerland)*, *13*(7), 381. DOI: 10.3390/socsci13070381

Sakagami, M., Pan, X., & Uttl, B. (2006). Behavioral inhibition and prefrontal cortex in decision-making. *Neural Networks*, *19*(8), 1255–1265. DOI: 10.1016/j.neunet.2006.05.040 PMID: 16952442

Sandini, G., Sciutti, A., & Morasso, P. (2024). Artificial cognition vs. artificial intelligence for next-generation autonomous robotic agents. *Frontiers in Computational Neuroscience*, *18*, 1349408. DOI: 10.3389/fncom.2024.1349408 PMID: 38585280

Santello, M., Toni, N., & Volterra, A. (2019). Astrocyte function from information processing to cognition and cognitive impairment. *Nature Neuroscience*, *22*(2), 154–166. DOI: 10.1038/s41593-018-0325-8 PMID: 30664773

Saw, S. N., Yan, Y. Y., & Ng, K. H. (2025). Current status and future directions of explainable artificial intelligence in medical imaging. *European Journal of Radiology*, *183*, 111884. DOI: 10.1016/j.ejrad.2024.111884 PMID: 39667118

Schemmer, M., Hemmer, P., Kühl, N., Benz, C., & Vössing, M. (2023). Appropriate reliance on AI advice: Conceptualization and the effect of explanations. In *Proceedings of the 28th International Conference on Intelligent User Interfaces* (pp. 410–422). DOI: 10.1145/3581641.3584066

Schmidgall, S., Ziaei, R., Achterberg, J., Kirsch, L., Hajiseyedrazi, S., & Eshraghian, J. (2024). Brain-inspired learning in artificial neural networks: A review. *APL Machine Learning*, *2*(2), 021501. DOI: 10.1063/5.0186054

Scholkmann, F., Kleiser, S., Metz, A. J., Zimmermann, R., Mata Pavia, J., Wolf, U., & Wolf, M. (2014). A review on continuous wave functional near-infrared spectroscopy and imaging instrumentation and methodology. *NeuroImage*, *85*, 6–27. DOI: 10.1016/j.neuroimage.2013.05.004 PMID: 23684868

Schulman, J., Wolski, F., Dhariwal, P., Radford, A., & Klimov, O. (2017). Proximal policy optimization algorithms. *arXiv preprint arXiv:1707.06347*. https://arxiv.org/abs/1707.06347

Schultz, W. (2024). The dopamine reward prediction error signal. SSRN *Electronic Journal*. DOI: 10.2139/ssrn.4816666

Schultz, W. (2016). Dopamine reward prediction error coding. *Dialogues in Clinical Neuroscience*, *18*(1), 23–32. DOI: 10.31887/DCNS.2016.18.1/wschultz

Schultz, W., Dayan, P., & Montague, P. R. (1997). A neural substrate of prediction and reward. *Science*, *275*(5306), 1593–1599. DOI: 10.1126/science.275.5306.1593

Sculley, D., Holt, G., Golovin, D., Davydov, E., Phillips, T., Ebner, D., Chaudhary, V., Young, M., Crespo, J.-F., & Dennison, D. (2015). Hidden technical debt in machine learning systems. *Advances in Neural Information Processing Systems*, *28*, 2503–2511.

Seaver, N. (2021). Care and scale: Decorrelative ethics in algorithmic recommendation. *Cultural Anthropology*, *36*(3), 509–537. DOI: 10.14506/ca36.3.11

Semmelrock, H., Kopeinik, S., Theiler, D., Ross-Hellauer, T., & Kowald, D. (2023). Reproducibility in machine learning-driven research. *arXiv preprint arXiv:2307.10320.*

Seok, H., Lee, D., Son, S., Choi, H., Kim, G., & Kim, T. (2024). Beyond von Neumann architecture: Brain-inspired artificial neuromorphic devices and integrated computing. *Advanced Electronic Materials*, *10*(8), 2300839. DOI: 10.1002/aelm.202300839

Seth, A. K., & Friston, K. J. (2016). Active interoceptive inference and the emotional brain. *Philosophical Transactions of the Royal Society B: Biological Sciences, 371*(1708), 20160007.

Shafik, W., Singh, R., & Kumar, V. (2024). Artificial intelligence transparency and explainability in sustainable healthcare. In R. Singh, W. Shafik, D. Crowther, & V. Kumar (Eds.), *Transforming healthcare sector through artificial intelligence and environmental sustainability* (Approaches to Global Sustainability, Markets, and Governance). Springer. https://doi.org/DOI: 10.1007/978-981-97-9555-0_9

Sheth, A., & Roy, K. (2024). Neurosymbolic value-inspired artificial intelligence (why, what, and how). *IEEE Intelligent Systems, 39*(1), 5–11. DOI: 10.1109/MIS.2023.3344353

Shi, F., Zhou, F., Liu, H., Chen, L., & Ning, H. (2022). Survey and tutorial on hybrid human-artificial intelligence. *Tsinghua Science and Technology, 28*(3), 486–499. DOI: 10.26599/TST.2022.9010022

Shin, J. H., Kwon, J., Kim, J. U., Ryu, H., Ok, J., Joon Kwon, S.,…Kim, T.-i. (2022). Wearable EEG electronics for a Brain–AI Closed-Loop System to enhance autonomous machine decision-making. *npj Flexible Electronics, 6*(1), 32.

Shirolapov, I. V., Zakharov, A. V., Shishkina, A. A., Sergeeva, M. S., Komarova, Y. S., Romanchuk, N. P., Bannov, V. M., Kuznetsova, O. G., & Khivintseva, E. V. (2024). Efficiency of computerized cognitive training for prevention of cognitive impairments and stimulation of neuroplasticity. *Advances in Gerontology = Uspekhi Gerontologii, 37*(3), 221–229. *Scopus*. Advance online publication. DOI: 10.34922/AE.2024.37.3.007

Shmueli, E. (2010). To explain or to predict? *Statistical Science, 25*(3), 289–310. DOI: 10.1214/10-STS330

Shneiderman, B. (2022). *Human-centered AI*. Oxford University Press. DOI: 10.1093/oso/9780192845290.001.0001

Silver, D., Huang, A., Maddison, C. J., Guez, A., Sifre, L., Van Den Driessche, G., ... & Hassabis, D. (2016). Mastering the game of Go with deep neural networks and tree search. *nature, 529*(7587), 484-489.

Simon, H. A. (1955). A behavioural model of rational choice. *The Quarterly Journal of Economics, 69*(1), 99–118. DOI: 10.2307/1884852

Sinaga, Y. K., Manugeren, M., & Barus, E. (2025). Systematic Literature Review on Social Epistemology of Science: Knowledge Production in Scientific Communities. *IDEAS: Journal on English Language Teaching and Learning. Linguistics and Literature, 13*(2), 7149–7159.

Singh, R., Crowther, D., & Ravesangar, K. (2024). Harnessing human capital for a sustainable future: HR's pivotal role toward net zero policies. In R. Singh & D. Crowther (Eds.), *Transition towards a sustainable future* (Approaches to Global Sustainability, Markets, and Governance). Springer. https://doi.org/DOI: 10.1007/978-981-97-5756-5_2

Singh, R., Ravesangar, K., Narayanan, S., & Singh, K. K. H. (2024). HRM practices in creating net zero: Supporting organizational sustainability. In R. Singh, S. Khan, A. Kumar, S. Luthra, & H. Chokshi (Eds.), *Net zero economy, corporate social responsibility and sustainable value creation* (CSR, Sustainability, Ethics & Governance). Springer. https://doi.org/DOI: 10.1007/978-3-031-55779-8_1

Singh, R., Joshi, A., Dissanayake, H., Iddagoda, A., Khan, S., Félix, M. J., & Santos, G. (2025). Integrating Industry 4.0, circular economy, and green HRM: A framework for sustainable transformation. *Sustainability (Basel), 17*(7), 3082. DOI: 10.3390/su17073082

Singh, R., Khan, S., Dsilva, J., Akram, U., & Haleem, A. (2024). Modelling the organisational factors for implementation of corporate social responsibility: A modified TISM approach. *Global Journal of Flexible Systems Managment, 25*(2), 283–301. DOI: 10.1007/s40171-024-00388-x

Sloane, M., Moss, E., Awomolo, O., & Forlano, L. (2020). Participation is not a design fix for machine learning. In *Proceedings of the Workshop on Participatory Approaches to Machine Learning, ICML 2020*. DOI: 10.48550/arXiv.2007.02423

Sloman, S. A. (1996). The empirical case for two systems of reasoning. *Psychological Bulletin, 119*(1), 3–22. DOI: 10.1037/0033-2909.119.1.3 PMID: 8711015

Srinivasan, K., Eysenbach, B., Ha, S., Tan, J., & Finn, C. (2020). Learning to be safe: Deep rl with a safety critic. *arXiv preprint arXiv:2010.14603.* https://arxiv.org/abs/2010.14603

Stammer, W., Friedrich, L., & Baier, D. (2021). *Revising neuro-symbolic concepts by interacting with their explanations*. CVPR Neuro-Symbolic XIL.

Stephan, K. E., Iglesias, S., Heinzle, J., & Diaconescu, A. O. (2015). Translational perspectives for computational neuroimaging. *Neuron, 87*(4), 716–732. DOI: 10.1016/j.neuron.2015.07.008

Steyvers, M., Tejeda, H., Kerrigan, G., & Smyth, P. (2022). Bayesian modeling of human–AI complementarity. *Proceedings of the National Academy of Sciences of the United States of America, 119*(11), e2111547119. DOI: 10.1073/pnas.2111547119 PMID: 35275788

Sullivan, E. V., Lannoy, S., Le Berre, A. P., Fama, R., & Pfefferbaum, A. (2023). Alcohol drinking and alcohol use disorder across the ages: Dynamic effects on the brain and function.

Sun, R. (2006). *Cognition and multi-agent interaction: From cognitive modeling to social simulation*. Cambridge University Press.

Sutton, R. S., & Barto, A. G. (2018). *Reinforcement learning: An introduction* (2nd ed.). MIT Press.

Sutton, R. T., Pincock, D., Baumgart, D. C., Sadowski, D. C., Fedorak, R. N., & Kroeker, K. I. (2020). An overview of clinical decision support systems: Benefits, risks, and strategies for success. *NPJ Digital Medicine, 3*(1), 17. DOI: 10.1038/s41746-020-0221-y PMID: 32047862

Takahashi, J. S. (2017). Transcriptional architecture of the mammalian circadian clock. *Nature Reviews. Genetics, 18*(3), 164–179. DOI: 10.1038/nrg.2016.150

Tenenbaum, J. B., Kemp, C., Griffiths, T. L., & Goodman, N. D. (2011). How to grow a mind: Statistics, structure, and abstraction. *Science, 331*(6022), 1279–1285. DOI: 10.1126/science.1192788 PMID: 21393536

Teso, S., Alkan, Ö., Stammer, W., & Daly, E. (2023). Leveraging explanations in interactive machine learning: An overview. *Frontiers in Artificial Intelligence, 6*, 113–129. DOI: 10.3389/frai.2023.1066049 PMID: 36909207

Thagard, P. (2003). Why cognitive science needs philosophy and vice versa. *Topics in Cognitive Science, 1*(2), 237–254. DOI: 10.1111/j.1756-8765.2009.01016.x PMID: 25164930

Theodorakopoulos, L., Theodoropoulou, A., & Halkiopoulos, C. (2025). Cognitive bias mitigation in executive decision-making: A data-driven approach integrating big data analytics, AI, and explainable systems. *Electronics (Basel), 14*(19), 3930. DOI: 10.3390/electronics14193930

Thoelen, G. T. (2023). *Essays in the Neurophysiology of Economic Decision-Making and Preference Formation* (Doctoral dissertation, The Claremont Graduate University).

Topol, E. (2019). *Deep medicine: How artificial intelligence can make healthcare human again*. Basic Books.

Topol, E. J. (2019). High-performance medicine: The convergence of human and artificial intelligence. *Nature Medicine*, *25*(1), 44–56. DOI: 10.1038/s41591-018-0300-7 PMID: 30617339

Tripathi, D. R., Urkude, A. S., & Parkhi, V. P. (2025). Explainable artificial intelligence (XAI): Interpretable models for ethical decision-making. *International Journal for Research in Applied Science and Engineering Technology*, *13*(1), 102–110. DOI: 10.22214/ijraset.2025.74356

Tversky, A., & Kahneman, D. (1981). The framing of decisions and the psychology of choice. *science, 211*(4481), 453-458.

Vaccaro, K., Karahalios, K., Sandvig, C., Hamilton, K., & Langlotz, C. P. (2019). Contestability in algorithmic systems. In *Companion Publication of the 2019 ACM Conference on Computer Supported Cooperative Work and Social Computing* (pp. 523–527). DOI: 10.1145/3311957.3359435

Van Den Bos, R., Homberg, J., & De Visser, L. (2013). A critical review of sex differences in decision-making tasks: Focus on the Iowa Gambling Task. *Behavioural Brain Research*, *238*, 95–108. DOI: 10.1016/j.bbr.2012.10.002

Vasuki, M., Kumar, A. D., Celestin, M., & Alghazali, T. A. H. (2025). Mathematical Logic as the Engine of Autonomous Decision Intelligence in Digital Systems. *International Journal of Applied and Advanced Scientific Research*, *10*(2), 97–108.

Vaswani, A., Shazeer, N., Parmar, N., Uszkoreit, J., Jones, L., Gomez, A. N., Kaiser, Ł., & Polosukhin, I. (2017). Attention is all you need. In *Advances in Neural Information Processing Systems, 30*, 5998–6008.

Vaswani, A., Shazeer, N., Parmar, N., Uszkoreit, J., Jones, L., Gomez, A. N., Kaiser, Ł., & Polosukhin, A. (2017). Attention is all you need. *Advances in Neural Information Processing Systems*, *30*, 5998–6008.

Vattonen, A., & Reunanen, M. (2024). Preferences for AI explanations based on cognitive style and socio-cultural factors. *Proceedings of the ACM on Human-Computer Interaction, 8*(CSCW1), 1–26.

Veltri, D., Kamath, U., & Shehu, A. (2018). Deep learning improves antimicrobial peptide recognition. *Bioinformatics (Oxford, England)*, *34*(16), 2740–2747. DOI: 10.1093/bioinformatics/bty179 PMID: 29590297

Vigneashwaran, R., Rohan, M., Praveen, N., & Vijay, N. (2025). IoT-Based Sewage Atmosphere and Health Monitoring for Sanitation Workers with Alerting System. 2025 7th International Conference on Inventive Material Science and Applications (ICIMA), Wang, G., Bao, H., Liu, Q., Zhou, T., Wu, S., Huang, T.,…Zhang, Z. (2024). Brain-inspired artificial intelligence research: A review. *Science China. Technological Sciences*, *67*(8), 2282–2296.

Volkow, N. D., Michaelides, M., & Baler, R. (2019). The neuroscience of drug reward and addiction. *Physiological Reviews*, *99*(4), 2115–2140. DOI: 10.1152/physrev.00014.2018

Wang, D., Yang, Q., Abdul, A., & Lim, B. Y. (2023). Designing theory-driven user-centric explainable AI. In Proceedings of the 2023 CHI Conference on Human Factors in Computing Systems (Article 601). DOI: 10.1145/3544548.3580894

Wang, K., & Bai, D. (2023, August). Overview of Game Decision Intelligence. In *2023 IEEE 11th International Conference on Information, Communication and Networks (ICICN)* (pp. 835-843). IEEE. DOI: 10.1109/ICICN59530.2023.10392489

Wang, G., Bao, H., Liu, Q., Zhou, T., Wu, S., Huang, T., Yu, Z. F., Lu, C. W., Gong, Y. H., Zhang, Z. X., & He, S. (2024). Brain-inspired artificial intelligence research: A review. *Science China. Technological Sciences*, *67*(8), 2282–2296. DOI: 10.1007/s11431-024-2732-9

Wang, Y., Yan, T., Li, S., Cai, Y., Zhang, P., Cao, M., Wang, F., Zhan, X., & Wang, Z. (2025). Brain-inspired neural networks: Neuromorphic devices and their practical applications. *Nanotechnology*, *36*(47), 472001. DOI: 10.1088/1361-6528/ae1840 PMID: 41151104

Wassim, M. (2023). *Explaining by conversing: The argument for conversational XAI* [Harvard undergraduate thesis]. Harvard DASH. https://dash.harvard.edu/server/api/core/bitstreams/934a4899-db75-4d91-9ed4-a259ec234485/content

Weick, K. E. (1995). *Sensemaking in organisations*. Sage Publications.

Weick, K. E., Sutcliffe, K. M., & Obstfeld, D. (2005). Organising and the process of sensemaking. *Organization Science*, *16*(4), 409–421. DOI: 10.1287/orsc.1050.0133

Wei, J., Wang, X., Schuurmans, D., Bosma, M., Ichter, B., Xia, F., Chi, E., Le, Q. V., & Zhou, D. (2022). Chain-of-thought prompting elicits reasoning in large language models. *Advances in Neural Information Processing Systems*, *35*, 24824–24837. DOI: 10.52202/068431-1800

Whittington, J. C. R., & Bogacz, R. (2019). Theories of error back-propagation in the brain. *Trends in Cognitive Sciences*, *23*(3), 235–250. DOI: 10.1016/j.tics.2018.12.005 PMID: 30704969

Wiehler, A., Branzoli, F., Adanyeguh, I., Mochel, F., & Pessiglione, M. (2022). A neuro-metabolic account of why daylong cognitive work alters the control of economic decisions. *Current Biology*, *32*(16), 3564–3575. DOI: 10.1016/j.cub.2022.07.010

Wingerter, T. L., Straub, T., & Schweitzer, S. (2025). Mitigating Automation Bias in Generative AI Through Nudges: A Cognitive Reflection Test Study. *Procedia Computer Science*, *270*, 2106–2114. DOI: 10.1016/j.procs.2025.09.331

Wong, A., Otles, E., Donnelly, J. P., Krumm, A., McCullough, J., DeSmet, O., Passarello, B., & Singh, K. (2021). External validation of a widely implemented proprietary sepsis prediction model in hospitalized patients. *JAMA Internal Medicine*, *181*(8), 1065–1070. DOI: 10.1001/jamainternmed.2021.2626 PMID: 34152373

Work, E. S. (2025). Stanford Encyclopedia of Philosophy. https://iris.unito.it/retrieve/e27ce434-0eb2-2581-e053-d805fe0acbaa/einstein-philscience_sc.pdf

Wu, H., Shao, H., Gao, X. Z., Yan, M., Heng, J., & Li, Z. (2025). Brain-inspired Network with Distinct Brain Mechanisms for Short-term Load Forecasting. *IEEE Internet of Things Journal*.

Wu, S. Z. (2025). Emerging Next-Generation AI Techniques. In *The AI City* (pp. 237–245). Springer Nature Singapore. DOI: 10.1007/978-981-96-2560-4_15

Wu, S., Oltramari, A., Francis, J., Giles, C. L., & Ritter, F. E. (2025). Cognitive LLMs: Toward human-like artificial intelligence by integrating cognitive architectures and large language models for manufacturing decision-making. *Neurosymbolic Artificial Intelligence*, *1*, 29498732251377341. DOI: 10.1177/29498732251377341

Wu, Y., Zhao, R., Zhu, J., Chen, F., Xu, M., Li, G., Song, S., Deng, L., Wang, G., Zheng, H., Ma, S., Pei, J., Zhang, Y., Zhao, M., & Shi, L. (2022). Brain-inspired global-local learning incorporated with neuromorphic computing. *Nature Communications*, *13*(1), 65. DOI: 10.1038/s41467-021-27653-2 PMID: 35013198

Xiong, H., Chu, C., Fan, L., Song, M., Zhang, J., Ma, Y., ... & Jiang, T. (2023). The digital twin brain: A bridge between biological and artificial intelligence. Intelligent Computing, 2, 0055.

Xu, W., Dainoff, M. J., Ge, L., & Gao, Z. (2021). *Transitioning to human interaction with AI systems: New challenges and opportunities for HCI professionals to enable human-centered AI* (Version 4). arXiv. DOI: 10.48550/ARXIV.2105.05424

Xu, B., & Poo, M. M. (2023). Large language models and brain-inspired general intelligence. *National Science Review*, *10*(10), nwad267. DOI: 10.1093/nsr/nwad267 PMID: 37942481

Xu, B., Zhang, D., Wang, Y., Deng, L., Wang, X., Wu, C., & Song, A. (2021). Decoding Different Reach-and-Grasp Movements Using Noninvasive Electroencephalogram. *Frontiers in Neuroscience*, *15*, 684547. DOI: 10.3389/fnins.2021.684547 PMID: 34650398

Xu, W. (2019). Toward human-centered AI: A perspective from human-computer interaction. *Interactions (New York, N.Y.)*, *26*(4), 42–46. DOI: 10.1145/3328485

Yair, O., Ben-Chen, M., & Talmon, R. (2019). Parallel Transport on the Cone Manifold of SPD Matrices for Domain Adaptation. *IEEE Transactions on Signal Processing*, *67*(7), 1797–1811. DOI: 10.1109/TSP.2019.2894801

Yalawar, M. S., Vijaya Babu, K., Mahender, B., & Singh, H. (2023). A brain-inspired cognitive control framework for artificial intelligence dynamic system. International Conference on Communications and Cyber Physical Engineering 2018, Ye, P., Wang, X., Zheng, W., Wei, Q., & Wang, F.-Y. (2022). Parallel cognition: Hybrid intelligence for human-machine interaction and management. *Frontiers of Information Technology & Electronic Engineering*, *23*(12), 1765–1779.

Yang, X., Yan, J., Wang, W., Li, S., Hu, B., & Lin, J. (2022). Brain-inspired models for visual object recognition: An overview. *Artificial Intelligence Review*, *55*(7), 5263–5311. DOI: 10.1007/s10462-021-10130-z

Yordanova, Z. (2026). Merging AIoE technology With Brain-Inspired Computing in Education. In *AI-Augmented Creativity in Learning Analytics* (pp. 273–292). IGI Global Scientific Publishing.

Zaimoglu, A. K., Pratt, L., & Fisher, B. (2023). Epistemological role of human reasoning in data-informed decision-making. *Frontiers in Communication*, *8*, 1250301. DOI: 10.3389/fcomm.2023.1250301

Zakharov, A. V., Bratchenko, I. A., Bratchenko, L. A., Neupokoeva, A. V., Khivintseva, E. V., Shirolapov, I. V., & Zakharov, V. P. (2025). Deep learning of surface-enhanced Raman spectroscopy data for multiple sclerosis diagnostics. *The European Physical Journal. Special Topics*, *234*(15), 4789–4797. Advance online publication. DOI: 10.1140/epjs/s11734-024-01449-z

Zakharov, A. V., Bulanov, V. A., Khivintseva, E. V., Kolsanov, A. V., Bushkova, Y. V., & Ivanova, G. E. (2020). Stroke Affected Lower Limbs Rehabilitation Combining Virtual Reality With Tactile Feedback. *Frontiers in Robotics and AI*, *7*, 81. https://www.frontiersin.org/articles/10.3389/frobt.2020.00081. DOI: 10.3389/frobt.2020.00081 PMID: 33501248

Zamfirescu-Pereira, J. D., Wong, R. Y., Hartmann, B., & Yang, Q. (2023, April). Why Johnny can't prompt: how non-AI experts try (and fail) to design LLM prompts. In *Proceedings of the 2023 CHI conference on human factors in computing systems* (pp. 1-21). DOI: 10.1145/3544548.3581388

Zander, T. O., & Kothe, C. (2011). Towards passive brain–computer interfaces: Applying brain–computer interface technology to human–machine systems in general. *Journal of Neural Engineering*, *8*(2), 025005. DOI: 10.1088/1741-2560/8/2/025005 PMID: 21436512

Zeng, Y., Zhao, D., Zhao, F., Shen, G., Dong, Y., Lu, E., & Zhao, Y. (2023). BrainCog: A spiking neural network based, brain-inspired cognitive intelligence engine for brain-inspired AI and brain simulation. *Patterns (New York, N.Y.)*, *4*(8), 100789. DOI: 10.1016/j.patter.2023.100789 PMID: 37602224

Zhang, B., Li, L., Chen, Y., Zhao, Q., Zhu, Z., Yi, X., & Xie, X. (2025). Knowing Your Uncertainty—On the application of LLM in social sciences. *arXiv preprint arXiv:2512.05461*.

Zhang, K., Yang, Z., & Başar, T. (2021). Multi-agent reinforcement learning: A selective overview of theories and algorithms. *Handbook of reinforcement learning and control*, 321-384.

Zhang, C., & Ma, Y. (Eds.). (2012). *Ensemble Machine Learning*. Springer. DOI: 10.1007/978-1-4419-9326-7

Zhang, H., Jiao, L., Yang, S., Li, H., Jiang, X., Feng, J., Zou, S., Xu, Q., Gu, J., Wang, X., & Wei, B. (2024). Brain–computer interfaces: The innovative key to unlocking neurological conditions. *International Journal of Surgery*, *110*(9), 5745–5762. DOI: 10.1097/JS9.0000000000002022 PMID: 39166947

Zhang, M. W., Ying, J., Wing, T., Song, G., Fung, D. S., & Smith, H. E. (2018). Cognitive biases in cannabis, opioid, and stimulant disorders: A systematic review. *Frontiers in Psychiatry*, *9*, 376. DOI: 10.3389/fpsyt.2018.00376

Zhang, R., Deng, H., & Xiao, X. (2024). The insular cortex: An interface between sensation, emotion and cognition. *Neuroscience Bulletin*, *40*(11), 1763–1773. DOI: 10.1007/s12264-024-01211-4 PMID: 38722464

Zhang, Y., Liao, Q. V., & Bellamy, R. K. E. (2020). Effect of confidence and explanation on accuracy and trust calibration in AI-assisted decision making. In *Proceedings of the 2020 ACM Conference on Fairness, Accountability, and Transparency* (pp. 295–305). DOI: 10.1145/3351095.3372852

Zhang, Y., Liao, Q. V., & Bellamy, R. K. E. (2024). From explainable to interactive AI: A literature review on current trends in human-AI interaction. *International Journal of Human-Computer Studies, 182*, 103140. DOI: 10.1016/j.ijhcs.2023.103140

Zhao, L., Zhang, L., Wu, Z., Chen, Y., Dai, H., Yu, X., Liu, Z., Zhang, T., Hu, X., Jiang, X., Li, X., Zhu, D., Shen, D., & Liu, T. (2023). When brain-inspired ai meets agi. *Meta-Radiology, 1*(1), 100005. DOI: 10.1016/j.metrad.2023.100005

About the Contributors

Ionica Oncioiu holds a Ph.D. in economy and accounting. Her research interests include the development of SMEs innovation, Project Management, Accounting Information Systems, Asset Management and E-Commerce Marketing. She had more than 20 years of experience in this area and has published 15 text books and more than 90 papers in scholarly peer reviewed international journals. She has previously published three book: in 2013 called "Business Innovation, Development, and Advancement in the Digital Economy" (IGI Global Book), in 2016 called "Driving Innovation and Business Success in the Digital Economy", in 2017 called "Ethics and Decision-Making for Sustainable Business Practices", in 2018 called "Network Security and Its Impact on Business Strategy ", in 2019 called "Throughput Accounting in a Hyperconnected World", in 2020 called "Improving Business Performance Through Innovation in the Digital Economy" and is interested in continuing writing in the accounting, finance, big data and robotics area. From 2020 to the present, he has published over 6 books under the IGI publishing house.

***

Ashraf Alam is currently serving as an Assistant Professor in the Department of Education at NIT Calicut, Kerala, India. He has been recognised among the World's Top 2% Scientists by Stanford University and Elsevier. His academic contributions include more than 80 Scopus-indexed publications and over 100 conference presentations delivered at national and international academic forums. His work has appeared in reputed research publication platforms including Elsevier, Springer, Taylor & Francis, CRC Press, Routledge, Atlantis Press, Springer Nature, Palgrave Macmillan, and IEEE. He earned his Ph.D. from IIT Kharagpur, where his research engaged with foundational concerns in educational policy, curriculum design, educational technology, philosophy of education, innovation in education, educational administration, governance, and pedagogical transformation. His work is grounded in a transdisciplinary orientation that draws from educational sciences, digital pedagogy, curriculum theory, learning analytics, sustainability education, instructional design, artificial intelligence, ethics of technology, and socio-technical studies. Dr. Ashraf's scholarship strives to interrogate and expand the boundaries of educational thought by examining the epistemological, ethical, and social dimensions of digital and post-digital pedagogies. His work critically engages with the changing nature of knowledge production, learner agency, pedagogical mediation, and institutional transformation in technologically saturated educational environments. He is particularly invested in advancing inclusive digital learning, gamified instructional strategies, AI-mediated educational ecosystems, and ethically governed technology integration in education. His research agenda is animated by the pursuit of design-oriented, ethically attuned, and contextually responsive frameworks that can support meaningful educational reform. Through his work, he seeks to contribute to the development of pedagogical models that are not only technologically innovative but also socially just, epistemically inclusive, developmentally sensitive, and aligned with the broader goals of sustainable and human-centred education.

Firdaus Alamsyah is a researcher and academic practitioner with research interests in explainable artificial intelligence, decision intelligence, and human-centered AI systems. His work focuses on integrating transparent AI techniques with cognitive computing approaches to support interpretable and trustworthy decision-making. He has been involved in research and academic projects related to intelligent systems, assistive technologies, and AI-driven decision support. His current interests include brain-inspired mechanisms, human–AI collaboration, and the ethical deployment of AI in complex decision-making environments.

Al-E'mari earned her Ph.D. in Cybersecurity from Universiti Sains Malaysia (USM) in 2022. She is the Chair of the Information Security Department and an Assistant Professor at the University of Petra (UoP). With expertise in Blockchain, Deep Learning, and Network Security, she has contributed significantly to cybersecurity research. She previously earned her Bachelor's and Master's in Computer Science from Yarmouk University, Jordan, shaping her strong foundation in security and technology.

Muralidhar L B is a professionally qualified dental surgeon. He holds a Ph.D. from Canara Bank School of Management Studies, Bangalore University, after his MBA from Christ University. He is qualified in management at UGC - NET and KSET. He has 18 years of total teaching experience spanning across dental education, healthcare management and general management and 4 years of corporate experience. He has worked as a research associate in the Institute for Social and Economic Change, Bangalore in a project entitled as "Management By and For Common Man: An Exploration of the Concept and Direction for Future"post Ph.D. He had secured 28th rank in the S.S.L.C. examination for Karnataka State in 1993. His 32 patents have been published, and 2 design patents have been granted. He has 31 research publications to his credit, 12 conference proceedings publications, 13 books and 26 book chapters. He has 37 research publications indexed in Scopus, 5 in ABDC and 1 in Web of Science database. He has facilitated SWAYAM MOOC on "Social Enterpreneurship and Sustainability" through, Ministry of Education, Govt of India, Indian Institute of Management, Bangalore, Jain Deemed to be University He is supervising Ph.D. research of 4 research scholars at JAIN (Deemed to be University). He works as an assistant professor in the Department of Business Management Studies, School of Commerce, Jain Deemed to be a University in Bangalore.

Kundan Sagar Bedmutha is an undergraduate student pursuing a Bachelor of Technology in Artificial Intelligence and Data Science at Vishwakarma University, India, and is expected to graduate in 2027. His academic interests focus on Artificial Intelligence, Machine Learning, Data Science, and Computer Science. He has actively worked on projects involving AI-driven decision support systems, intelligent web-based applications, predictive modeling, and data analytics. His areas of interest include AI commercialization frameworks, intelligent personalization systems, and applied machine learning for real-world problem solving. He is particularly interested in bridging theoretical AI concepts with practical system implementation and scalable software architectures.

Annmol Damien is a Research Scholar at the School of Business and Management, CHRIST (Deemed to be University), Bengaluru, specializing in Neurofinance. Her doctoral research investigates how risk tolerance, motivation, and emotional states influence financial investment decisions using EEG-based neural measurements and experimental task designs. With a strong academic background in finance and marketing, complemented by professional experience in sales and market research, she brings an interdisciplinary perspective that integrates neuroscience, psychology, and financial behavior. She has presented her work at international conferences and has published a book chapter in neurofinance, reflecting her emerging contributions to this growing field.

Fadi Fataftah is an information technology professional, researcher, and educator with expertise in cybersecurity, cloud computing, healthcare IT, virtual care, artificial intelligence, and emerging digital technologies. He holds a PhD in

Information Technology from North-West University, South Africa, where his research focused on risk assessment for big data in cloud computing environments. He currently works in healthcare technology and virtual care within a Canadian hospital setting, focusing on clinical workflow analysis, virtual care implementation, testing, training, change-impact review, and secure digital health service delivery. He is also a sessional faculty member at University Canada West, teaching in Computer Science, information systems, and emerging technologies. His publications include peer-reviewed journal articles, IEEE conference papers, and book chapters related to cybersecurity, cloud security, artificial intelligence, business intelligence, digital transformation, cyber insurance, edge intelligence, digital twins, and healthcare technology.

Swapna H. R holds two Ph.D. degrees in Management — one from CMR University, Bangalore, and another from Rayalaseema University, Kurnool. She is an MBA in Finance from VTU University. She has over 18 years of teaching experience in the areas of Finance, Analytics, Statistics, Entrepreneurship, and Human Resource Management, along with 4 years of industry experience. Her research expertise spans Statistics, Behavioural Finance, Employee Engagement, Artificial Intelligence, Entrepreneurship, research on millennials, customer engagement, social media, and the health sector. She has published over 100 research articles in National and International Journals, of which 46 have been indexed in Scopus, Web of Science (WoS), Emerald, Springer, and ABDC. She has presented papers at National and International Conferences, including in the USA, Malaysia, and Spain, and has received several Best Paper Presenter Awards. She has authored over 25 textbooks in the fields of Statistics, Entrepreneurship, Business Analytics, Healthcare, and Event Management. Her book Business Analytics received the Golden Book Award of 2022. She also received the Excellence Researcher Award on May 30, 2025. Another feather in her cap is her achievement in the field of Intellectual Property Rights, with 14 patents published, 2 granted, and 4 more filed. She has also curated and developed a SWAYAM course titled "Microfinance and Self-Help Groups," contributing to the promotion of inclusive education and financial literacy at the national level. She serves as an editor for several Scopus-indexed journals under IGI Global, Public Administration and Policy (Emerald Publishing), and PubMed-indexed medical journals. She is an active member of ENACTUS, an international social entrepreneurship forum linked with several NGOs, where she contributes to sustainability initiatives and drives impactful community projects. She passionately mentors students to develop sustainable business solutions and promotes social innovation and environmental responsibility. Dr. Swapna has also been at the forefront of social service initiatives such as organizing blood donation camps in association with the Rotary Club and TTK, supporting the Ministry of Education's

One-on-One COVID-19 emotional support drive, and leading environmental awareness campaigns like "One Earth."

Rajkumar Jagdale is an Assistant Professor and Head of the Department of Computer Science at the Faculty of Science and Technology, Vishwakarma University, Pune. He earned his Ph.D. in Computer Science and IT from Dr. Babasaheb Ambedkar Marathwada University, Chh. Sambhajinagar, Maharashtra, India.With over six years of teaching and research experience, complemented by two years in the industry as Data Scientist, Dr. Jagdale specializes in Sentiment Analysis, Opinion Mining, Text Mining, ML, Big Data Analytics, and RPA. He has made significant academic contributions, with 13 publications in reputable international journals and conferences, and holds one patent related to his research. Dr. Jagdale's work has received notable recognition, including prestigious scholarships such as the JASSO Scholarship to attend a Spring School on "Smart Materials and AI" at Tokushima University, Japan, and the DST Inspire Fellowship from the Department of Science and Technology, Government of India. Additionally, he was awarded a Gold Medal for his exemplary performance in the M.Sc. Computer Science program. He reviewed conference papers and journals and He is member of CSI, ISCA and IAENG.

Sheeba K P., M. Sc, PhD (pursuing). She is having a academic Experience of 5 years and 8 years in health care industries like hospitals, health insurance organizations and TPA companies both in abroad and in India. She is a post graduate in Hospital Administration from Rajiv Gandhi University of Health Sciences, currently pursuing her PhD in management. She is a member of associate ship in general insurance from Insurance Institute of India (III). She has attended many workshops and Faculty Development Programs, seminars, and orientation programs. . She has presented more than 3 research articles at National and International conferences. She has published few research articles in UGC care listed journals. She is currently working as Assistant Professor in Department of Business Management Studies, School of Commerce, Jain Deemed to be University, Bengaluru, India.

Jyoti Kumari is an Assistant Professor in the Department of Computer Science and Engineering, specializing in Machine Learning and Artificial Intelligence. Her research primarily focuses on AIML (Artificial Intelligence and Machine Learning), data analytics, and related cutting-edge areas. She has authored and published numerous research papers, chapters, patents, copyrights and books in the field of AIML, contributing valuable insights to the academic and scientific community. Dedicated to shaping the next generation of technologists, she is committed to preparing students for successful careers in technology and research. Through her teaching, she instills a strong foundation in core concepts, emphasizes hands-on practical experience, and cultivates an innovative mindset. Her pedagogical approach goes beyond traditional instruction by inspiring students to explore novel ideas, adopt emerging techniques, and stay aligned with the latest industry trends and technological advancements.

Jain Mathew serves as Professor and Dean of the School of Business and Management at CHRIST University, Bangalore. With over 35 years of experience, he is recognized for his expertise in Finance and Human Resource Management. He has supervised17 PhD and 20 MPhil scholars, contributing significantly to academia. Dr. Mathew's research spans leadership taxonomy, market volatility, and social media technology. His projects have received funding from the Indo-Korean Science & Technology Institute and Liverpool Hope University. He has published over 60 research articles and delivered 40 high-impact conference presentations. His accolades include the Manubhai M. Shah Memorial Research Award and Best Paper Award at the International Business Research Conference in Dubai. Dr. Mathew also has served as a doctoral committee member at Karunya University and VIT University. He has contributed to Boards of Studies at Mount Carmel College and Kristu Jayanti University. Currently, Dr. Mathew is a member of CHRIST Universitys Finance Committee and Executive Council. A passionate educator and researcher, he continues to inspire future generations through his unwavering dedication to academic excellence.

Madhu S has 32+ years of experience in academia and industry. A prolific author and researcher, he has authored 110+ books, holds 27 design patent, 6 utility patent published, published 30+ articles and book chapters, and has won multiple awards. He serves as a consultant, keynote speaker, jury member, resource person, and academic mentor, guiding hundreds of students and contributing widely to research and innovation. BoS and BoE member for various universities.

Yousef K. Sanjalawe is an Assistant Professor specializing in Cybersecurity and Cloud Computing at the University of Jordan (JU), Jordan. He earned his PhD in Cloud Computing from Universiti Sains Malaysia (USM) in 2020, where his research focused on enhancing cloud service broker selection using optimization algorithms. Dr. Sanjalawe has extensive experience in academia, research, and professional training. He has supervised Ph.D. students in Cybersecurity, Cloud Computing, IoT, Fog Computing, Optimization, and Artificial Intelligence (AI). From 2023 to 2024, he served as the Chair of the Cybersecurity Department at the American University of Madaba (AUM), Jordan, and from 2022 to 2024, he was an Assistant Professor in Cybersecurity at AUM.

Rubee Singh is an Assistant Professor of HR at the Institute of Business Management, GLA University, India. She is also a Post-Doctoral (D.Litt.) Researcher at Kumaun University, Nainital, India. Dr. Singh earned her Ph.D. in Management from Noida International University, and an MBA in Human Resource Management from Dr. A.P.J. Abdul Kalam Technical University, India. A prolific academic, she has authored and edited over 30 academic books and published more than 100 research papers in reputed journals indexed in SCI, ABDC-A, Scopus, and UGC. Her research excellence is reflected in her Google Scholar citation count exceeding 2000 with an H-index of 28. Her core research interests include Net Zero initiatives, Circular Economy, Digital Circular Economy, Sustainability, Digital HRM, Sustainable HRM, Industry 4.0, Artificial Intelligence, and Innovation Management. Dr. Singh has been the recipient of several prestigious awards, including the Best Researcher Award (for three consecutive years: 2022–23, 2023–24, and 2024–25) by GLA University, the Iconic Faculty Award (2019), the International Young Researcher Award (2021), and the Young Scientist Award (2020).

Nidhi Tripathi is serving as an Assistant Professor at CMP Degree College, a constituent college of the University of Allahabad, since February 2018. She is involved in teaching and mentoring undergraduate and postgraduate students in the field of zoology and biological sciences. Her research focuses on developmental neurotoxicity, particularly the effects of prenatal exposure on brain development using albino rat models. She was awarded the Junior Research Fellowship (JRF) and Senior Research Fellowship (SRF) during her research tenure. Her academic interests include neurobiology, developmental biology and toxicology.

Pradeep Yadav is a distinguished academician with over 20 years of teaching experience and expertise in Computer Science and Engineering. He holds a Ph.D. in Computer Engineering and has an extensive research portfolio, including many international journal papers, multiple conference proceedings, and several books. His research interests span machine learning, artificial intelligence, and data mining. Currently, Dr. Yadav serves as an Associate Professor and Head of Department of CSE at ITM Gwalior, MP, India. He has also contributed significantly to the field through patents and copyrights and actively engages as a reviewer, conference chair and editor.

Index

A

B

C

D

E

F

G

H

M

N

P

Q

R

S

T

W

X

www.ingramcontent.com/pod-product-compliance
Lightning Source LLC
LaVergne TN
LVHW080912300926
849162LV00003B/654